CW01371263

*Modern Advanced
Mathematics for Engineers*

Modern Advanced Mathematics for Engineers

Vladimir V. Mitin
Dmitri A. Romanov
Michael P. Polis

A Wiley-Interscience Publication
JOHN WILEY & SONS, INC.
New York / Chichester / Weinheim / Brisbane / Singapore / Toronto

This text is printed on acid-free paper. ∞

Copyright © 2001 by John Wiley & Sons, Inc. All rights reserved.

Published simultaneously in Canada.

No part of this publication may be reproduced, stored in a retrieval system or transmitted in any form or by any means, electronic, mechanical, photocopying, recording, scanning or otherwise, except as permitted under Sections 107 or 108 of the 1976 United States Copyright Act, without either the prior written permission of the Publisher, or authorization through payment of the appropriate per-copy fee to the Copyright Clearance Center, 222 Rosewood Drive, Danvers, MA 01923, (978) 750-8400, fax (978) 750-4744. Requests to the Publisher for permission should be addressed to the Permissions Department, John Wiley & Sons, Inc., 605 Third Avenue, New York, NY 10158-0012, (212) 850-6011, fax (212) 850-6008, E-Mail: PERMREQ @ WILEY.COM.

For ordering and customer service, call 1-800-CALL-WILEY.

Library of Congress Cataloging in Publication Data is available.

ISBN 0-471-41770-X

Printed in the United States of America

10 9 8 7 6 5 4 3 2 1

*To my granddaughter,
Christina, cats, and pets
who by their warm and
quiet attitude encouraged
me to complete this work.
V. V. M.*

*To Lioudmila and Paul,
ever patient and inspiring.
D. A. R.*

*To my wife and family
for their support over the
years: Claudette,
Melanie, Pascal, and
Karine.
M. P. P.*

Preface

This book is written with the aim of providing students with a strong foundation in modern applied mathematics. It is intended for use as a text for an introductory one-semester graduate course for engineering students.

All those who deal with engineering education are well aware of the precarious state of basic mathematical skills and conceptual comprehension which is typical of modern students. Most beginning engineering graduate students haven't had a good experience with mathematics. As a result, in many specialized engineering courses, such as Control Systems, Solid State Electronics, or Communication Theory, instructors have to begin with a review of the mathematics they to use in the course. This, in turn, leads to a fragmentation of mathematical knowledge in the student's head. Vector analysis stays forever uniquely associated with electromagnetic waves, set operations with computer architecture, and the Laplace transform with circuits and signal processing. The mathematical methods encountered in each course do not evolve into unified patterns which the future engineer would be able to recognize and use universally. This smokestack mentality inflicts incalculable damage to engineering practice, especially in many areas closely connected to research and development.

We strongly believe that this situation can be and should be radically improved. We think that it is truly possible to offer a beginning graduate student a concise yet comprehensive course which summarizes, unifies, and completes his/her mathematical knowledge by constructing a comprehensive system of operating mathematics and setting up patterns that can be recalled for use in a wide range of engineering disciplines. This book is our response to this challenge.

The main difficulty in developing such a course lies in the wide diversity of ideas and concepts of modern mathematics which are in use in contemporary engineering analysis. Although many excellent textbooks exist, in our opinion, they take too narrow an approach, and none provides the full range of necessary topics. This self-narrowing stems from an inclination to either use an operator approach or to employ a discrete-mathematics framework. Another weak point of existing texts on engineering mathematics is their formidable volume (cf. two of the most popular texts on Advanced Engineering Mathematics, Kreyszig [1] and Wylie and Barrett [2], are each well over a thousand pages). Although quite suitable for a systematic in-depth study, such books are not very helpful for a one-semester introductory course that most university programs can only allow themselves.

To avoid both the Scylla of narrowing and the Charybdis of voluminosity, in this course we develop all the topics from one general concept, that of mappings, their properties and manifestations. This unification of concepts and terminology saves time and effort which we then use to tailor several levels of explanation, addressing various ranges of students' background. The review of undergraduate topics, such as basic calculus, determinants, or Gaussian elimination, evolves gradually through exercises which involve more and more mathematical background.

Intended to be as concise as possible, the text has several salient features which need some explanation. First, in the methods of delivery, emphasis is more on student understanding than on mathematical rigor. We strongly believe that basic ideas of modern mathematics can be understood and successfully used by engineers without delving too far into technical details and terminological subtleties. With this in mind we have mostly forgone the theorem-proof format for a more informal style. We must confess that it was done with certain relish; in this sense, as in many others, this is a book written by engineers for engineers. However, we have not eliminated all the theorems and have not presented the applied mathematics as merely a bag of tricks. The important theorems we retained are used as pivotal points in the exposition of particular concepts. They play an important role in summarizing concepts and making the student consciously realize that any established method or technique has well-defined limitations.

The second feature of the text is our approach to examples and exercises. It is our intention to give the student a can-do frame of mind about modern mathematics and confidence in using it. Accordingly, the assortment of examples ranges from simple ones, included only as illustrations, through ones that are both difficult and involved. In all cases, we tried our best to discuss all the details and not to leave anything significant behind "it is easy to show".

In our choice of the examples and problems we tried to resist following the widely held opinion that in a book for engineers all the examples must be very tangible and preferably taken directly from engineering practice. Nowadays, the average engineering student is expected to change fields of specialization several times during

---

[1] E. Kreyszig, *Advanced Engineering Mathematics*, Wiley, New York, 1993.
[2] C.R. Wylie, L.C. Barrett, *Advanced Engineering Mathematics*, McGraw-Hill, New York, 1995.

his/her career. Thus, those students need much more general ("convertible") concepts and methods than given by particular examples doomed to become quickly obsolete. Even now, all the systems of "masses and springs", "inductors and capacitors", and "connected pendula on moving platforms", that invariably pass from one book to another, have little in common with modern engineering which functions on a much higher level of abstraction.

On the other hand, given the diversity of modern engineering disciplines, any *real* problem from one field would require far too much space to explain to somebody from another field. Thus, we intentionally restrict our goal to providing the patterns of proper mathematical structures, referring to engineering disciplines but not relying on them. We merely state that a particular topic is important and discuss the basic concepts and related mathematical methods, leaving the student to joyfully recognize the described pattern in a particular field of his/her future specialization and to make use of the acquired methods and skills.

The main body of the text consists of eleven chapters. Each chapter corresponds to approximately four one-hour lectures of the course. The first part, Chapters 1 through 5, develops the concepts of sets, mappings, and linearity, which are fundamental for this course. The second part, Chapters 6 through 8, concentrates on the concepts of distance, inner product, and orthogonality. Finally, the third part, Chapters 9 – 11, demonstrates various applications of the basic concepts and methods developed in the first two parts of the text.

In more detail, the eleven chapters cover the following topics.

Chapter 1 has a twofold goal. First, set theory is introduced giving the principle notations, general philosophy, and *modus operandi* being used throughout the book. Second, this theory is used as the model *mathematical structure* to be referred to in the subsequent chapters. We consider the very basics of set theory, which are necessary to understand any subject in modern applied mathematics.

Chapter 2 is the foundation of the entire book. We introduce the mathematical concept of relation in terms of set theory, as it is most appropriate in computer applications and design of algorithms. To this end, we discuss in detail the properties of ordered n-tuples and cartesian product sets. Emphasis is given to binary relations; we give an expanded classification of them and show how to specify particular relations in scientific and engineering applications. The terminology elaborated here is then used throughout the book and serves as the framework unifying the diverse topics.

The main emphasis of Chapter 3 is to show the basic similarity of mathematical structures. Here, we treat mathematical logic in terms of mappings and functions, and develop methods of truth-value calculations resembling advanced calculus and demonstrating the overriding unity of the methods being in widest use in engineering. We pay special attention to carefully delineate the basic concept of mathematical logic and to the interrelations between logical objects, logical statements and truth values. Having obtained basic formulas for finding truth values of complex logical statements, we close the chapter by discussing the general structure of mathematical logic, giving an axiomatic description of Boolean algebra, and showing its application to mathematical logic and set theory.

In Chapter 4, we endeavor to describe the essentials of the algebraic structures and methods most needed in applications. We start with a classification of algebraic structures with emphasis on group notion and group properties. The concept of a linear space and its ramifications form the major part of this chapter. We concentrate on discussing properties of linear subspaces and on choosing a convenient basis in a finite-dimensional linear space. The space of column vectors is established as a representation of any linear space of given dimension.

In Chapter 5 we introduce matrices as a natural tool for dealing with linear mappings between linear spaces, and discuss in detail properties of matrices, operations with matrices, and the structure of matrix algebra. We pay special attention to those topics in matrix theory which are in most use in various applications. We proceed to diagonalization of square matrices and go through the calculation of functions of matrices. We also consider the virtues and limitations of LU decomposition as applied to solving systems of linear algebraic equations.

Chapter 6 is devoted to introducing and discussing the concept of metric mapping and metric distance. This is the second basic concept of the course, complementary to that of a linear space. We discuss in detail different ways to introduce a convenient and useful definition of distance. Then, we introduce the important concepts of continuity and passing to the limit. We use these notions to describe the global structure of a closed set and to familiarize the student with continuous mappings. We introduce the Lipschitz condition and pay considerable attention to contraction mappings which are so important in applications.

Chapters 7 and 8 constitute the central part of the text. They complete the exposition of the basic theory, and open the way for considering applications in the third part of the book.

In Chapter 7 we unite the concepts of metric mapping and linearity to define normed mappings and Banach spaces. We discuss in detail some important examples of finite-dimensional and infinite-dimensional Banach spaces, and introduce the notion of infinite basis. The continuity of the norm mapping and the closedness of finite-dimensional subspaces are established as important concepts used in optimization procedures. We also introduce the notion of inner product and describe the properties of finite- and infinite-dimensional Hilbert spaces. The Projection Theorem provides a basis for the approximation theory we develop further in Chapter 9.

Chapter 8 deals mainly with decomposition of functions over orthonormal bases in a Hilbert space. It includes some digression to the theory of functions of a complex variable and to the related calculation techniques. Making use of the Gram-Schmidt orthonormalization procedure, we present a method of constructing an orthonormal basis in an infinite-dimensional Hilbert space. Then, we use the Projection Theorem to decompose a projection of an arbitrary element of a Hilbert space over an orthonormal basis. The Generalized Fourier series arises as a corollary of this theorem. We discuss properties of Fourier coefficients and applications of classical Fourier series. We also introduce other systems of orthonormal functions on finite intervals.

The application-oriented third part of the book starts with Chapter 9, which is devoted to operator equations. We consider methods of approximate solution for a general problem, while the exercises emphasize solving differential equations. We discuss three generic problems related to operator equations, (1) the root-finding problem, (2) the eigenvalue problem, and (3) the fixed-point problem. Having proved the Banach Fixed-Point Theorem, we convert it into the method of successive approximations. Concurrently, we introduce the residual principle, and compare the virtues and disadvantages of successive and residual approximations. We then apply both of these methods to approximately solving ordinary differential equations. We discuss in detail incompletely specified problems involving uncertainty and applications of the least-square method and pseudo-inversion to analysis of experimental data.

In Chapter 10, the Fourier integral is heuristically introduced as the limit of a Fourier series. Then, the Fourier transform is presented as a full-fledged bijection mapping, with a variety of properties important in applications. We introduce the delta-function as an integral operator related to the Fourier transform, and develop this topic by considering *distributions* as a specific kind of mapping and briefly discuss the algebra of distributions. The Laplace transform is then introduced as a natural overgrowth of the Fourier integral and Fourier transforms. Examples of particular transforms and related calculation techniques based on the theory of functions of a complex variable are considered in detail. We also compare the properties of Fourier and Laplace transforms and discuss their advantages in applications. Another important part of this chapter is the discussion of signal processing.

Chapter 11 covers Partial Differential Equations; it is intended to demonstrate the application of the general approach to a very specific area and to give the student a model example of applying the general theory to practical problems. Drawing from all of the previous chapters, we show how to reduce a problem to a known format and to apply proper means to solve it. In this way, we start with constructing special linear mappings of the variables and the unknown function, which reduce a second-order equation to its canonical form. Then, we consider the three basic (canonical) types of second-order equations on examples taken from engineering applications. We discuss the most useful methods of solving these equations and typical properties of solutions.

This text was developed for use in a required graduate engineering mathematics course and has evolved over a period of eight consecutive semesters of teaching at Wayne State University with an enrollment of 65 – 85 students each semester. Students appeared both to grasp and to appreciate the consolidation of concepts and unification in the methods of explanation. To improve the scope of the course and its important details, we extensively used feedback from students and alumni, as well as selective questioning among high-technology industrial companies and governmental laboratories. We greatly appreciate their help and criticism. We also use this opportunity to express our sincere gratitude to Dr. Robert Barnard, Professor Emeritus in the Department of ECE at Wayne State University, for discussions which greatly enlightened us on the essence of mathematical approaches and mathematical mindset. We are grateful to Rustam Bashirov, who diligently read the final version of our manuscript and helped us to eradicate many typos and to clarify some obscure passages. We have received much help from the stern but encouraging critiques on the part of the reviewers choosen by John Wiley & Sons.

We conclude by noting that this book (or any book) is not meant as a substitute for the knowledgeable and enthusiastic person teaching the course. The aim is to guide both teachers and students through the study of modern applied mathematics.

Vladimir V. Mitin
Dmitri A. Romanov
Michael P. Polis

Detroit, Michigan
Rochester, Michigan

Contents

	Dedication	v
	Preface	vii
1	The Basics of Set Theory	1
	1.1 Introduction	1
	1.2 Logical Connectives — Notation	2
	1.3 Introduction of Sets	2
	1.4 Specifications of Sets	3
	1.5 Examples of Sets	4
	1.6 Basic Sets of Numbers	4
	1.7 Paradoxes	5
	1.8 Axioms of Set Theory	6
	1.9 Sample Problems	13
	1.10 Summary of Chapter 1	15
	1.11 Problems	15
	1.12 Further Reading	16
2	Relations and Mappings	19
	2.1 Relations – The Naive Approach	19
	2.2 Ordered n-tuple	20

2.3	An n-tuple Versus a Set	21
2.4	Cross-Product Set and Its Properties	22
2.5	Sample Problem I	25
2.6	n-ary Relation	27
2.7	Binary Relations — Terminology	29
2.8	Mappings — Classification	30
2.9	Sample Problem II	36
2.10	Summary of Chapter 2	38
2.11	Problems	39
2.12	Further Reading	39

3 Mathematical Logic — 41

3.1	Logical Objects and Logical Statements	41
3.2	Logical-Statement and Truth-Value Mappings	42
3.3	Logical Connectives and Logical Formulas	44
3.4	Adequate Set of Connectives	44
3.5	Truth Values of Complex Logical Formulas	45
3.6	Tautologies and Contradictions	49
3.7	Sample Problems I	50
3.8	Fundamental Tautologies. De Morgan Laws. Proof Formats	52
3.9	Properties of Connectives. An Axiomatic Approach to Mathematical Logic	57
3.10	Sample Problems II	58
3.11	Boolean Algebra. The Analogy Between Mathematical Logic and Set Theory	59
3.12	Sample Problems III	61
3.13	Summary of Chapter 3	63
3.14	Problems	64
3.15	Further Reading	65

4 Algebraic Structures: Group through Linear Space — 67

4.1	Classification of Algebraic Structures	67
4.2	Algebra of Complex Numbers	73
4.3	Geometric Representation of Complex Numbers	75
4.4	Linear Space: Axioms and Examples	77
4.5	Linear Combination and Span. Linear Mappings	80
4.6	Linear Subspaces	80
4.7	Linear Independence	82

	4.8	Basis and Dimension of a Linear Space	82
	4.9	Decomposition as Mapping. Isomorphism of Finite-Dimensional Linear Spaces	85
	4.10	Sample Problems	86
	4.11	Summary of Chapter 4	90
	4.12	Problems	92
	4.13	Further Reading	92
5	Linear Mappings and Matrices		95
	5.1	Linear Mappings and Column Vectors	95
	5.2	Matrices	95
	5.3	The Matrix Product	97
	5.4	Matrix Algebra	99
	5.5	Transpose of a Matrix	101
	5.6	Fundamental Subspaces of Column Vectors Associated With a Given Matrix	102
	5.7	The Echelon Matrix. LU-decomposition	103
	5.8	The Fundamental Theorem of Linear Algebra	106
	5.9	Sample Problem I	108
	5.10	Eigenvalues, Eigenvectors, and the Characteristic Equation	110
	5.11	Matrix Diagonalization	111
	5.12	Functions of Matrices	112
	5.13	Sample Problem II	117
	5.14	Multiple Eigenvalues. The Jordan Form	118
	5.15	Summary of Chapter 5	122
	5.16	Problems	123
	5.17	Further Reading	124
6	Metrics and Topological Properties		127
	6.1	Necessity of Distance Concept	127
	6.2	Metric Mapping and Distance: Definition and Axioms	127
	6.3	Metric Spaces	129
	6.4	Sample Problem	132
	6.5	Proximity. Limits in Metric Spaces	133
	6.6	Cauchy Sequences	134
	6.7	Open Ball: A Geometric Interpretation	134
	6.8	Examples of Cauchy Sequences	135
	6.9	Open and Closed Sets. Completeness vs. Closedness	137

6.10	Continuous mapping	140
6.11	Lipschitz (Bounded) Mapping	141
6.12	Example of a Contraction Mapping	141
6.13	Summary of Chapter 6	142
6.14	Problems	143
6.15	Further Reading	144

7 **Banach and Hilbert Spaces** — 147
- 7.1 Introduction: The Great Alliance of Linearity and Metrics — 147
- 7.2 Norm Mapping. Banach Space — 147
- 7.3 Sample Problem I — 149
- 7.4 Finite-Dimensional Subspaces in Banach Space — 150
- 7.5 Inner-Product Mapping. Hilbert Space — 152
- 7.6 Sample Problems II — 154
- 7.7 The Cauchy-Schwarz Inequality — 155
- 7.8 Inner Product and Matrices. Hermitian Matrices — 157
- 7.9 Continuity of the Inner-Product Mapping — 159
- 7.10 Orthogonality — 160
- 7.11 Sample Problem III — 160
- 7.12 Eigenvectors of Hermitian Matrices — 162
- 7.13 The Projection Theorem — 163
- 7.14 Summary of Chapter 7 — 167
- 7.15 Problems — 168
- 7.16 Further Reading — 169

8 **Orthonormal Bases and Fourier Series** — 171
- 8.1 Orthonormal Basis — 171
- 8.2 The Gram - Schmidt Orthonormalization — 172
- 8.3 Sample Problem I — 176
- 8.4 Sample Problem II — 178
- 8.5 Bessel's Inequality — 180
- 8.6 The Projection Decomposition — 181
- 8.7 Sample Problem III — 184
- 8.8 Complete Orthonormal Sequences — 186
- 8.9 The Fourier Series — 190
- 8.10 Properties of the Fourier Series — 191
- 8.11 Sample Problem IV — 192
- 8.12 Sample Problem V — 194

8.13	Summary of Chapter 8	196
8.14	Problems	197
8.15	Further Reading	197

9 Operator Equations 199
9.1	Inverse Mappings and Operator Equations	199
9.2	Three Generic Problems and Their Equivalence	200
9.3	Fixed-point Problem. The Contraction Mapping Theorem	201
9.4	Successive Approximations	204
9.5	Application to Differential Equations	207
9.6	Sample Problem I	212
9.7	Residual Principle and Residual Approximation	215
9.8	Incompletely Specified Equations. Least-Square Approximation	221
9.9	Problem of Uniqueness. Pseudo-Inverse Mapping	226
9.10	Sample Problem II	228
9.11	Summary of Chapter 9	230
9.12	Problems	231
9.13	Further Reading	232

10 Fourier and Laplace Transforms 235
10.1	Fourier Integral	235
10.2	The Fourier Transform as a Mapping	237
10.3	Digression. Properties of the δ-function	239
10.4	Back to Fourier Transform Mapping	241
10.5	Properties of the Fourier Transform	241
10.6	Application to Linear Systems. The Convolution Product	246
10.7	Sample Problem I: Smoothing Operator	250
10.8	Residue Theory	251
10.9	From Fourier Transform to Laplace Transform	256
10.10	Properties of the Laplace Transform	258
10.11	Basic Laplace Transforms	262
10.12	The Laplace Transform Table	264
10.13	Sample Problems II	265
10.14	Laplace Transform of the Convolution Product	267
10.15	Applications of the Laplace Transform to Solving Differential Equations	268

xviii   CONTENTS

    10.16 Sample Problem III   269
    10.17 Summary of Chapter 10   269
    10.18 Problems   270
    10.19 Further Reading   271

11 Partial Differential Equations   273
    11.1 A Short Introduction   273
    11.2 Definitions   274
    11.3 Canonical Forms   274
    11.4 The Parabolic Case   280
    11.5 Canonical Forms — A Summary   281
    11.6 Sample Problem I   282
    11.7 The Wave Equation. D'Alembert's Formula   285
    11.8 The Diffusion Equation. Transform Methods   288
    11.9 The Elliptic Case: Poisson's Equation   292
    11.10 A Very Brief Comment in Defense of the Laplace Transform   299
    11.11 Sample Problem II   300
    11.12 Summary of Chapter 11   302
    11.13 Problems   303
    11.14 Further Reading   303

Topic Index   306

1
The Basics of Set Theory

1.1 INTRODUCTION

The goal of this chapter is twofold. First, we introduce sets, their basic properties and the elementary operations on sets, as a necessary foundation of any mathematical construction and, in particular, of all those fields of modern mathematics to be considered further in this book. Second, in the course of getting acquainted with sets and related notions and concepts, we will develop specific language and begin customizing methods of proof and inference to be used throughout the book.

The concept of set is of fundamental importance in modern mathematics, all fields of which successfully exploit the language of set theory. The set-oriented approach has proved useful in computer-aided modeling and in investigating problems in engineering and science. Indeed, such inherent features of the set approach as categorizing, classification and subordination of objects are unavoidable in any problem setting. It has been even argued that all mental operations are nothing but the recognition of, and operation on, sets. What is particularly useful in this regard is the generalization of categorizing from objects to processes. Realizing that one may classify processes, it is possible to explicitly search for models and categories that exhibit similarities and to derive solutions from this identification of similarities. For instance, oscillation is a process-based category, of which there are various types, from simple to complex, but all of which have certain features in common. Thus, classifying a process as oscillation may well allow one to begin taking steps toward explicit modeling of the phenomenon.

From the standpoint of constructing mathematical concepts, set theory is significant because it reveals the role of paradoxes or contradictions in developing a consistent theory. Having started with the intuitive perception of a set as a collection of elements, set theory revealed certain paradoxes which necessitated its reformulation on an axiomatic basis. In the course of this reformulation it was shown, however, that any formal mathematical theory based on a finite number of axioms will always include assertions which cannot be either proved or disproved, that is, the cost of consistency is incompleteness, the notion having become the cornerstone of modern mathematical thinking. These profound consequences and ramifications are, however, well beyond our humble goals.

We consider in this chapter only those few features of set theory which are useful (and indispensable) in those branches of modern mathematics to be dealt with in this book. We will start with "naive" set theory, go through paradoxes to an axiomatic framework describing basic properties of sets and operations on sets, and consider pertinent examples.

1.2 LOGICAL CONNECTIVES — NOTATION

In mathematics it is particularly necessary to be very cautious concerning the statements one makes and the exact meaning of the words one uses. This is the reason for the common trend to completely eradicate words from mathematical statements so as not to be confused with possible connotations. Thus, precisely defined symbols are used instead of words. We introduce the following notations to be used as shorthand signs for English connectives. The fundamental role of these *logical connectives* in the structure of Mathematical Logic will be discussed in Chapter 3. These basic connectives are the following:

$\quad \land \quad$ and
$\quad \lor \quad$ or
$\quad \sim \quad$ not
$\quad \rightarrow \quad$ if...then
$\quad \leftrightarrow \quad$ if and only if

In addition, there are special notations for general statements, so called *quantifiers*:

$\quad \forall \quad$ for all
$\quad \exists \quad$ there exists

1.3 INTRODUCTION OF SETS

An "Intuitive" Definition of a Set
The concept of set is one of the basic concepts not only in mathematics but in general

knowledge systems as well. It relies on a basic mental operation. As usual with such fundamental concepts, it is hard (or even hardly possible) to give a satisfactory definition of set. We start with a quite naive one.

A *set* is defined as a collection of objects which can be treated as an entity. This definition implies that the objects have some classifying attributes, and all the objects in the set have the same attributes. Note also, that *object* does not necessarily mean *material* object. We may well talk about the set of transistors in a given circuit, and about the set of all operating frequencies of this circuit.

An object in such a collection is called an *element* or a *member* of the set. Sets are usually denoted with capital letters, such as A, B, and so on, and their elements with corresponding lower-case letters, a_i, b_j, and so on, i and j being identifying indices for different elements. Then, s being an element of a set S is denoted as

$$s \in S, \qquad (1.1)$$

which reads "s is an element of S or s belongs to (is contained in) S". This notation can be used in the opposite direction as well,

$$S \ni s, \qquad (1.2)$$

which means "S contains s".

If, on the contrary, an element s' does not belong to the set S (s' is not an element of S, $\sim (s' \in S)$), this fact is expressed as

$$s' \notin S. \qquad (1.3)$$

This notation, of course, also can be used in the opposite direction,

$$S \not\ni s', \qquad (1.4)$$

S does not contain s'.

The signs $\in$, $\ni$, $\notin$, and $\not\ni$ allow one to distinguish the elements of a given set from all the other objects in the universe.

1.4 SPECIFICATIONS OF SETS

As we mentioned, defining a set as a collection of its elements implies ways to specify this collection. There are two principal ways to write such a specification:
1) Explicit, by listing all the element of the set in braces: $D = \{d_1, d_2, d_3, ...\}$;
2) Implicit, by indicating the distinctive attribute of the elements belonging to the set and putting this description in braces, too.

For instance, we can specify the same set D of natural numbers less then 5 in either of these ways:

$$D = \{1, 2, 3, 4\} \qquad (1.5)$$

4 THE BASICS OF SET THEORY

or

$$D = \{d \mid d \text{ is a natural number} \wedge d < 5\}. \tag{1.6}$$

In Eq. (1.6), we denote an element of the set with some variable, d, and put the predicate description of this d after the dividing sign |.

1.5 EXAMPLES OF SETS

1) The set of all components in a given TV-set;
2) The set of TV-channels at a given location (in this regard, one can distinguish "basic set", "cable set" and "premium set");
3) The set of positions of a simple electric switch (a very simple set, containing just two elements);
4) The set of positive even integers less then 10: $\{2, 4, 6, 8\}$;
5) The set of books in a given public library: a large but finite set;
6) The set of live tigers in that public library: the *null* or *empty* set (hopefully);
7) The set of integers greater than 4: $\{5, 6, 7, 8, 9, 10, ...\}$, an infinite but *countable* set, that is, a set whose elements can be indexed by integer numbers. The ellipsis "..." in the set description stays for all remaining elements of the set;
8) The set of two elements $C = \{A, B\}$, where A is the set of all even integers and B is the set of all odd integers. In this interesting case the elements of set C are sets themselves (which is not forbidden!), so that $A \in C$; $B \in C$. It is notable that any single element of either set A or set B does not belong to the set C. For instance, the even number 2 belongs to the set A, but $2 \notin C$, because the set A participates in the set C only as a whole entity.

1.6 BASIC SETS OF NUMBERS

Among all sets of numbers, there are several basic sets, which are of special importance and which therefore deserve special attention and special notations. These special sets are the following.
1) The set of natural numbers,

$$\mathcal{N} = \{1, 2, 3, 4, ...\}. \tag{1.7}$$

2) The set of integer numbers,

$$\mathcal{I} = \{.... -4, -3, -2, -1, 0, 1, 2, 3, 4, ...\}. \tag{1.8}$$

3) The set of real numbers, $\mathcal{R}$. This is an example of an *uncountable set*, that is, the elements of which cannot be indexed with integer numbers. Thus, this set cannot be described by just listing its elements, even with ellipses, as in the two preceding

cases. Moreover, even the implicit specification of the set of real numbers, that is, the answer to the question "what is a real number", faces tremendous difficulties, which we do not intend to discuss. We merely assume that it is possible to indicate for any number that it belongs to this set:

$$1 \in \mathcal{R}; \quad \frac{2}{3} \in \mathcal{R}; \quad \sqrt{2} \in \mathcal{R}, \text{ etc.} \tag{1.9}$$

Note that, among these three special sets, each set includes all the elements of the preceding one. We will discuss this relation in more detail a bit later, see Eq. (1.19). In Chapter 4, we will add one more step to this ladder by considering *complex numbers*.

For further reference (and use), it is also worthwhile to introduce two kinds of specific uncountable subsets of the set $\mathcal{R}$, the so-called *intervals* of real numbers.

A *"closed interval"* is a subset of $\mathcal{R}$ which consists of all real numbers between two given numbers r_1 and r_2, including the numbers r_1 and r_2 themselves. The closed interval is denoted as

$$[r_1, r_2] = \{r | r \in \mathcal{R} \land r_1 \leq r \leq r_2\}. \tag{1.10}$$

An *"open interval"* is a subset of $\mathcal{R}$ which consists of all real numbers between two given numbers r_1 and r_2 but does not include the numbers r_1 and r_2 themselves, denoted as

$$(r_1, r_2) = \{r | r \in \mathcal{R} \land r_1 < r < r_2\}. \tag{1.11}$$

A hybrid of these two basic intervals is an interval between given numbers r_1 and r_2 which includes, for instance, r_1 but does not include r_2. Such monstrous interval is denoted as

$$[r_1, r_2) = \{r | r \in \mathcal{R} \land r_1 \leq r < r_2\}. \tag{1.12}$$

In practice, this notation is most often used for intervals extending to infinity, like $[r_1, \infty)$.

1.7 PARADOXES

As history has proved, having started with a *naive* definition of set as a collection of elements and going further and further into set theory, one sooner or later inevitably faces inherent and unsolvable contradictions, the so-called paradoxes of set theory. Maybe, the most famous among them is *Russel's paradox*, which naturally arises in the following way.

As already mentioned, a set can contain other sets as its elements. Now, consider the set whose elements are *by definition* all the possible sets. Then, this set must be an element of itself! This situation, though it seems weird, is by no means contradictory. However, only slightly changing the specification of the set, we get a contradiction.

6   THE BASICS OF SET THEORY

Let the set contain as its elements all the possible sets which *are not* elements of themselves. Is this new set an element of itself?

There is a simple illustration of this situation. Imagine, there is a regiment, and there is a barber in this regiment. The duty of this barber is to shave all those, and only those, soldiers who don't shave themselves. The question is whether the barber should shave himself. Both the positive and negative answer to the question are equally wrong. If the barber shaves himself, then he belongs to those people who shave themselves. Therefore, he shouldn't shave himself. On the other hand, if he does not shave himself, he belongs to those who do not shave themselves. Thus, he has to shave himself.

To avoid such unpleasant situations, rigorous foundations of set theory have been laid in the beginning of 20th century (by German mathematicians Zermelo and Fraenkel). Although we do not intend to go into all the intricacies and ramifications of set theory, we do want to be confident in the basics of the mathematical structures we will consider in the following chapters.

First of all, it is assumed that not every arbitrary collection of elements constitutes a set. A more general name for such a collection of objects is a *class*. Then, classes are of two types: a class that can be an element of a class is called a *set* and a class that cannot is called a *nonset*. All sets satisfy the six axioms we are about to introduce, and are elements of the *universal set* $\mathcal{U}$, the set containing as its elements all the elements of all the possible sets and all these possible sets as well. In this book we will consider only classes which are sets. (In these terms the solution of Russel's paradox is that the barber is a nonset from the point of view of shaving...)

1.8   AXIOMS OF SET THEORY

Although we are trying our best to avoid a very stiff formal narration, in this section we must be extremely accurate. So, the basics of set theory are to be formulated in a highly mathematized form of *axioms* (statements taken for granted), *theorems* (statements which can be proven on the basis of axioms and preceding theorems), and *corollaries* (statements being direct consequences of either a given axiom or a given theorem). To make the emerging notions and concepts more concrete, they will be illustrated on the simple example of the set of natural numbers less than 5 which was introduced in Eq. (1.5).

The six basic axioms of Set Theory are the following:

1) *The Axiom of Extension*: two sets, A and B, are equal if and only if they consist of the same elements:

$$A = B \longleftrightarrow \forall x \left((x \in A \to x \in B) \wedge (x \in B \to x \in A) \right), \qquad (1.13)$$

that is, regardless of how the set is specified, only its real content matters. In our basic example, the sets

$$D = \{1, 2, 3, 4\}, \quad D' = \{3, 2, 4, 1\}, \quad \text{and} \quad D'' = \{x \mid x \in \mathcal{N} \land x < 5\} \quad (1.14)$$

are the same,

$$D = D' = D''. \quad (1.15)$$

Moreover, if in specifying a set an element is counted twice or more, it does not change the set. Thus, the set

$$\tilde{D} = \{1, 1, 2, 3, 3, 3, 4\} \quad (1.16)$$

is still the same set D,

$$\tilde{D} = D. \quad (1.17)$$

2) *The Containment Axiom* (or Axiom of Subset Existence): If all the elements of a class A turn out to be elements of a set B, then A is also a set; it is called a subset of B. This relation is denoted as $A \subset B$,

$$A \subset B \leftrightarrow \forall a \, (a \in A \rightarrow a \in B). \quad (1.18)$$

Referring to the basic example, the set $A = \{1, 3\}$ is a subset of the example set D of Eq. (1.5), $A \subset D$. It is also instructive to illustrate the notion of subset by sets of points in a plane, as it is shown in Fig. 1.1. The area A in this figure is a part of a broader area B. Apparently, all points of the area A being elements of the set A belong to the area B, that is, to the set B, as well. Thus, $A \subset B$.

Applying the Containment Axiom to the special sets of numbers, the set of natural numbers, $\mathcal{N}$ (Eq. (1.7)), the set of integer numbers, $\mathcal{I}$ (Eq. (1.8)), and the set of real numbers, $\mathcal{R}$ (Eq. (1.9)), their relations are expressed as

$$\mathcal{N} \subset \mathcal{I} \subset \mathcal{R}, \quad (1.19)$$

the set of natural numbers is a subset of the set of integer numbers, which, in its turn, is a subset of the set of real numbers.

The Containment Axiom has two obvious consequences:

1) *Corollary*: Any set is a subset of itself,

$$A \subset A. \quad (1.20)$$

This directly follows from the definition (1.18). If $A \subset B$ and $A \neq B$, the set A is sometimes called a *proper subset* of the set B.

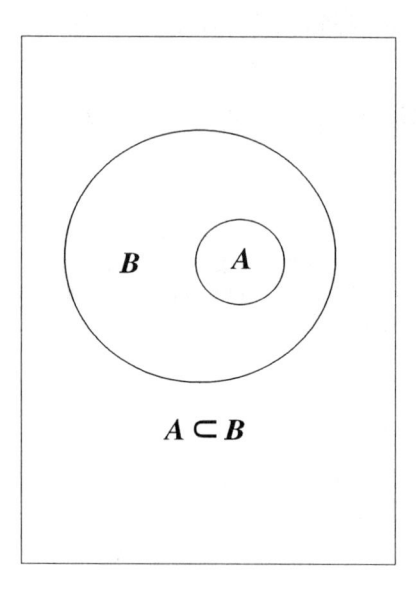

Fig. 1.1 Subset of a set of points in plane.

2) *Theorem 1.1*
The statement of two sets being equal is equivalent to the statement that *each* of these sets is a subset of the other,

$$A = B \leftrightarrow \Big((A \subset B) \wedge (B \subset A)\Big). \tag{1.21}$$

To prove this theorem, we make use of the statement of the Containment Axiom, Eq. (1.18), and rewrite the right-hand side of (1.21) as

$$\forall a\, ((a \in A \to a \in B) \wedge (a \in B \to a \in A)). \tag{1.22}$$

But this is exactly the statement of the Axiom of Extension, Eq. (1.13), establishing $A = B$. Thus, the theorem is proven.

3) *The Axiom of Union* (the union of sets is also a set): for any two sets, A, and B, the class consisting of all the elements belonging either to A or to B is also a set,

$$\forall (A, B)\, \exists\, C (x \in C \leftrightarrow x \in A \vee x \in B). \tag{1.23}$$

The set C in this expression is called *union* of the sets A and B and denoted as

$$C = A \cup B. \tag{1.24}$$

For example, let

$$A = \{1, 4\} \subset D, \quad B = \{2\} \subset D. \tag{1.25}$$

(The latter set consists of one element only). Then,

$$A \cup B = \{1, 2, 4\}. \tag{1.26}$$

In Fig. 1.2 the union of two sets is illustrated on the example of plane areas. Here, the sets A and B are sets of points in the plane which happen to have some points in common. The union set, $A \cup B$, is the area including all points of either set.

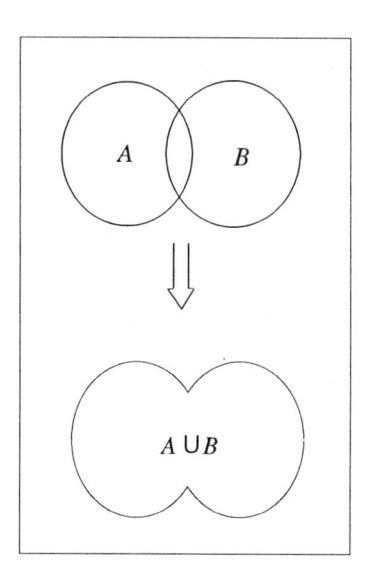

Fig. 1.2 Union of two sets.

4) *The Axiom of Intersection* (the intersection of sets is also a set): for any two sets, A and B, the class consisting of all elements belonging to *both* A and B is also a set,

$$\forall (A, B) \; \exists \, F \; (x \in F \; \leftrightarrow \; x \in A \land x \in B). \tag{1.27}$$

The set F is called the *intersection* of the sets A and B and denoted as

$$F = A \cap B. \tag{1.28}$$

For example, the sets

$$A = \{1, 4\} \subset D \quad \text{and} \quad E = \{3, 4\} \subset D \tag{1.29}$$

have their intersection as

$$\{1, 4\} \cap \{3, 4\} = \{4\}. \tag{1.30}$$

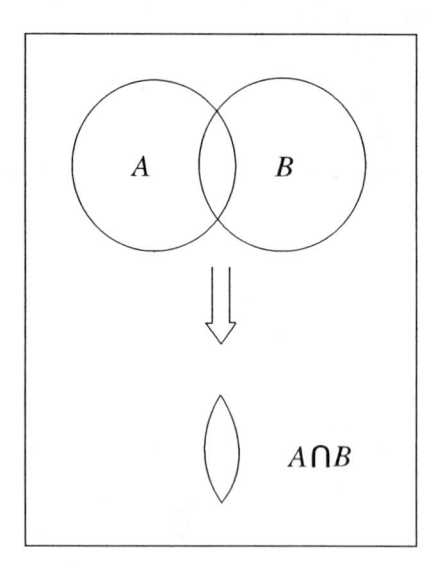

Fig. 1.3 Intersection of two sets.

In Fig. 1.3, for the same sets A and B as in Fig. 1.2, the area common to both sets is just their intersection, $A \cap B$.

Some keen person might raise a doubt at this point. Let again, as in example of Eq. (1.25), $A = \{1, 4\}$, $B = \{2\}$. These two sets evidently have no common elements! So, their intersection should be a set containing no members at all. Thus, for the sake of consistency, we must take into consideration such a special set. It is the so-called *empty set*, which is the subject of the next axiom.

5) *The Empty Set.* There exists a special set, denoted $\emptyset$ and called the *empty, null,* or *void* set, which contains no elements at all.

For the example we just considered,

$$\{1, 4\} \cap \{2\} = \emptyset. \tag{1.31}$$

(In fact, we already met what we hope is an empty set in Section 1.5: the set of live tigers in a public library.)

As we will see shortly, this set $\emptyset$ plays a crucial role in the whole structure of Set Theory. Meanwhile, we illustrate its properties by two theorems.

1) *Theorem 1.2*
The empty set is a subset of any set,

$$\forall A; \quad \emptyset \subset A. \tag{1.32}$$

The Containment Axiom, Eq. (1.18), requires that if $\emptyset \subset A$, then, for any element x,

$$x \in \emptyset \ \rightarrow \ x \in A. \tag{1.33}$$

But $\emptyset$ is just the *absence* of elements, and this absence is certainly included in any set A!

2) *Theorem 1.3*
The empty set is unique.

To prove this statement, we assume the opposite to be true and demonstrate that this assumption leads to a contradiction. Suppose there are two *different* empty sets: $\emptyset$ and $\emptyset'$. According to Theorem 1.2, the $\emptyset$ is a subset of any set; in particular,

$$\emptyset \subset \emptyset'. \tag{1.34}$$

On the other hand, as the $\emptyset'$ is also a quite legal empty set, the same theorem requires it also be a subset of any set, and, in particular,

$$\emptyset' \subset \emptyset. \tag{1.35}$$

Applying to these two statements Theorem 1.1, that is, putting them in the right-hand side of Eq. (1.21), we come to the conclusion that

$$\emptyset = \emptyset'. \tag{1.36}$$

Thus, our original assumption that the sets $\emptyset$ and $\emptyset'$ were different is wrong; they are the same set, that is, the empty set is unique.

Important notice
Though the set $\emptyset$ is empty, that is, containing no elements, it can be, in its capacity of a set, an element of some other set. For instance, the set $\{\emptyset\}$ is a set containing just one element, that is, the empty set; the set $\{\emptyset, \{\emptyset\}\}$ consists of two elements, one of which is the empty set, the other one is a one-element set.

These five axioms look very natural and almost trivial. This is so because we intuitively use them from childhood, operating with numbers in the elementary arithmetic. There is, however, an additional, more specific axiom, whose importance we will completely understand only in Chapter 3.

6) *Power Set Axiom*: For any set A, there exists a special class, the collection of all subsets of the set A. This collection is also a set; it is denoted as $\mathcal{P}(A)$ and is called *the Power Set of the set A*,

$$\forall A \, \exists \, \mathcal{P}(A) \, (\forall B \, ((B \in \mathcal{P}(A)) \leftrightarrow (B \subset A))), \tag{1.37}$$

that is, a set B is an element of the power set $\mathcal{P}(A)$ if and only it is a subset of the set A itself. In particular, for any set A, the empty set $\emptyset$ and the set A itself are evidently elements of the power set $\mathcal{P}(A)$.

For example, we construct explicitly the power set for the example set of Eq. (1.5), $D = \{1, 2, 3, 4\}$. This power set has the form:

$$\mathcal{P}(D) = \{\emptyset, \{1\}, \{2\}, \{3\}, \{4\}, \{1,2\}, \{1,3\}, \{1,4\}, \{2,3\}, \{2,4\}, \{3,4\},$$
$$\{2,3,4\}, \{1,3,4\}, \{1,2,4\}, \{1,2,3\}, \{1,2,3,4\}\} \tag{1.38}$$

(note that the set $\emptyset$ proudly leads this chain, and the set D itself is the last element in the chain). Thus, for the original set D of four elements, the power set $\mathcal{P}(D)$ consists of sixteen elements.

This observation can be generalized. What we intend to do here is to give an answer to the question: if a set A consists of n elements, how many elements are in its power set $\mathcal{P}(A)$?

To find this answer, we critically review the way in which we constructed the set $\mathcal{P}(D)$ in Eq. (1.38). In fact, we just listed the sets being all the possible *combinations* of $m = 0$, $m = 1$, $m = 2$, $m = 3$, and $m = 4$ elements out of the given $n = 4$ elements of the original set D. The numbers of these combinations are given by the binomial coefficients,

$$C_4^0 = 1, \quad C_4^1 = 4, \quad C_4^2 = 6, \quad C_4^3 = 4, \text{ and } C_4^4 = 1, \tag{1.39}$$

where

$$C_n^m = \frac{n!}{m!(n-m)!} \tag{1.40}$$

is the number of *combinations* of m elements out of n elements. Thus, in the general case of an n-element set A, the list of all combinations of m elements out of n for all $m \leq n$ will contain

$$\sum_{m=0}^{n} C_n^m \tag{1.41}$$

elements. Comparing this expression with the Newton's binomial formula,

$$(x+y)^n = \sum_{m=0}^{n} C_n^m x^m y^{n-m}, \tag{1.42}$$

we see that the expression (1.41) coincides with the right-hand side of the formula (1.42) at $x = y = 1$. Thus,

$$\sum_{m=0}^{n} C_n^m = (1+1)^n = 2^n. \tag{1.43}$$

This simple approach, however, works for finite sets only. It is quite different matter to deal with the power set of an infinite set, let alone *uncountable* sets. These much more complicated subjects originated the whole branch in the Set Theory, which (fortunately) lies quite beyond our scope in this book.

The six axioms lay a firm foundation to the concept of a set. By adhering to them, we will not face any paradoxes or other unpleasant situations. All sets satisfy these axioms and can be considered as elements of some general "universal" set $\mathcal{U}$. (To tell the whole truth, this optimistic conclusion is not completely right. As was shown, some paradoxes are unavoidable in any formal mathematical system. The real advantage of the axiomatization is that paradoxes are pushed quite far away, almost to an infinite distance, from those fields of mathematics we will deal with.) We can also consider the formulation of the set axioms from a different angle, that of setting up a system of *operations* on, or transformations of sets. This line of reasoning will be continued to much greater extent in the next chapter.

1.9 SAMPLE PROBLEMS

I. List the elements of the following sets:
1) $A = \{x \mid x \text{ is an integer number} \land 5 < x < 12\}$;
2) $B = \{x \mid x \text{ is a decimal digit}\}$;
3) $C = \{x \mid x = 3 \land x = 8\}$;
4) $D = \{x \mid x = 2 \lor x = 5\}$.

Solutions
1) $A = \{6, 7, 8, 9, 10, 11\}$;
2) $B = \{0, 1, 2, 3, 4, 5, 6, 7, 8, 9\}$;
3) $C = \emptyset$;
4) $D = \{2, 5\}$.

II. Find the power sets $\mathcal{P}(A \cap B)$, $\mathcal{P}(C \cup D)$, and $\mathcal{P}(C \cap D)$, where the sets A, B, C, and D are specified as follows:

$$A = \{n \mid\mid n \text{ is a natural number} \land n^2 + 20 = 9n\}; \qquad (1.44)$$

$$B = \{x \mid\mid x \text{ is a real number} \land \cos(\pi x) = 1\}; \qquad (1.45)$$

$$C = \{\text{vowels in the word "grade"}\}; \qquad (1.46)$$

$$D = \mathcal{P}(\{1\}). \qquad (1.47)$$

Solutions
First, we specify explicitly all the included sets:
The elements of the set A are determined by the condition:

$$n^2 + 20 = 9n. \qquad (1.48)$$

This is square equation for n,

$$n^2 - 9n + 20 = 0, \qquad (1.49)$$

and its two roots are the following:

$$n_{1,2} = 9/2 \pm \sqrt{81/4 - 20} = \frac{9 \pm 1}{2}, \qquad (1.50)$$

14 THE BASICS OF SET THEORY

that is,

$$n_1 = 5; \quad n_2 = 4. \quad (1.51)$$

Therefore, the set A is explicitly

$$A = \{4, 5\}. \quad (1.52)$$

The set B is specified by the equation:

$$\cos(\pi x) = 1, \quad (1.53)$$

which has the following solution:

$$x = 2k, \quad (1.54)$$

where k is an integer. Thus, x is any even integer number, and the set B is explicitly specified as

$$B = \{0, \pm 2, \pm 4, \pm 6, \pm 8...\}. \quad (1.55)$$

The word "grade" contains two vowels, "a" and "e". Thus, the explicit specification of the set C is

$$C = \{a, e\}. \quad (1.56)$$

The power set $\mathcal{P}(\{1\}) = \{\emptyset, \{1\}\}$, it contains $2^1 = 2$ elements. So, the set D is a set whose elements are sets,

$$D = \{\emptyset, \{1\}\}. \quad (1.57)$$

Now, using the explicit specifications of Eqs. (1.52), (1.55), (1.56), and (1.57), we find the unions and intersections,

$$A \cap B = \{4, 5\} \cap \{0, \pm 2, \pm 4, \pm 6, ...\} = \{4\}, \quad (1.58)$$

$$C \cup D = \{a, e\} \cup \{\emptyset, \{1\}\} = \{\emptyset, a, e, \{1\}\}, \quad (1.59)$$

and

$$C \cap D = \{a, e\} \cap \{\emptyset, \{1\}\} = \emptyset. \quad (1.60)$$

Finally, proceeding as in the example for Power Set Axiom (Eq. (1.19)), we find all the required power sets:

$$P((C \cap D)) = P(\emptyset) = \{\emptyset\}; \quad (1.61)$$

it contains $2^0 = 1$ element.

$$P(A \cap B) = P(\{4\}) = \{\emptyset, \{4\}\}; \quad (1.62)$$

it contains $2^1 = 2$ elements.

$$P(C \cup D) = P(\{\emptyset, a, e, \{1\}\})$$
$$= \{\emptyset, \{\emptyset\}, \{a\}, \{e\}, \{\{1\}\}, \{\emptyset, a\}, \{\emptyset, e\}, \{\emptyset, \{1\}\}, \{a, e\}, \{a, \{1\}\},$$
$$\{e, \{1\}\}, \{a, e, \{1\}\}, \{\emptyset, e, \{1\}\}, \{\emptyset, a, \{1\}\},$$
$$\{\emptyset, a, e\}, \{\emptyset, a, e, \{1\}\}\};\qquad(1.63)$$

it contains $2^4 = 16$ elements, as required by formula (1.43).

1.10 SUMMARY OF CHAPTER 1

1) A set is a collection of elements which can be treated as an entity.
2) The two basic methods of set specification are the following:
    a) explicit specification, by listing elements of the set (with the use of an ellipsis, "...", to indicate the omitted elements, when it is necessary or convenient);
    b) implicit specification, by indicating classifying attributes of the elements of the given set.
3) Special sets of numbers: the set of natural numbers, $\mathcal{N}$, the set of integer numbers, $\mathcal{I}$, and the set of real numbers, $\mathcal{R}$.
4) Perceiving any collection of elements as a set leads to Paradoxes of Set Theory and necessitates a more rigorous axiomatic approach.
5) The six Axioms of Set Theory are:
    a) The Axiom of Extension,
    b) The Containment Axiom,
    c) The Axiom of Union,
    d) The Axiom of Intersection,
    e) The Axiom of Empty Set,
    f) The Power Set Axiom.
They describe the basic properties of sets, operations on sets, and relations between sets. Adhering to these axioms effectively eliminates paradoxes and provides consistency of the theory.

1.11 PROBLEMS

I. Consider the following two sets:
1) $A = \{x \mid (x \in \mathcal{I}) \wedge (x^4 + 2 = 4.5x^2)\}$;
2) $B = \{y, z, u, v\}$, where the *real* numbers y, z, u, and v satisfy the following

16 THE BASICS OF SET THEORY

system of linear equations:
$$\begin{cases} y + 2z + 2u + v = 15 \\ 2y + 5z + 6u + 3v = 42 \\ 3y + 6z + 7u + 5v = 56 \\ 4y + 8z + 7u + 3v = 53 \end{cases}$$

Find $A \cup B$ and $A \cap B$.

II. Consider the following two sets:
1) $C = \{f(\pi/2), f(\pi)\}$, where the function $f(x)$ of a real variable x is the solution to the following differential equation and initial conditions:
$$\frac{d^2 f}{dx^2} + 4f = 0; \quad f(0) = 1; \quad \left.\frac{df}{dx}\right|_{x=0} = 0.$$

2) $D = \{g(\pi/2), g(\pi)\}$, where the function $g(x)$ is the solution to the following differential equation and initial conditions:
$$\frac{d^2 g}{dx^2} - 4g = 0; \quad g(0) = 0; \quad \left.\frac{dg}{dx}\right|_{x=0} = 1.$$

Find the power sets $\mathcal{P}(C \cup D)$ and $\mathcal{P}(C \cap D)$.

III. Consider the following sets:
1) $A = \{3, 4, 5\}$;
2) $B = \{n \mid n \text{ is a natural number} \wedge n^2 + 30 = 11n\}$;
3) $C = \{x \mid x \text{ is a real number} \wedge \sin \pi x = 0\}$;
4) $D = \{\text{vowels in the word "real"}\}$;
5) $E = \{n \mid n \text{ is a natural number} \wedge n^2 - 2n = 2\}$.

Find the power sets $\mathcal{P}(A \cup B), \mathcal{P}(B \cap A), \mathcal{P}(A \cap E), \mathcal{P}(D \cup B), \mathcal{P}(B \cap C), \mathcal{P}(D \cup E), \mathcal{P}(A \cap C)$, and $\mathcal{P}(D \cap E)$. Are some of these sets equal?

1.12 FURTHER READING

More detailed consideration of the basics of set theory, given in the same vein as in this chapter, can be found in the following books for students in mathematics:

P. R. Halmos, *Naive Set Theory*, Van Nostrand, New York, 1960;

J. D. Monk, *Introduction to Set Theory*, McGraw-Hill, New York, 1969;

K. Devlin, *The Joy of Sets: Fundamentals of Contemporary Set Theory*, 2nd ed., Springer-Verlag, New York, 1995.

The first is a celebrated introduction to the subject, it is well written and containing a good dose of humor; the two others contain a more formal axiomatic treatments of set theory.

For a deeper understanding and special topics in set theory the following books are recommended:

P. C. Suppes, *Axiomatic Set Theory*, Dover, New York, 1972.

Though being a graduate course for mathematics majors, it is written in an informal style and avoids excessive symbolism. These features make it attractive and useful for a "casual" reader.

R. L. Wilder, *Introduction to the Foundations of Mathematics* (2nd ed.), Wiley, New York, 1965.

An exhaustive discussion of set theory paradoxes and their role in the development of an axiomatic theory.

F. R. Drake, D. Singh, *Intermediate Set Theory*, Wiley, Chichester, 1996.

An intermediate-level course, mostly devoted to tracing the implications and applications of set theory in abstract mathematics.

D. Van Dale, H. C. Doets, and H. De Swart, *Sets: Naive, Axiomatic, and Applied*, Pergamon, Oxford, 1978.

A book which emphasizes a number of topics illustrating the actual use of set theory in "everyday" mathematics.

Those, who have become seriously interested in the basics of set theory, or are just curious as to how much can be derived from those basics, are referred to, for instance,

A. Levy, *Basic Set Theory*, Springer-Verlag, Berlin, 1979.

This is very advanced and detailed treatment of the axiomatic foundations.

A. S. Kechris, *Classical Descriptive Set Theory*, Springer-Verlag, New York, 1995.

This is also a very advanced treatment stressing applications of set theory in the foundations of various fields of modern mathematics.

2
Relations and Mappings

2.1 RELATIONS – THE NAIVE APPROACH

In Chapter 1 we have set out the basics of a consistent set theory and introduced two elementary operations on sets: union and intersection. Sets, as collections of elements, provide the raw material upon which a mathematical theory can be built. If these elements are supposed to constitute not only a loose collection but some structure (which is always true as long as the theory somehow represents the real world), this structure should be characterized by the *relations* between the elements. Relations are of fundamental importance in both theory and applications (particularly, in computer applications, which juxtapose various data arrays). As is the case with all mathematical concepts, the mathematical concept of relation is based on the common notion of relationship between objects in the real world. It generalizes and purifies this notion, making it well defined and unambiguous. Consider first some examples of relations that we can use without particular regard to their deep meaning.

Examples of relations:
1) We can compare the elements of a set by certain of their characteristics. In our example set $D = \{1, 2, 3, 4\}$ of Eq. (1.5) we can compare the numbers by their values, for instance, in relations $1 < 3$, $2 < 4$ and so on.

2) We can combine objects of different sets. For example, we combine the elements of sets "colors" V_1 and "animals" V_2:

$$V_1 = \begin{pmatrix} white \\ black \\ grey \\ red \end{pmatrix} ; \quad V_2 = \begin{pmatrix} cat \\ dog \\ horse \\ cow \end{pmatrix}, \quad (2.1)$$

to get relations of the kind "cat is white", "dog is black", and so on.

3) A relation may connect more than two objects. We can formulate "three-object" relations on the same set $D = \{1, 2, 3, 4\}$ in the form of statements $1 + 3 = 4$, $2 + 1 = 3$, and so on.

4) Many-object relations can involve objects of different sets as well. We add to sets V_1 and V_2 of example 2 (Eq. (2.1)) the set of attitudes,

$$V_3 = \begin{pmatrix} hate \\ like \\ disgust \\ \cdot \\ \cdot \\ \cdot \end{pmatrix}, \quad (2.2)$$

and construct such five-object relations as "grey dog hates white cat", which involves elements of all three sets.

In principle, we can construct a relationship which holds for n objects. Note, that when making an assertion that a relationship holds among these n objects, it is in general necessary to specify not only the objects, but also an *ordering* of them. For instance, in Example 1) the assertion $2 < 4$ is as possible as $4 < 2$, but the first assertion is true, the other one is false. In Example 2) "cat is red" is unusual, but "red is cat" is just meaningless! Thus, the order of the elements is an essential feature of a relation, and this feature should be properly represented in a mathematically correct description that we are about to set out.

2.2 ORDERED N-TUPLE

The examples just considered lead us to the mathematical concept of "ordered n-tuple" or "ordered n-element tuple" as a specific structure of n elements in which the order of the elements matters. We define such an n-tuple as a sequence of n objects a_j and denote it as $(a_1, a_2, a_3, ..., a_n)$. Here we use parentheses instead of braces to emphasize the difference between ordered n-tuples and sets. The objects a_j are called components of the n-tuple. In the general case they may belong to different sets: $a_1 \in V_1; a_2 \in V_2; a_3 \in V_3; ...; a_n \in V_n$. However, they also may be taken from the same set, or even be the same element of a set. It may happen, for instance, that $V_1 = V_2$ or even $a_1 = a_2$.

In contrast to sets, a repetition of the same element in the specification of an n-tuple makes sense, because in an n-tuple not only the *presence* but also the *position* of an element is significant. Two n-tuples are equal if and only if they contain the same elements in the same order:

$$(a_1, a_2, a_3, ..., a_n) = (b_1, b_2, b_3, ..., b_n) \leftrightarrow a_1 = b_1 \wedge a_2 = b_2 \wedge ... \wedge a_n = b_n. \quad (2.3)$$

For example, the pairs $(1, 3)$, $(2, 4)$, or (cat, white) are simple two-tuples, $(1, 3, 4)$ or $(3, 2, 2)$ are three-tuples. Note, however, that the components of an n-tuple might be quite different in nature. For instance, $(1, 3, \text{cat})$, or even (nightingale, $\emptyset$, bomber) are still legal three-tuples.

Note also that if all the components of an n-tuple are real numbers, we can represent this n-tuple as a point in the n-dimensional space with the cartesian coordinates being the components of this n-tuple. For example, the two-tuple $(1, 3)$ is represented by the point in (x, y)-plane with coordinates $x = 1$, $y = 3$.

We emphasize once more that the essential feature distinguishing an n-tuple from a set is the ordering of the elements: $(1, 3)$ and $(3, 1)$ are different two-tuples, while $\{1, 3\}$ is the same set as $\{3, 1\}$.

2.3 AN N-TUPLE VERSUS A SET

According to its definition, an n-tuple seems to be very different from a set (due to ordering). Moreover, we have pointed out that this difference is a salient feature. If so, it should be possible to develop a new self-consistent and non-contradictory system of axioms being able to handle "n-tuple theory". This sounds disappointing, since we have just put so much time and effort into establishing a system of axioms of *set* theory. The fact is, however, that new complications will not arise, because an n-tuple can be completely represented by some specific (actually, *very* specific) set. We prove this important statement by constructing this specific set explicitly.

Given an n-tuple $(a_1, a_2, a_3, ..., a_n)$, we consequently construct the following sets having the components of the n-tuple as their elements:

The one-element set consisting of the first component of the n-tuple — $\{a_1\}$;

The two-element set consisting of the first two components of the n-tuple — $\{a_1, a_2\}$;

The three-element set consisting of the first three components of the n-tuple — $\{a_1, a_2, a_3\}$ ...

and so on, up to the last (n-th) n-element set, $\{a_1, a_2, a_3, ..., a_n\}$, including as its elements all the components of the n-tuple.

Each of these constructed sets, by definition, has no ordering of its elements. But putting all these sets together, the order of the components of the initial n-tuple is preserved. It is clear that in this collection of n sets having different number of elements there is only one one-element set, and the single element of this set is exactly the first component of the n-tuple. Then, there is only one two-element set, and this set contains the first and the second components of the n-tuple. The first component we know already from the one-element set; thus, the two-element set provides us with the second component of the n-tuple. Going further in the same manner, we restore from these n sets the order of the n components of the initial n-tuple. Thus, the "superset"

$$\{\{a_1\}, \{a_1, a_2\}, \{a_1, a_2, a_3\}, ..., \{a_1, a_2, a_3, ..., a_n\}\} \tag{2.4}$$

containing n element-sets represents the initial n-tuple in full, with all its components and their ordering.

The bottom line is that we may treat n-tuples as sets and apply to them the axioms and all the concepts and constructions of set theory.

Examples of the set representation of n-tuples
Let $a_1 = 1, a_2 = 2, a_3 = 3$. Two different three-tuples, $(1, 2, 3)$ and $(3, 2, 1)$, are constructed from these elements. The first 3-tuple is represented by set $\{\{1\}, \{1, 2\}, \{1, 2, 3\}\}$, the second one is represented by set $\{\{3\}, \{3, 2\}, \{3, 2, 1\}\}$. The sets are clearly different; this corresponds to different ordering in the original three-tuples.

Another example is the four-tuple $D = (1, 2, 3, 4)$, constructed on the basis of our sample set $D = \{1, 2, 3, 4\}$, Eq. (1.5). The set representation of this four-tuple is

$$D = \{\{1\}, \{1, 2\}, \{2, 3, 1\}, \{2, 3, 1, 4\}\}, \tag{2.5}$$

where we intentionally reshuffled the elements of the included sets to stress the insignificance of their ordering *inside* the sets.

The four-tuple $D' = (1, 3, 2, 4)$ differs from D in the positions of the elements 3 and 2, and this alteration immediately manifests itself in the set representation of this four-tuple,

$$D' = \{\{1\}, \{3, 1\}, \{2, 1, 3\}, \{1, 2, 3, 4\}\}. \tag{2.6}$$

2.4 CROSS-PRODUCT SET AND ITS PROPERTIES

Consider now several sets, $V_1, V_2, V_3, ..., V_n$. Let us construct all the possible n-tuples $(v_1, v_2, v_3, ..., v_n)$, in which the first component is one of the elements of the set V_1, $v_1 \in V_1$, the second component is one of the elements of the set V_2, $v_2 \in V_2$, the third component is one of the elements of the set V_3, $v_3 \in V_3$, and so on, up to the last, the n-th component being one of the elements of the set V_n, $v_n \in V_n$. Let us then carefully collect all these n-tuples and consider this collection as a new set. This set of n-tuples is called a *product* (or *cartesian product*, or *cross product*) of the sets $V_1, V_2, V_3, V_4, ... V_n$. It is denoted as $V_1 \times V_2 \times V_3 \times V_4 \times ... \times V_n$. In particular, if all the components of n-tuples are taken from the same set V, the product set,

$$V \times V \times V \times V \times ... \times V, \tag{2.7}$$

is often denoted as V^n

If each of the sets $V_1, V_2, V_3, V_4, \ldots V_n$ contains a finite number of elements, say, m_1 elements in the set V_1, m_2 elements in the set V_2, m_3 elements in the set $V_3, \ldots, m_n$ elements in the set V_n, then the number of elements of the product set, $V_1 \times V_2 \times V_3 \times V_4 \times \ldots \times V_n$, is equal to the *product* of all these numbers, $m_1 \cdot m_2 \cdot m_3 \cdots m_n$.

Example of a product set.
Let
$$V_1 = \{1, 2\}; \quad V_2 = \{3, 4\}; \quad V_3 = \{0\}; \quad V_4 = \emptyset. \tag{2.8}$$
The cross-product set $V_1 \times V_2$ contains $2 \times 2 = 4$ elements:
$$V_1 \times V_2 = \{(1,3), (1,4), (2,3), (2,4)\}. \tag{2.9}$$

Set $V_1 \times V_3$ should contain $2 \times 1 = 2$ elements, and this set is
$$V_1 \times V_3 = \{(1,0), (2,0)\}. \tag{2.10}$$

It is important to note that
$$V_1 \times V_4 = V_2 \times V_4 = V_3 \times V_4 = \emptyset. \tag{2.11}$$

Indeed, an element of the set $V_1 \times V_4$ should have been a 2-tuple (v_1, v_4) with its second component taken from the elements of the set V_4. But there are no elements in the set V_4! Thus, there are no such 2-tuples, and the set $V_1 \times V_4$ is *empty*.

Generalizing this statement, if among several sets at least one is the empty set, then the cross product of these sets is also the empty set. (This is quite consistent with the previous statement that the number of elements in the cross-product set is the product of the numbers of elements in each set: if one of the set contains *zero* elements, then the cross product set also must contain *zero* elements).

If V_1 and V_2 are sets of *real* numbers, their cartesian product $V_1 \times V_2$ can be represented as a set of points in the cartesian plane. For instance, the set $V_1 \times V_2$ of the previous example, Eq. (2.9), is represented in Fig. 2.1.

The cartesian product of the same two sets taken in inverse order, the set
$$V_2 \times V_1 = \{(3,1), (4,1), (3,2)(4,2)\} \tag{2.12}$$

is represented in Fig. 2.2.

The obvious difference between these figures clearly shows that the cross-product operation, $\times$, is generally not commutative, that is,
$$V_1 \times V_2 \neq V_2 \times V_1. \tag{2.13}$$

The cartesian-plane diagrams can also be used to easily visualize cross-products of *infinite* sets such as *intervals* of real numbers. These cross-products are just areas in the plane. For instance, Fig. 2.3 shows $V_{13} \times V_{24}$ and $V_{24} \times V_{13}$, where $V_{13} = [1, 3]$

24 RELATIONS AND MAPPINGS

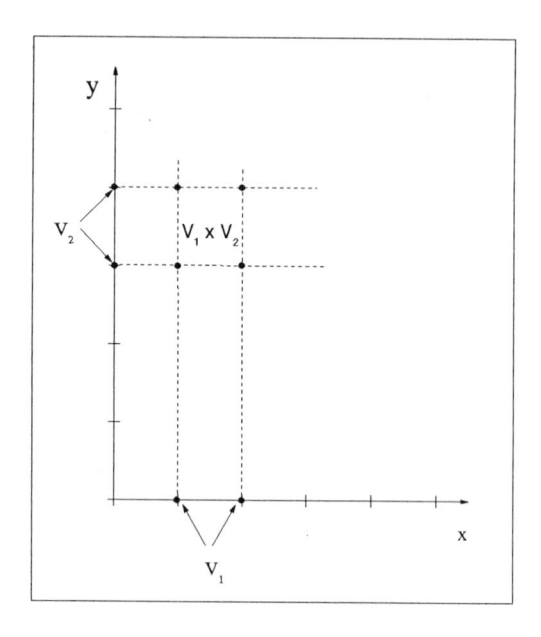

Fig. 2.1 Cross-product of two sets of real numbers: $V_1 = \{1, 2\}$, $V_2 = \{3, 4\}$; $V_1 \times V_2 = \{(1, 3), (1, 4), (2, 3), (2, 4)\}$.

(that is, the closed interval of real numbers between 1 and 3, see Section 2.6) and $V_{24} = [2, 4]$.

As two further important properties of the cross-product set, we mention (without proof) that the operation of cartesian product distributes over operations of union and intersection of sets, introduced in Chapter 1 (Section 1.8):

$$A \times (B \cup C) = (A \times B) \cup (A \times C), \tag{2.14}$$
$$A \times (B \cap C) = (A \times B) \cap (A \times C); \tag{2.15}$$

and

$$(B \cup C) \times A = (B \times A) \cup (C \times A), \tag{2.16}$$
$$(B \cap C) \times A = (B \times A) \cap (C \times A). \tag{2.17}$$

It is easy to illustrate these relations on the cartesian plane. Let the participating sets in relations (2.14) – (2.16) be closed intervals of real numbers,

$$A = [1, 2], \quad B = [1, 3], \quad C = [2, 4]. \tag{2.18}$$

Then,

$$B \cup C = [1, 4], \quad B \cap C = [2, 3], \tag{2.19}$$

and we see in Fig. 2.4 that indeed relations (2.14) – (2.16) hold.

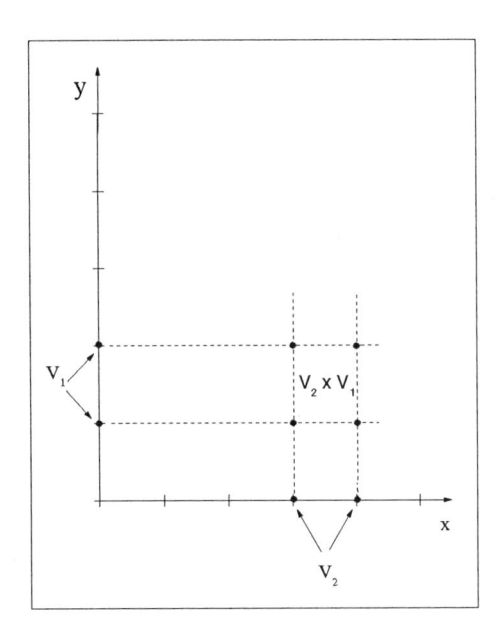

Fig. 2.2 Cross product of the same two sets as in Fig. 2.1 but taken in inverse order: $V_2 \times V_1 = \{(3,1), (3,2), (4,1), (4,2)\}$.

2.5 SAMPLE PROBLEM I

Consider again the sets of Problem II, Chapter 1:
1) $A = \{n \mid n \text{ is a natural number} \wedge n^2 + 20 = 9n\}$;
2) $B = \{x \mid x \text{ is a real number} \wedge \cos(\pi x) = 1\}$;
3) $C = \{\text{vowels in the word "grade"}\}$;
4) $D = P(\{1\})$.

Find the power sets $\mathcal{P}((A \cap B) \times C)$, $\mathcal{P}((A \cap B) \times (C \cup D))$, and $\mathcal{P}((A \cup B) \times (C \cap D))$.

Solution
(1) We take the explicit specifications of all the required sets from the solution to Problem II given in Chapter 1, and write down all the required unions and intersections:

$$A \cap B = \{4, 5\} \cap \{0, \pm 2, \pm 4, \pm 6, ...\} = \{4\};$$
$$C \cup D = \{a, e\} \cup \{\emptyset, \{1\}\} = \{\emptyset, a, e, \{1\}\};$$
$$C \cap D = \{a, e\} \cap \{\emptyset, \{1\}\} = \emptyset;$$
$$A \cup B = \{4, 5\} \cup \{0, \pm 2, \pm 4, \pm 6, ...\}$$
$$= \{0, \pm 2, \pm 4, 5, \pm 6, \pm 8....\}. \tag{2.20}$$

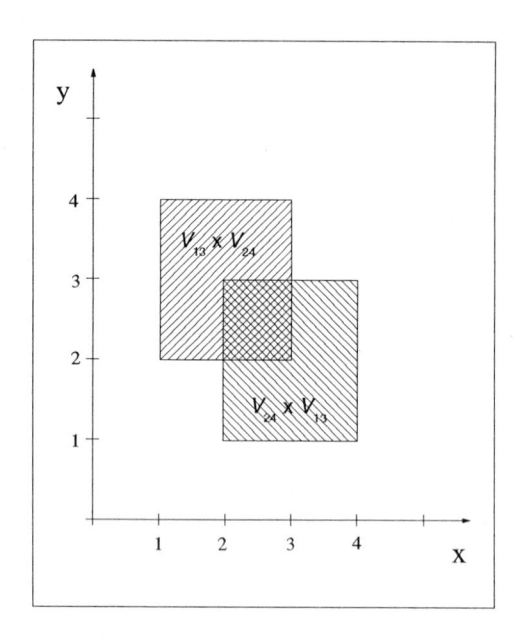

Fig. 2.3 Cross products of the closed intervals of real numbers, $V_{13} = [1, 3]$ and $V_{24} = [2, 4]$.

(2) We write explicitly all the cartesian product sets:

$$(A \cap B) \times C = \{4\} \times \{a, e\} = \{(4, a), (4, e)\};$$
$$(A \cap B) \times (C \cup D) = \{4\} \times \{\emptyset, a, e, \{1\}\}$$
$$= \{(4, \emptyset), (4, a), (4, e), (4, \{1\})\};$$
$$(A \cup B) \times (C \cap D) = (A \cup B) \times \emptyset = \emptyset. \tag{2.21}$$

(3) Now we can find all the required power sets:

$$P((A \cup B) \times (C \cap D)) = P(\emptyset) = \{\emptyset\}, \tag{2.22}$$

it contains $2^0 = 1$ element.

$$P((A \cap B) \times C) = P(\{(4, a), (4, e)\})$$
$$= \{\emptyset, \{(4, a)\}, \{(4, e)\}, \{(4, a), (4, e)\}\}, \tag{2.23}$$

it contains $2^2 = 4$ elements.

$$P((A \cap B) \times (C \cup D)) = P(\{(4, \emptyset), (4, a), (4, e), (4, \{1\})\}$$
$$= \{\emptyset, \{(4, \emptyset)\}, \{(4, a)\}, \{(4, e)\}, \{(4, \{1\})\}, \{(4, \emptyset), (4, a)\}, \{(4, \emptyset), (4, e)\},$$
$$\{(4, \emptyset), (4, \{1\})\}, \{(4, a), (4, e)\}, \{(4, a), (4, \{1\})\},$$

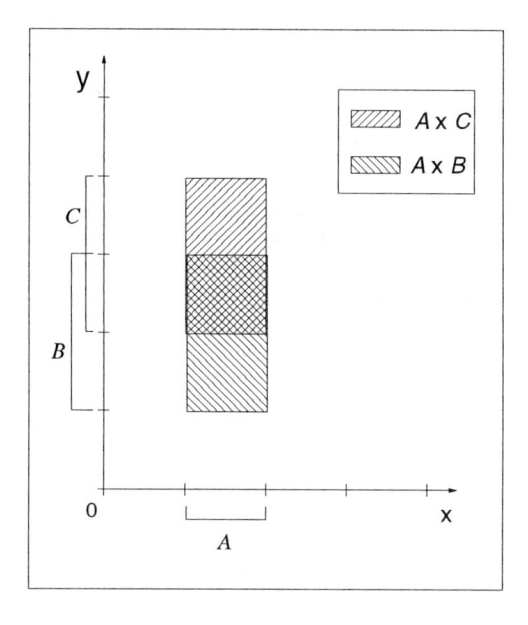

Fig. 2.4 Distribution of cross product and operations of union and intersection of sets on an example of intervals of real numbers: $A = [1, 2]$, $B = [1, 3]$, $C = [2, 4]$; $B \cup C = [1, 4]$, $B \cap C = [2, 3]$.

$$\{(4, e), (4, \{1\})\}, \{(4, a), (4, e), (4, \{1\})\},$$
$$\{(4, \emptyset), (4, e), (4, \{1\})\}, \{(4, \emptyset), (4, a), (4, \{1\})\},$$
$$\{(4, \emptyset), (4, a), (4, e)\}, \{(4, \emptyset), (4, a), (4, e), (4, \{1\})\}\}, \quad (2.24)$$

it contains $2^4 = 16$ elements, just as expected.

2.6 *N*-ARY RELATION

Now we are in a position to give a mathematically rigorous definition of an *n-ary relation* in terms of an *n*-tuple.

Definition. A subset of a given cross-product set $V_1 \times V_2 \times V_3 \times V_4 \times \ldots \times V_n$,

$$R \subseteq V_1 \times V_2 \times V_3 \times V_4 \times \ldots \times V_n, \quad (2.25)$$

is called an *n-ary relation in the given cross-product set*.

This definition means that there is some criterion or rule for selecting elements of the subset R out of all the elements of the cross-product set $V_1 \times V_2 \times \ldots \times V_n$. However, once the selection has been completed and the subset R explicitly defined,

28 RELATIONS AND MAPPINGS

the selecting rule can be put aside.

Note on terminology
Here, the notation of the subset R relates to the word *Relation*. This R should not be confused with the set of real numbers of Chapter 1 (Section 1.6), which is denoted $\mathcal{R}$.

Then, there are special terms for special kinds of relations:
If the subset $R = \emptyset$, it is called the *empty (or void) relation*;
If $R = V_1 \times V_2 \times V_3 \times V_4 \times \ldots \times V_n$, that is, it includes all the n-tuples of the product set, it is called the *universal* relation;
If the cross-product set is a set of *one-tuples*, that is, just elements, its subset $R \subseteq V$ is called a *unary relation*;
If the cross-product set is a set of two-tuples, the subset $R \subseteq V_1 \times V_2$ is called a *binary relation*;
In the same way one can introduce *ternary relation, quaternary relation,* and so on.

Equality of relations.
Two relations,
$R_1 \subseteq V_1 \times V_2 \times V_3 \times V_4 \times \ldots \times V_n$
and
$R_2 \subseteq W_1 \times W_2 \times W_3 \times W_4 \times \ldots \times W_m$,
are equal, if (and only if) the following conditions are met:
a) $n = m$;
b) $V_1 = W_1, V_2 = W_2, V_3 = W_3, \ldots, V_n = W_n$;
c) The sets R_1 and R_2 are equal in the sense of Section 2.8, that is, they consist of the same elements.

Note, that satisfying the "natural" condition c) is not sufficient for the equality of the two relations. For example, in the simplest case of unary relations, the two relations expressed by the conditions "John is the bravest man in this room" and "John is the bravest man in the world" are definitely not the same, though they refer to the same person. We will also clearly see this insufficiency in the following example.

Example of Different Relations:
Let
$$V_1 = \{1, 2, 3, 4\}, \quad V_2 = V_1; \tag{2.26}$$
Consider the binary relation in the cross-product set $V_1 \times V_2 = V_1^2$, defined as
$$R = \{(v_1, v_2) || v_1 \in V_1 \land v_2 \in V_2 \land v_1 < v_2\}. \tag{2.27}$$
The explicit specification of set R is
$$R = \{(1,2), (1,3), (1,4), (2,3), (2,4), (3,4)\}. \tag{2.28}$$
Then, if we take the sets
$$V_1 = \{1, 2, 3\} \text{ and } V_2 = \{1, 2, 3, 4\} \tag{2.29}$$

and consider the binary relation in the cross-product set $\tilde{R} \subset V_1 \times V_2$, defined by the same conditions of Eq. (2.27), we obtain, as the explicit specification of the set $\tilde{R}$, the same set R of Eq. (2.28). Despite this, these two relations are not the same, because the original cross-product sets $V_1 \times V_2$ are different!

2.7 BINARY RELATIONS — TERMINOLOGY

Among various kinds of relations, binary relations are of special importance both in theory and applications. It is not surprising that they have their own extensive terminology. At this stage we merely introduce all the corresponding notions and notations, which we will extensively use throughout this book.

For a binary relation B in a cross-product set $V \times W$,

$$B \subseteq V \times W, \qquad (2.30)$$

the set V is called the *domain* and the set W the *codomain* of the relation B. Then, for any two-tuple $(v, w) \in B$, its second component, w, is called the *value* of the first component, v, under the mapping B. To emphasize w being the value of v, sometimes it is denoted as $b(v)$ where small b corresponds to capital B of the relation.

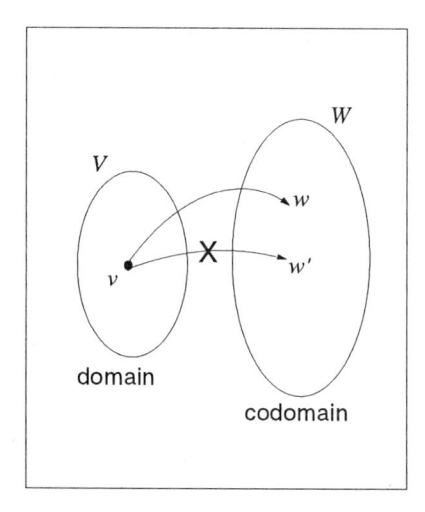

Fig. 2.5 Due to the definition of mapping, it is forbidden for an element of the domain to have two different values in the codomain.

The general concept of binary relation does not impose any restrictions on the content of the set B. So, in principle, any element of the domain might meet any element of the codomain in a two-tuple belonging to some binary relation. There is, however, a class of binary relations which is so important for the real world that it is given a special name. Namely, if each element v of the domain corresponds to a unique value w in the codomain (see Fig. 2.5), the binary relation $B \subseteq V \times W$ is called a *mapping* from the set V to the set W. As it will be shown in this book, most subjects of modern mathematics can be understood as describing properties of specific mappings between various domains and codomains. Thus, the mapping-based viewpoint is one that we will use in the following to underline the deep inherent similarity of the seemingly very different mathematical structures to be considered.

2.8 MAPPINGS — CLASSIFICATION

Mappings are so important and so frequently encountered that they are meticulously classified, so that different kinds of mapping are given special terms. These designations are based on two different viewpoints.

Local viewpoint, or *elemental* point of view, which is concerned with what happens to individual elements of the domain when they are mapped. Again, the definitive feature of a mapping is a unique value for any element of the domain,

$$b(v) \neq b(v') \longrightarrow v \neq v'. \tag{2.31}$$

This does not imply, however, the uniqueness of the element of the domain which produces a given value. In fact, a binary relation can be a mapping and, at the same time, have several elements of its domain corresponding to the same element in the codomain (for an illustration, see Fig. 2.6). But in many instances the inverse condition to that of Eq. (2.31) is also satisfied, that is,

$$v \neq v' \longrightarrow b(v) \neq b(v'), \tag{2.32}$$

for any value in the codomain its original element in the domain is unique, see Fig. 2.6. The mappings which meet this additional condition, are called *one-to-one mappings*, or *injections*.

The feature of being an injection is definitely attractive in a mapping because it establishes some equality between the domain and the codomain. This is, however, quite local and thus a very illusive equality. Indeed, under a given *mapping*, even if it is an *injection*, not every element of the codomain is necessarily a value of some element of the domain; some elements of the codomain might be not involved in the mapping at all. This observation induces us to consider another dimension of mapping, complementary to the one-element approach discussed so far.

Global viewpoint, or *whole-set-based* point of view. Here, the focus is switched from tracing transformations of individual elements of the domain under a given mapping to considering sets of such elements.

Consider some subset, $\tilde{V}$, of the domain V, and trace what happens with its elements under a mapping B. The set of values of all the elements of $\tilde{V}$ constitute

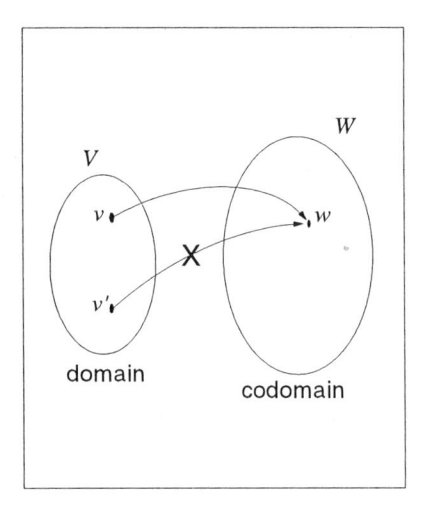

Fig. 2.6 Due to the definition of injection, each value must correspond to unique element of the domain.

some subset of the codomain W. We denote this subset $b(\tilde{V})$. It is called the *image* (or, more precisely, the *direct image*) of the subset $\tilde{V}$ under the mapping B.

In the case when the subset $\tilde{V}$ is the entire domain, $\tilde{V} = V$, its direct image, $b(V)$ is called the *range* of the mapping B. Generally, the range is some subset of the codomain, see Fig. 2.7. This means that some elements of the codomain have no connection with the domain whatsoever. To deal with such mavericks is rather uncomfortable. A nice situation occurs when the range covers all of the codomain. If this is the case, that is,

$$b(V) = W, \qquad (2.33)$$

such a mapping B is called an *onto mapping* or a *surjection*.

The mapping being a surjection establishes a "global" equality of the domain and codomain. Note, however, that the qualities of a mapping of being an injection and being a surjection are of different nature and have nothing in common (see the following examples of different combinations of these qualities). A mapping being both an injection and a surjection is called a *bijection*.

The mapping being an injection and/or a surjection directly relates to the general concept of *inverse* mapping, that is, to the problem of restoring the original element of the domain by its value in the codomain. In particular, if the mapping in question, B, is an *injection*, then definitely each element w *of the range* of B corresponds to that unique element of the domain whose value it is, $w = b(v)$. Thus, the binary relation

$$B^{-1} = \{(b(v), v)\} \subset b(V) \times V \qquad (2.34)$$

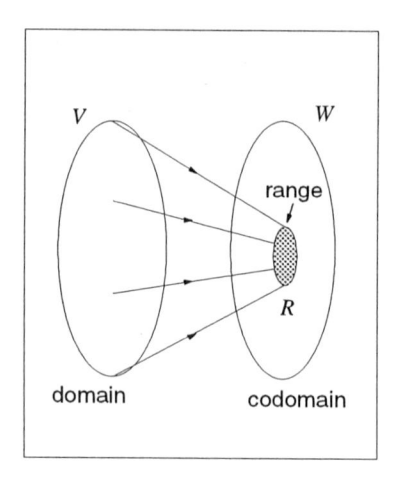

Fig. 2.7 "Global" view on a mapping: all the elements of the domain are mapped into range R being a subset of codomain.

is a mapping. This mapping, $v = b^{-1}(w)$, is called the inverse mapping with respect to the mapping B.

On the other hand, if the mapping B is not an injection, for any element w of its range there exists a set of elements in the domain which are mapped into this element w,

$$\tilde{V}(w) = \{v|b(v) = w\}. \qquad (2.35)$$

But this means that for any subset $\tilde{W}$ of the range, $\tilde{W} \subset b(V)$, one can find the subset U of the domain, $U \subset V$, the values of whose elements constitute the set $\tilde{W}$. This subset U is just the *union* of the sets $\tilde{v}(w)$ for all the elements $w \in \tilde{W}$. In our newly introduced terms, the set $\tilde{W}$ is the direct image of the set U under mapping B. Correspondingly, the subset U of the domain V is called the *inverse image* of the subset $\tilde{W}$ under mapping B and is denoted as $b^{-1}(\tilde{W})$. Note, that inverse image always exists whether the "local" inverse of mapping B exists or not.

Finally, only a mapping which is a bijection is *globally* invertible, that is, *each* element of the codomain corresponds to *just one* element of the domain.

Examples: visual demonstration of binary relations
Let both the domain and the codomain be the same set of real numbers, $V = \mathcal{R}$, $W = \mathcal{R}$. Then the cross-product set; $V \times W$ can be represented as the cartesian plane, V being the x-axis, and W being the y-axis. Consider the following relations F_i represented by *lines* in this plane.
1) $F_1 \subset V \times W$, in which the components of two-tuples are connected by the expression $v^2 + w^2 = 1$. This is the unit circle, Fig. 2.8. This relation is not a mapping, because each $v < 1$ corresponds to *two* values in W.
2) $F_2 \subset V \times W$, the components of two-tuples of F_2 being connected though $w = v^2$.

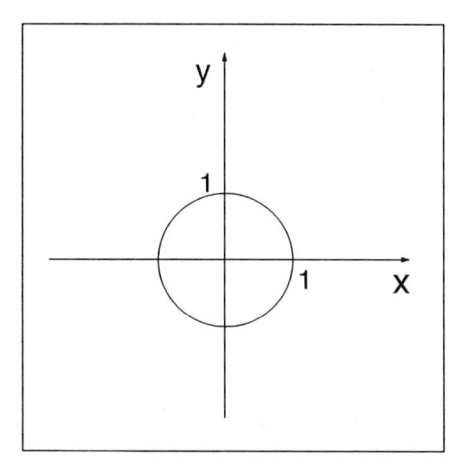

Fig. 2.8 Relation $F_1 \subset V \times W$, $V = W = \mathcal{R}$, determined by condition $v^2 + w^2 = 1$. The relation is not a mapping.

This is a parabola, Fig. 2.9. The mapping F_2 is not an injection, because the number v and $-v$ correspond to the same value w. Also, this mapping is not a surjection because the range covers only the positive half-axis.

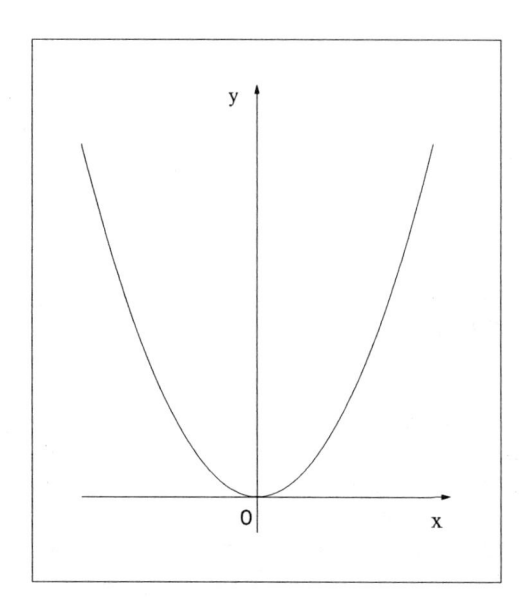

Fig. 2.9 Relation $F_2 \subset V \times W$, $V = W = \mathcal{R}$, determined by condition $w = v^2$. This relation is a mapping, though neither injection nor surjection.

34   RELATIONS AND MAPPINGS

3) $F_3 \subset V \times W$, the determining expression being $w = e^v$, Fig 2.10. This mapping still is not a surjection, because the values again fall in the positive half-axis, but this mapping is definitely an injection and, by the way, the inverse mapping (from the range) is defined as the *natural logarithm*, $v = \ln w$.

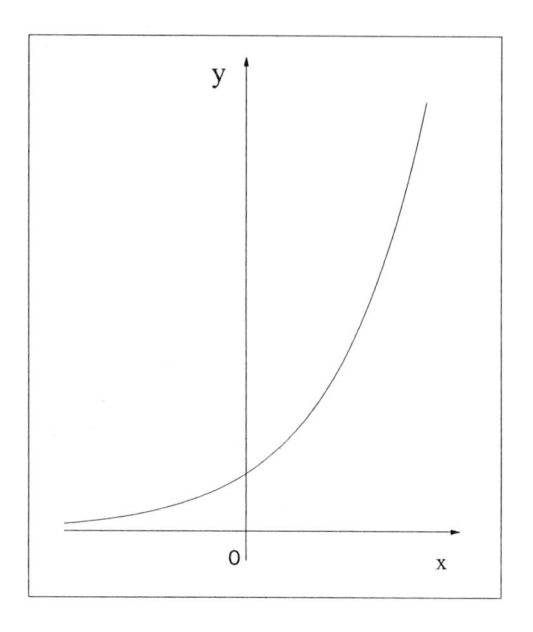

Fig. 2.10   Relation $F_3 \subset V \times W$, $V = W = \mathcal{R}$, determined by the condition $w = e^v$. The relation is a mapping. It is an injection but not a surjection.

4) $F_4 \subset V \times W$, $w = v^2 + v^3$, Fig. 2.11. Now, the mapping F_4 is finally a surjection, the entire y-axis is involved. However, it is not an injection, because on some interval as many as *three* points of the domain correspond to the same value in the codomain.

5) $F_5 \subset V \times W$, $w = v$, Fig. 2.12. This mapping is both an injection and a surjection, and thus presents an example of a bijection. The above five examples relate to the specific case of mappings which should be familiar to you; ordinary real functions of real arguments from calculus. Given a variety of such functions, it is possible to represent all the basic properties of mappings. Even the concept of mapping itself can be thought of as a generalization of the notion of a function in calculus. The practical implication of this fact is that in handling any mapping-related problem it is always useful to invoke images of ordinary functions for heuristic help.

MAPPINGS — CLASSIFICATION   35

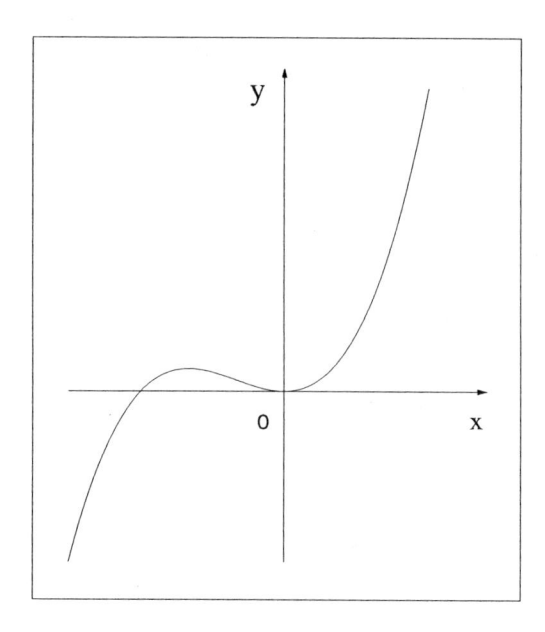

Fig. 2.11 Relation $F_4 \subset V \times W$, $V = W = \mathcal{R}$, determined by condition $w = v^2 + v^3$. The relation is a mapping. It is a surjection but not an injection.

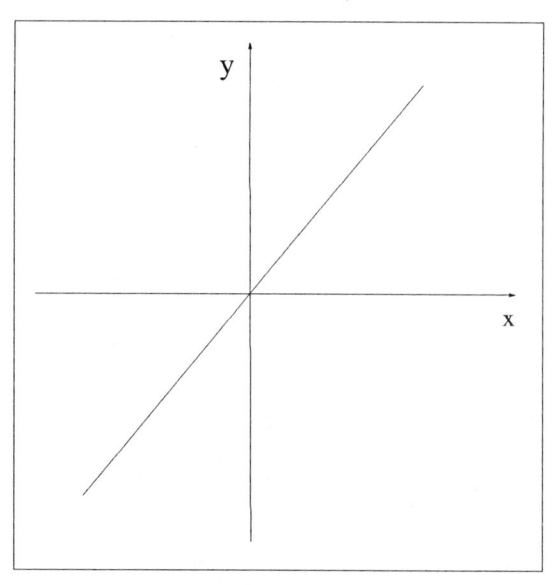

Fig. 2.12 Relation $F_5 \subset V \times W$, $V = W = \mathcal{R}$, determined by the condition $w = v$. This relation is a mapping being both an injection and a surjection, that is, this is a bijection.

36   RELATIONS AND MAPPINGS

2.9   SAMPLE PROBLEM II

Consider the following four sets:
1) $A = \{x \mid x \text{ is real number} \land x^3 - 20x = x^2\}$;
2) $B = \{y \mid y \text{ is integer number} \land \tan(\pi/y) = -1\}$;
3) $C = \{z \mid z \text{ is real number} \land \sin(\pi z/5) = 0\}$;
4) $D = \mathcal{P}(A \cap C)$.
   a) Find the power set $\mathcal{P}(B \times D)$;
   b) A binary relation $R \subset A \times (B \cup (A \cap C))$ is defined by the condition $\forall (r_1, r_2) \in R \ \ r_1 < r_2$. Present the relation as set of points in the cartesian plane. Is this a mapping? If yes, is it an injection? Is it a surjection?

Solution
a) We first specify all the sets explicitly, that is, by listing their elements:
1) Set A : the defining equation,

$$x^3 - x^2 - 20x = x \cdot (x^2 - x - 20) = x \cdot (x+4) \cdot (x-5) = 0, \qquad (2.36)$$

has roots $x_1 = 0$, $x_2 = 5$, and $x_3 = -4$. Thus,

$$A = \{-4, 0, 5\}. \qquad (2.37)$$

2) Set B: the defining equation,

$$\tan\left(\frac{\pi}{y}\right) = -1, \qquad (2.38)$$

has multiple roots defined by:

$$\frac{\pi}{y} = -\frac{\pi}{4} + \pi \cdot k, \qquad k = 0, \pm 1, \pm 2, \ldots. \qquad (2.39)$$

From the expression (2.39) we get

$$y = -\frac{4}{1 - 4 \cdot k}. \qquad (2.40)$$

Clearly, among these solutions the only integer number is $y = -4$. Thus,

$$B = \{-4\}. \qquad (2.41)$$

3) Set C: the equation

$$\sin\left(\frac{\pi \cdot z}{5}\right) = 0 \qquad (2.42)$$

also has multiple roots,

$$\frac{\pi \cdot z}{5} = \pi \cdot k, \qquad k = 0, \pm 1, \pm 2, \ldots, \qquad (2.43)$$

which gives
$$z = 5 \cdot k. \tag{2.44}$$
Thus,
$$C = \{..... -15, -10, -5, 0, 5, 10, 15, ...\}. \tag{2.45}$$
4) Set D: the set
$$A \cap C = \{-4, 0, 5\} \cap \{0, \pm 5, \pm 10, ...\} = \{0, 5\}. \tag{2.46}$$
Therefore,
$$D = \mathcal{P}(\{0, 5\}) = \{\emptyset, \{0\}, \{5\}, \{0, 5\}\}. \tag{2.47}$$
b) Now we can find the cross-product set $B \times D$:
$$\begin{aligned} B \times D &= \{-4\} \times \{\emptyset, \{0\}, \{5\}, \{0, 5\}\} \\ &= \{(-4, \emptyset), (-4, \{0\}), (-4, \{5\}), (-4, \{0, 5\})\}. \end{aligned} \tag{2.48}$$
c) The power set $\mathcal{P}(B \times D)$ is given by
$$\begin{aligned} \mathcal{P}(B \times D) = &\{\emptyset, \{(-4, \emptyset)\}, \{(-4, \{0\})\}, \{(-4, \{5\})\}, \{(-4, \{0, 5\})\}, \\ &\{(-4, \emptyset), (-4, \{0\})\}, \{(-4, \emptyset), (-4, \{5\})\}, \\ &\{(-4, \emptyset), (-4, \{0, 5\})\}, \{(-4, \{0\}), (-4, \{5\})\}, \{(-4, \{0\}), (-4, \{0, 5\})\}, \\ &\{(-4, \{6\}), (-4, \{0, 5\})\}, \\ &\{(-4, \{0\}), (-4, \{5\}), (-4, \{0, 5\})\}, \{(-4, \emptyset), (-4, \{5\}), (-4, \{0, 5\})\}, \\ &\{(-4, \emptyset), (-4, \{0\}), (-4, \{0, 5\})\}, \{(-4, \emptyset), (-4, \{0\}), (-4, \{5\})\}, \\ &\{(-4, \emptyset), (-4, \{0\}), (-4, \{5\}).(-4, \{0, 5\})\}\}, \end{aligned} \tag{2.49}$$
It has 16 elements.

d) Set
$$\begin{aligned} A \times (B \cup (A \cap C)) &= \{-4, 0, 5\} \times (\{-4\} \cup \{0, 5\}) \\ &= \{-4, 0, 5\} \times \{-4, 0, 5\} = A \times A \\ &= \{(-4, -4), (-4, 0), (-4, 5), (0, -4), (0, 0), (0, 5), \\ &\quad (5, -4), (5, 0), (5, 5)\}. \end{aligned} \tag{2.50}$$
The subset R contains all the pairs in which the first element is less than the second. It is
$$R = \{(-4, 0), (-4, 5), (0, 5)\}. \tag{2.51}$$
Treating the elements of R as pairs of coordinates (x, y), we obtain three points in the cartesian plane, see Fig. 2.13.

The relation B is not a mapping, because one element of the domain, (-4), corresponds to two different elements of the codomain, 0 and 5.

38 RELATIONS AND MAPPINGS

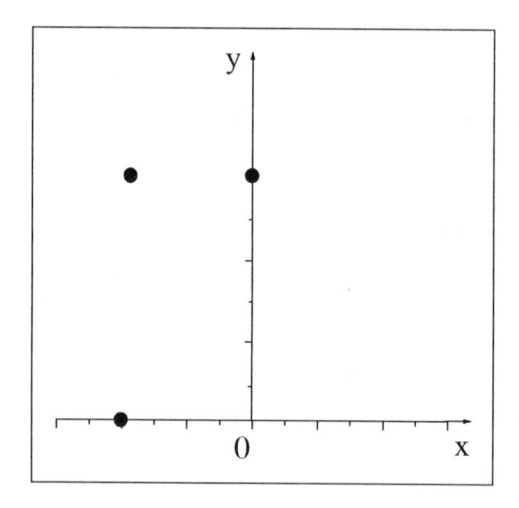

Fig. 2.13 The relation $R \subset A \times A$ determined by Eqs. (2.50) and (2.51)

2.10 SUMMARY OF CHAPTER 2

1) A *relation* between elements of a set or between elements of different sets implies *ordering* of these elements. This ordered structure requires a special description, because the concept of set by itself does not imply any order.

2) The required special mathematical construction representing ordering is provided by the *ordered n−tuple*, $(v_1, v_2, v_3, \ldots, v_n)$, in which components have prescribed positions. The n-tuple, however, can be represented by a certain intentionally constructed and very special set.

3) The set of all possible n−tuples with the first component being an element of set V_1, the second component being an element of set V_2, and so on, up to the last, n-th component being an element of set V_n is called *cross-product* set and is denoted as $V_1 \times V_2 \times \cdots \times V_n$.

4) Then, *n-ary relation* in the cross-product set $V_1 \times V_2 \times \cdots \times V_n$ is determined as a subset R of this cross-product set: $R \subset V_1 \times V_2 \times \ldots \times V_n$. Thus, relations can be completely described in set-theory language without any additional assumptions.

5) Most important in applications are *binary relations*. A *binary relation* is a relation between the elements of two sets, *domain V* and *codomain W*. The element $w \in W$ that corresponds to a given $v \in V$ is called the *value* of the element v under the given relation.

6. A special kind of binary relation is a *mapping*, for which any element of the domain corresponds to just one value in the codomain.

7) Further, special kinds of mappings are the *injection* (one-to-one mapping) and the *surjection* (onto mapping). A mapping which is an injection and a surjection is called

a *bijection*. In this case, the *inverse* mapping exists on the entire codomain.

2.11 PROBLEMS

I. Consider the following sets:
1) $E = \{e_1, e_2 \mid (e_1 \in \mathcal{N}) \land (e_2 \in \mathcal{N}) \land (e_1^2 + e_2^2 = 25)\}$;
2) $F = \{f \mid f \text{ is a digit in the number } 1999\}$;
3) $G = \{\emptyset\}$.
Find $\mathcal{P}(E \times F \times G)$.

II.
1) A binary relation $R \subset \mathcal{R} \times \mathcal{R}$ is determined by the condition

$$\forall (r_1, r_2) \in R, \quad r_2 = \int_{\pi/2}^{r_1} \cot(r_1') dr_1'.$$

Represent the relation as a set of points in the cartesian plane (r_1, r_2). Is this a mapping? If yes, is it an injection? Is it a surjection?
2) Answer the same questions for the binary relation $\tilde{R} \subset \mathcal{R} \times \mathcal{R}$ determined by the condition

$$\forall (r_1, r_2) \in \tilde{R}, \quad r_2 = \int_0^{r_1} (1 + r_1') e^{r_1'} dr_1'.$$

2.12 FURTHER READING

The material covered in this chapter is a classical extension of set theory. Therefore, it is treated, at least briefly, in a number of books, particularly, in the books we referred to in Chapter 1. In addition to those, the following texts could be recommended as treating details and particularities.

J. C. Abbott, *Sets, Lattices, and Boolean Algebras*, Allyn and Bacon, Boston, 1969.
This is a classical senior-level text, which we will also mention in the next chapter. Regarding the concepts of relations and mappings, they are established here as basic justification for the central ideas of linear algebra, functional analysis and topology.

S. Mac Lane, G. Birkhoff, *Algebra*, Chelsea, New York, 1988.
This is another classical book. Its first chapters provide a distinct but related development of the material of our Chapters 1 and 2, given on a much higher and more rigorous level.

K. Ciesielski, *Set Theory for the Working Mathematician*, Cambridge University Press, Cambridge, England, 1997.

This is a rather advanced text, tracing implications of the ideas of sets and relations in various fields of modern mathematics. However, with its emphasis on language and notations rather than on sophisticated theories, it can be helpful in satisfying the curiosity of non-mathematicians.

Among the many books addressing on different levels various implications of the concepts of sets and mappings in specific fields of computer engineering and computer science, the following two are closest to our approach.

D. F. Stanat, D. McAllister, *Discrete Mathematics in Computer Science*, Prentice-Hall, Englewood Cliffs, NJ, 1977.

Though being an undergraduate text, it is quite extensive and treats many additional, computer-oriented aspects of relations and mappings.

A. V. Aho, J. E. Hopcroft, J. D. Ullman, *The Design and Analysis of Computer Algorithms*, Addison-Wesley, Reading, MA, 1974.

The book presents and thoroughly analyzes many algorithms directly associated with sets and mappings.

3
Mathematical Logic

In this chapter we will use the general concepts of mapping, which we elaborated in Chapter 2, for certain particular mathematical structures being very important in both theory and applications.

3.1 LOGICAL OBJECTS AND LOGICAL STATEMENTS

When we defined relation as a subset R of n-tuples in the cross-product set $V_1 \times V_2 \times V_3 \times \ldots \times V_n$, we implicitly supposed that we have a procedure which separates all the n-tuples of the cross-product set into two sets, R and not R. In other words, for each n-tuple $(a_1, a_2, a_3, \ldots, a_n)$ we are able to decide whether the statement $(a_1, a_2, a_3, \ldots, a_n) \in R$ is true or false. Therefore, we have moved to the world of logical statements. Of course, this is not our fist experience in this world. We can certainly say that most sentences we read, write, tell, or hear, are logical statements.

To set the notion of logical statement properly (we mean, mathematically properly), we must clearly define the method by which we choose the logical statements among all the possible assertions. By definition, *logical statements* are assertions to which truth values can be assigned (i.e., about which we can say whether the assertion is true or false). In this sense, "I am a liar" is a logical statement, while "get lost" or even "I am lying" are not (the latter statement is a paradox, because if it is true, it must be false, and vice versa...). The things about which such assertions can be made are called *logical objects*. (*Subject* "I" is such *object* in the above example.) However, in particular cases the thing being a logical object and the statement being

41

a logical statement depend on the assumed conventions.

Example of logical and non-logical statements
"A dragon breaths fire" is not a logical statement in zoology, which does not recognize dragons. Nevertheless, it is a logical statement in the framework of fairy tales, with an assigned truth value. On the other hand, "Couzdra is very glockish" is not a logical statement in any way, because there is neither convention of what is "couzdra" nor any slightest idea of what "glockish" means.

We denote the set of all logical objects as $\mathcal{U}$ (akin to Universe), the set of all logical statements as $\mathcal{P}$ (do not confuse this $\mathcal{P}$ with power set of Chapter 1, $\mathcal{P}(A)$, which always has an argument, a set A). We then call the elements of the sets $\mathcal{U}$ and $\mathcal{P}$ object variable and statement variable, respectively.

3.2 LOGICAL-STATEMENT AND TRUTH-VALUE MAPPINGS

We can establish a consistent relation between the world of logical objects and logical statements, by generalization (or rectification) of our everyday experience. Let $V_1, V_2, \ldots V_n$ be subsets of the set of logical objects, $\mathcal{U}$, $V_1 \subset \mathcal{U}, V_2 \subset \mathcal{U}, \ldots, V_n \subset \mathcal{U}$.

Consider now some n-ary relation as a subset R in the cross-product set

$$W = V_1 \times V_2 \times V_3 \times \ldots \times V_n. \tag{3.1}$$

Of course, $W \subset \mathcal{U}$, because the elements of W, n-tuples $(v_1, v_2, \ldots, v_n)$, are logical objects (we definitely can make logical statements about them). Let us, then, define a specific mapping P from the set W into set $\mathcal{P}$ which transforms elements of set W into logical statements "this n-tuple belongs to R":

$$(v_1, v_2, \ldots, v_n) \rightarrow ((v_1, v_2, \ldots, v_n) \in R) \tag{3.2}$$

This mapping is called the n-ary *predicate function* corresponding to the n-ary relation R. In particular, if R is an unary relation, then the mapping P just transforms the elements of a set V into statements "this element belongs to a given subset $R \subset V$,"

$$v \rightarrow (v \in R) \tag{3.3}$$

and is called the *unary predicate function*. The statements being obtained as a result of predicate-function mapping,

$$p(a_1, a_2, a_3, \ldots, a_n) = ((a_1, a_2, a_3, \ldots, a_n) \in R), \tag{3.4}$$

are called *atomic statement variables* or just *atomic statements*. Note that they belong to a set of a nature different from that of the set R. As it is defined, the predicate-function mapping produces a "personal" logical statement for each of the n-tuples of the set W. So, this mapping is an injection. However, it is not a surjection, because the obtained atomic statements definitely do not exhaust all the possible logical statements.

Example of atomic statements
Let logical objects be pairs of real numbers, so that $W = \mathcal{R} \times \mathcal{R}$. Then, the binary predicate function $P : W \to \mathcal{P}$,

$$(a_1, a_2) \in W \to (a_1 \leq a_2) \in \mathcal{P}, \tag{3.5}$$

produces the atomic statement variable, "the number a_1 is less than or equal to the number a_2".

Important note on notation
Logical statements being elements of a set $\mathcal{P}$, it is natural to denote them with small letters. However, as we will see, there exists a fundamental analogy between *logical statements* in Mathematical Logic and *subsets* in Set Theory. For this reason, it is commonly accepted (and we will follow this convention) to denote logical statements with *capital* letters, for example, $A \in \mathcal{P}$, $B \in \mathcal{P}$, and so forth.

Truth Values
Since "logical statement" was defined to be an assertion to which definite truth values can be assigned, there exists a mapping B from the set $\mathcal{P}$ into a *set of truth values*, $\mathcal{B}$. Since a logical statement can only be either true or false, the set $\mathcal{B}$ consists of only two elements, $\mathcal{B} = \{0, 1\}$ (or $\{f, t\}$), where 0 (or f) stands for *false*, and 1 (or t) stands for *true*.

The system of logic in which the truth values are allowed to take only two values is sometimes referred as "two level" or "conventional" logic. This is distinct from the recently developed "fuzzy logic", where the truth value can be *any* real number in the interval $[0, 1]$. In what follows we will consider only conventional logic.

The mapping T from set $\mathcal{P}$ into the set of truth values $\mathcal{B}$,

$$A \in \mathcal{P} \to b = t(P) \in \mathcal{B} \tag{3.6}$$

is naturally called the *truth-value function*. This mapping transforms a given statement into the number 1 (if it is true) or the number 0 (if it is false). Obviously, this mapping is a surjection, because the set of *all* logical statements definitely contains true statements as well as false statements. (Moreover, the truth-value mapping is usually a surjection even when its domain is a *subset* of $\mathcal{P}$, unless we intentionally construct a set of logical statements containing only true or only false statements.) It is also obvious that the mapping T is not an injection: since all the infinite variety of logical statements correspond to only two elements of the truth-value set, there are many statements per each of these two elements.

From the truth-value-function standpoint, the distinctive feature of atomic statement variables with respect to all the logical statements is that atomic statements are supposed to have prescribed *a priori* truth values.

3.3 LOGICAL CONNECTIVES AND LOGICAL FORMULAS

The main aim of mathematical logic is to find truth values of complex statements. These complex statements are obtained step by step from the atomic statement variables by means of *logical connectives* and quantifiers. There are five basic connectives and two quantifiers, which we have already introduced in the very beginning of Chapter 1 and indiscriminately used as abbreviations. Now we are going to treat them more seriously.

Connectives:
$\wedge$ and;
$\vee$ or;
$\sim$ not;
$\rightarrow$ if ... then;
$\leftrightarrow$ if and only if.
Quantifiers:
"universal" $\forall$ means "for all";
"existential" $\exists$ means "for some" (or "there exist").

Note that the connectives connect logical statements, while the quantifiers act directly on logical objects. Given a set of logical objects, V, the statement $\forall v(A)(v \in V)$ means "for all logical objects of the set V the logical statement A holds". The statement $\exists v(A)(v \in V)$ means "there exists such a logical object in the set V for which logical statement A holds".

The whole set $\mathcal{P}$ of complex logical statements or *logical formulas* can be determined on the basis of atomic statements in an inductive manner:
1) Atomic statements are formulas;
2) If A and B are logical formulas, then $\sim A$, $A \wedge B$, $A \vee B$, $A \rightarrow B, A \leftrightarrow B$, $\forall v(A)(v \in V)$, and $\exists v(A)(v \in V)$ are also logical formulas.
3) All the logical formulas are generated by 1) and 2).

Even in situations when one starts with a few atomic statements, multiple use of the connectives unavoidably produces an infinite number of pertinent logical formulas. Having obtained this infinite number of logical formulas, the basic goal of mathematical logic is to provide means to obtain the truth value of any logical formula, being given the truth values of the atomic statements involved.

3.4 ADEQUATE SET OF CONNECTIVES

Before determining the rules of finding truth values of complex statements, it is worth trying to reduce the number of "indispensable" connectives. In reality, not all the connectives are independent.

a) *Quantifiers*

Though statement $\forall v(A)(v \in V)$ seems to be irreducible, it is not so. For instance, in the case of a finite set of logical objects, $V = \{v_1, v_2, v_3, \ldots, v_n\}$, the statement $\forall v(A)(v \in V)$ just means

$$A(v_1) \wedge A(v_2) \wedge \ldots \wedge A(v_n). \tag{3.7}$$

Although in the case of an infinite set V we would need to write in Eq. (3.7) an infinite chain of statements, and although in the case of uncountable set V we could not write a chain indexed by integer numbers, but in all cases the quantifier $\forall$ is reducible to multiple use of the connective $\wedge$.

A similar observation can be made for the statement $\exists v(A)(v \in V)$. For the same finite set V this statement means:

$$A(v_1) \vee A(v_2) \vee \ldots \vee A(v_n). \tag{3.8}$$

Thus, the quantifier $\exists$ can be successfully replaced by multiple use of the connective $\vee$.

b) *Connective* $\leftrightarrow$
The statement $A \leftrightarrow B$ can be rewritten as a complex statement $(A \to B) \wedge (B \to A)$. Thus, the connective $\leftrightarrow$ is also an excessive luxury.

c) *Connective* $\to$
Reducing this connective is not at all evident. In fact, this is the most subtle result in this chapter. We start with a simple example: the statement "if I am not shaven, I have a beard" certainly means the same as "I am shaven or I have a beard." Then, we generalize this observation to the equivalence of the complex statements $\sim A \to B$ and $A \vee B$ made from any two logical statements, A and B. We can also illustrate the equivalence of these two complex statements by means of a set analogy, see Fig. 3.1. Then, we denote the negation $\sim A$ as a statement C and come to equivalence of the statements $C \to B$ and $\sim C \vee B$. Thus, we always can express the connective $\to$ in terms of connectives $\sim$ and $\vee$.

Based on these considerations, we are left with three basic connectives: $\wedge$, $\vee$, and $\sim$. Actually, as we will shortly see, it is even sufficient to keep only $\sim$ and either $\wedge$ or $\vee$ to express all the other connectives. However, this "radical reduction" would lead to overcomplicated formulas.

3.5 TRUTH VALUES OF COMPLEX LOGICAL FORMULAS

Having established relations between logical connectives, we can now proceed to establishing truth values of complex logical formulas.

All *atomic logical statements*, A, B, C, and so forth, have their assigned truth values $t(A)$, $t(B)$, $t(C)$, and so on. Then, the statement $\sim A$ obviously has the truth value "opposite" to that of a statement A:

$$t(\sim A) = 1 - t(A). \tag{3.9}$$

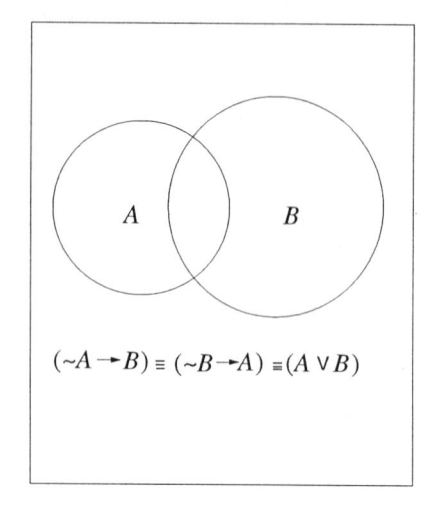

Fig. 3.1 Illustration to the reduction of the set of connectives to the adequate set.

$$(\sim A \to B) \equiv (\sim B \to A) \equiv (A \vee B)$$

As for the statement $A \wedge B$, it must be false if at least one of the statements A and B is false, and it is true only if both of the statements A and B are true. We can express this as

$$t(A \wedge B) = \min\{t(A), t(B)\}. \tag{3.10}$$

Since $t(A)$ and $t(B)$ can only take values 0 and 1, this minimum value is 1 if $t(A) = t(B) = 1$ and it is 0 otherwise. In more handy form, this result is written as

$$t(A \wedge B) = t(A)t(B). \tag{3.11}$$

In a similar fashion, the statement $A \vee B$ is true if at least one of the statements A or B is true, and it is false only if both of the statements are false. Thus,

$$t(A \vee B) = \max\{t(A), t(B)\}, \tag{3.12}$$

which for the allowed binary truth-values gives the symmetric formula

$$t(A \vee B) = 1 - (1 - t(A))(1 - t(B)). \tag{3.13}$$

For convenience, the formula (3.13) can be further simplified to

$$t(A \vee B) = t(A) + t(B) - t(A)t(B). \tag{3.14}$$

The truth-value formulas (3.9), (3.11), and (3.13) cover all the three basic connectives. Now we can construct truth values of logical formulas with the other, reducible, connectives and quantifiers.

Connective $\to$

$$\begin{aligned}
t(A \to B) &= t(\sim A \vee B) \\
&= 1 - (1 - t(\sim A))(1 - t(B)) \\
&= 1 - (1 - (1 - t(A)))(1 - t(B)) \\
&= 1 - t(A) + t(A)t(B) \\
&= 1 - t(A)(1 - t(B)).
\end{aligned} \qquad (3.15)$$

Connective $\leftrightarrow$

$$\begin{aligned}
t(A \leftrightarrow B) &= t((A \to B) \wedge (B \to A)) \\
&= (1 - t(A)(1 - t(B)))(1 - t(B)(1 - t(A))) \\
&= 1 - t(A)(1 - t(B)) - t(B)(1 - t(A)) \\
&\quad + t(A)t(B)(1 - t(A))(1 - t(B)).
\end{aligned} \qquad (3.16)$$

The last term of the last expression is zero, because for both possible values $t(A) = 0$ and $t(A) = 1$ one has

$$t^2(A) = t(A), \qquad (3.17)$$

and

$$t(A)(1 - t(A)) = t(A) - t^2(A) = 0. \qquad (3.18)$$

So,

$$t(A \leftrightarrow B) = 1 - t(A) - t(B) + 2t(A)t(B), \qquad (3.19)$$

which we can also write as

$$t(A \leftrightarrow B) = 1 - (t(A) - t(B))^2. \qquad (3.20)$$

Finally, we obtain truth values of formulas with quantifiers by means of generalization.

Quantifier $\forall$
We have reduced this quantifier to the multiple use of connective $\wedge$. Therefore, according to Eq. (3.10),

$$t(\forall v(A)(v \in V)) = \min_{v \in V} \{t(A(v))\}. \qquad (3.21)$$

For a countable set $V = \{v_i\}$ this can be further simplified by use of Eq. (3.11)

$$\begin{aligned}
t(\forall v_i(A)) &= t(A(v_1) \wedge A(v_2) \wedge A(v_3) \wedge \ldots) \\
&= t(A(v_1)) \cdot t(A(v_2)) \cdot t(A(v_3)) \cdot \ldots \\
&= \prod_i t(A(v_i)).
\end{aligned} \qquad (3.22)$$

Quantifier ∃

This quantifier was reduced to the multiple use of connective ∨. According to Eq. (3.12), this means,

$$t(\exists v(A)(v \in V)) = \max_{v \in V}\{t(A(v))\}. \qquad (3.23)$$

This formula also can be further simplified in the case of a finite set $V = \{v_j\}$, by means of Eq. (3.13). However, the derivation is a bit more complicated than that for the quantifier ∀. We do it by *induction*, starting with formula (3.13). Then, for three statements, A, B, and C, we obtain, by using this formula twice,

$$\begin{aligned} t(A \vee B \vee C) &= 1 - (1 - t(A \vee B))(1 - t(C)) \\ &= 1 - (1 - (1 - (1 - t(A))(1 - t(B))))(1 - t(C)) \\ &= 1 - ((1 - t(A))(1 - t(B))(1 - t(C)). \end{aligned} \qquad (3.24)$$

Repeating this argument for four, five, six and so on statements, we eventually get the generalization:

$$t(\exists v_j(A)) = t(A(v_1) \vee A(v_2) \vee A(v_3) \vee \ldots) = 1 - \prod_j(1 - t(A(v_j))). \qquad (3.25)$$

Combining all these newly obtained formulas in one table yields Table 4.1, which from now on will be our basic reference point concerning logical formulas.

Table 4.1. Basic Truth-Value Formulas

$t(\sim A) = 1 - t(A)$

$t(A \wedge B) = t(A)t(B)$

$t(A \vee B) = 1 - (1 - t(A))(1 - t(B))$

$t(A \to B) = 1 - t(A) + t(A)t(B))$

$t(A \leftrightarrow B) = 1 - t(A) - t(B) + 2t(A)t(B))$

$t(\forall v_i(A)) = \prod_i t(A(v_i))$

$t(\exists v_j(A)) = 1 - \prod_j(1 - t(A(v_j)))$

3.6 TAUTOLOGIES AND CONTRADICTIONS

The truth value of a complex logical formula is basically determined by two factors, the truth values of atomic statements participating in the formula and the structure of the formula, that is, the connectives tying the atomic statements together. However, in some interesting cases, the truth value is determined entirely by the structure, being completely independent of the truth values of the component statements.

Consider, for instance, the statement $A \wedge (\sim A)$. Referring to Table 4.1, its truth values is

$$t(A \wedge (\sim A)) = t(A)t(\sim A)$$
$$= t(A)(1 - t(A)) = t(A) - t^2(A) = 0, \qquad (3.26)$$

whatever the truth value of A is, that is,

$$t(A \wedge (\sim A)) = 0 \qquad (3.27)$$

always.

On the other hand, using the formulas of Table 4.1 for the statement $A \vee (\sim A)$, we obtain:

$$t(A \vee (\sim A)) = t(A) + t(\sim A) - t(A)t(\sim A)$$
$$= t(A) + 1 - t(A) - t(A)(1 - t(A))$$
$$= t(A) + 1 - t(A) - t(A) + t^2(A) = 1, \qquad (3.28)$$

again, whatever $t(A)$ is. So,

$$t(A \vee (\sim A)) = 1 \qquad (3.29)$$

always.

Of course, such cases are of special interest and of great value, because they provide convenient formats to be successfully used everywhere regardless of particular content of the formula, that is, of the meanings of its parts. A formula whose truth value is always 1, independent of the truth values of the components, is called a *tautology*; a formula whose truth value is always 0 is called a *contradiction*. In these terms, the statement $A \wedge (\sim A)$ is a contradiction, while the statement $A \vee (\sim A)$ is a tautology.

When a statement $A \to B$, where A and B are in turn some complex statements, happens to be a tautology, A is said to *logically imply* B. This fact is denoted as $A \Rightarrow B$. In this situation, the expression for the truth value of Eq. (3.15) reads

$$t(A \to B) = 1 - t(A) + t(A)t(B)) = 1. \qquad (3.30)$$

50 MATHEMATICAL LOGIC

It imposes definite restriction on the truth values of the incorporated statements A and B,

$$t(A) \cdot (1 - t(B)) = 0. \tag{3.31}$$

Thus, it should be either $t(A) = 0$ or $t(B) = 1$. We can express this as

$$A \Rightarrow B \text{ if and only if } t(A) \leq t(B). \tag{3.32}$$

The statement A logically implies the statement B only when $t(A)$ is less than or equal to $t(B)$.

This criterion leads sometimes to some psychogenic confusion. Indeed, it means that any *a priori* false statement implies everything, "if two times two is five, then witches exist". However, despite its seemingly disturbing form, this rule of the game is quite consistent with the structure of logic and, therefore, with the interrelations of things in the real world.

When a statement $A \leftrightarrow B$ is a tautology, A is said to be *logically equivalent* to B, which is denoted as $A \equiv B$. In this case we also can find some restriction on the truth values of the statements A and B. From Eq. (3.19),

$$t(A \leftrightarrow B) = 1 - t(A) - t(B) + 2t(A)t(B)) = 1, \tag{3.33}$$

which means

$$\begin{aligned} & t(A) + t(B) - 2t(A)t(B) \\ & = t^2(A) + t^2(B) - 2t(A)t(B) \\ & = (t(A) - t(B))^2 = 0. \end{aligned} \tag{3.34}$$

The only way to satisfy this equation is for A and B to have equal truth values:

$$A \equiv B \text{ if and only if } t(A) = t(B), \tag{3.35}$$

two logical formulas are logically equivalent if and only if they have equal truth values.

3.7 SAMPLE PROBLEMS I

In these problems we use the rules of truth-value calculations to find the truth-value of a complex formula and to prove this formula to be a tautology or contradiction.

Classify the following logical formulas, that is, determine whether a given formula is a tautology, a contradiction, or neither.

I. $(P \land \sim P) \to Q$.

Solution
The truth value of the left-hand side is found as

$$t(P \land \sim P) = \min\{t(P), 1 - t(P)\} = 0. \tag{3.36}$$

It is certainly less or at least equal to the truth value of the right-hand side, whatever the latter truth-value is,

$$t(P \wedge \sim P) \leq t(Q). \tag{3.37}$$

Thus, the given formula is a tautology.

II. $(P \vee \sim Q) \to Q$.

Solution
The truth value of the left-hand side is

$$t(P \vee \sim Q) = \max\{t(P), 1 - t(Q)\}. \tag{3.38}$$

Thus, if $t(P) = 1$, the truth value of the left-hand side is 1 regardless of $t(Q)$. This means, that if $t(Q) = 1$, the statement of the formula is true, but if $t(Q) = 0$, the statement is false. So, the truth value of the given formula depends on the truth values of the included statements; this formula is neither a tautology nor a contradiction.

III. $P \to (P \vee \sim Q)$.

Solution
The truth value of the right-hand side,

$$t(P \vee \sim Q) \geq t(P). \tag{3.39}$$

Thus, whatever the truth value $t(P)$ is, the inequality

$$t(\text{left-hand side}) \leq t(\text{right-hand side}) \tag{3.40}$$

holds. Thus, the given formula is a tautology.

IV. $(P \wedge Q) \to P$.

Solution
The truth value of the left-hand side is

$$t(P \wedge Q) = \min\{t(P), t(Q)\} \leq t(P). \tag{3.41}$$

Thus, the truth value of the left-hand side is always less than that of the right-hand side, independent of the truth values of the included statements. So, the formula is a tautology.

V. $((P \wedge Q) \leftrightarrow P) \leftrightarrow (P \leftrightarrow Q)$.

Solution
The truth value of the right-hand side is

$$t((P \leftrightarrow Q)) = 1 \quad \text{if} \quad t(P) = t(Q), \tag{3.42}$$

and

$$t((P \leftrightarrow Q)) = 0 \quad \text{otherwise.} \tag{3.43}$$

For the left-hand side we have:

$$t((P \wedge Q) \leftrightarrow P) = \begin{cases} 1 & \text{if } t(P) = t(Q) \\ 1 & \text{if } t(P) = 0, \ t(Q) = 1 \\ 0 & \text{otherwise} \end{cases} . \tag{3.44}$$

Thus, the truth value of the left-hand side sometimes is equal to that of the right-hand side, sometimes not, depending on the truth values of the included statements. This means that the formula is neither a tautology nor a contradiction.

VI. $((A \rightarrow (B \vee (\sim C))) \wedge (\sim A) \wedge (B \rightarrow C)$

Solution
By using the truth-value table (Table 3.1), we get:

$$\begin{aligned} &t((A \rightarrow (B \vee (\sim C))) \wedge (\sim A) \wedge (B \rightarrow C)) \\ &= t(A \rightarrow (B \vee (\sim C)))t(\sim A) \cdot t(B \rightarrow C)) \\ &= (1 - t(A) + t(A)t(B \vee (\sim C)))t(\sim A)t(B \rightarrow C) \\ &= (t(\sim A) - t(A)t(\sim A) + t(A)t(\sim A)t(B \vee (\sim C)))t(B \rightarrow C) \\ &= t(\sim A)t(B \rightarrow C). \end{aligned} \tag{3.45}$$

This truth-value clearly depends on the truth values of A, B, and C. For instance, if $t(A) = t(B) = t(C) = 1$, then

$$t(\sim A) \cdot t(B \rightarrow C) = 0. \tag{3.46}$$

If, on the other hand, $t(A) = t(B) = t(C) = 0$, then

$$t(\sim A) \cdot t(B \rightarrow C) = 1. \tag{3.47}$$

Thus, the given formula is neither a tautology nor a contradiction.

3.8 FUNDAMENTAL TAUTOLOGIES. DE MORGAN LAWS. PROOF FORMATS

There are some fundamental tautologies constituting the basis of logical inference. At first sight they seem quite natural and even trivial, but this very fact shows their consistency with the order of the real world. On the other hand, these tautologies being provable through the established truth-value calculation rules demonstrates in turn the consistency of mathematical logic with that real world. We consider only a few of such fundamental tautologies.

1)
$$A \Rightarrow A. \tag{3.48}$$

According to Eq. (3.32), this logical implication presumes the truth value of the left-hand side to be not greater than the truth value of the right-hand side. But it is quite clear that $t(A) \leq t(A)$. So, the statement $A \to A$ is really a tautology, and the implication of Eq. (3.48) holds.

2)
$$(A \wedge B) \Rightarrow A. \tag{3.49}$$

We use here the criterion of Eq. (3.32) again. In this case, according to the formula of Eq. (3.11), the truth value of the left-hand side is

$$t(A \wedge B) = t(A)t(B) \leq t(A), \tag{3.50}$$

because $t(B)$ is either 1 or 0. Thus, the statement $(A \wedge B) \to A$ is a tautology, that is, the logical implication of Eq. (3.49) holds.

3)
$$A \Rightarrow (A \vee B). \tag{3.51}$$

Here, we use the criterion (3.32) and the truth-value formula of Eq. (3.14) to obtain

$$\begin{aligned} t(A \vee B) &= t(A) + t(B) - t(A)t(B) \\ &= t(A) + t(B)(1 - t(A)) \\ &= t(A) + \text{"something positive or zero"} \geq t(A). \end{aligned} \tag{3.52}$$

Thus, the truth value of the left-hand side in formula (3.51) is always less then that of the right-hand side, which establishes the logical implication of this formula.

As a general commentary to formulas (3.49) and (3.51), one can say that adding something to a logical formula via the connective $\wedge$ always *lessens* its truth value, while adding something via the connective $\vee$ always *increases* the truth value.

The following two important tautologies are less trivial and more informative and useful. They reveal immanent and intimate relations between the connectives $\wedge$ and $\vee$. For their indispensable service, they deserve a special name; they have been named after their author.

De Morgan laws

$$\sim (A \wedge B) \equiv (\sim A) \vee (\sim B); \tag{3.53}$$

$$\sim (A \vee B) \equiv (\sim A) \wedge (\sim B). \tag{3.54}$$

54 MATHEMATICAL LOGIC

To prove, for instance, the first of these statements, we need to show that in Eq. (3.53) the truth value of the right-hand side equals that of the left-hand side. According to formulas (3.9) and (3.11), the truth value of the left-hand side is just

$$t(\sim (A \wedge B)) = 1 - t(A \wedge B) = 1 - t(A)t(B). \tag{3.55}$$

As for the truth value of the right-hand side, we use the formulas of Eqs. (3.9) and (3.13) to write it as

$$t((\sim A) \vee (\sim B)) = 1 - (1 - t(\sim A))(1 - t(\sim B))$$
$$= 1 - (1 - (1 - t(A)))(1 - (1 - t(B))) = 1 - t(A)t(B). \tag{3.56}$$

Comparing (3.55) and (3.56), we see that the truth values of the left-hand side and the right-hand side of Eq.(3.53) are equal. Thus, the first of the De Morgan tautologies is duly established.

The proof of the second of the De Morgan tautologies is a slight variation of the above argumentation; the reader is invited to prove it by him(her)self.

Having recalled that logical quantifiers $\forall$ and $\exists$ are generalizations of connectives $\wedge$ and $\vee$, respectively, we can correspondingly generalize the De Morgan tautologies.

In generalized form the first De Morgan law reads,

$$\sim \forall v(A)(v \in V) \equiv \exists v(\sim A)(v \in V), \tag{3.57}$$

that is, the statements "not for all objects $v \in V$ a statement A holds" and "among objects $v \in V$ there exists at least one for which the statement "not A" holds" are logically equivalent.

The generalization of the second De Morgan law results in the form

$$\sim \exists v(A)(v \in V) \equiv \forall v(\sim A)(v \in V), \tag{3.58}$$

that is, the statements "there is no object among $v \in V$, for which a statement A holds" and "for all the objects $v \in V$ the statement "not A" holds" are equivalent.

We can illustrate the generalized form of the first De Morgan tautology, formula (3.57), with the following "tangible" example: the statements "not all pigs have wings" and "there is at least one wingless pig" mean the same. Correspondingly, for the second De Morgan tautology, formula (3.58), the pig-wing illustration reads: the statements "there is no pig having wings" and "all the pigs are wingless" mean the same.

Proof formats

The following tautologies are much more sophisticated. They belong to the so-called *proof formats*, that is, standard schemes of mathematical reasoning. We will use these schemes in our further considerations. On the other hand, these methods of inferring are essentially the same as the common sense uses in everyday life. In mathematical logic they just appear in explicit form.
1)

$$A \wedge (A \to B) \Rightarrow B. \tag{3.59}$$

This proof format bears the latin name *modus ponens*, which means "proposing mode". For example, if in Eq. (3.59) the statement A is "it is raining" and the statement B is "cats and dogs get wet", then the complex statement "it is raining; and if it is raining, then cats and dogs get wet" logically implies that cats and dogs are really wet.

To prove the statement of Eq. (3.59), we make use of the truth-value formulas of Table 4.1. and the equality of Eq. (3.17) to compute the truth value of (3.59):

$$t(A \wedge (A \to B)) = t(A)t(A \to B) = t(A)(1 - t(A) + t(A)t(B))$$
$$= t(A) - t^2(A) + t^2(A)t(B) = t(A) - t(A) + t(A)t(B)$$
$$= t(A)t(B) \leq t(B). \tag{3.60}$$

Thus, the truth value of the left-hand side of Eq. (3.59) is always less than that of the right-hand side of this equation. So, the tautology (3.59) is proved.

2)
$$(A \to B) \wedge (\sim B) \Rightarrow (\sim A). \tag{3.61}$$

This also gives the latin name *modus tollens*, which means "removing mode". As can be seen, it is a counterpart of *modus ponens*. For instance, in the example of wetting cats and dogs we just used for *modus ponens*, the *modus tollens* would state: the complex statement "if it is raining, then cats and dogs get wet, and the cats and dogs are not wet" which logically implies that it is not raining.

The proof of the logical implication of Eq. (3.61) is as follows. We use the truth-value formulas of Eqs. (3.11) and (3.15) to find the truth value of the left-hand side of Eq. (3.61) in the form,

$$t((A \to B) \wedge (\sim B)) = t(A \to B)t(\sim B)$$
$$= (1 - t(A) + t(A)t(B))t(\sim B)$$
$$= (1 - t(A))t(\sim B) + t(A)t(B)t(\sim B). \tag{3.62}$$

The last term in this expression is zero. Moreover, making use of formula (3.9), we rewrite (3.62) as

$$t((A \to B) \wedge \sim B) = t(\sim A)t(\sim B) \leq t(\sim A). \tag{3.63}$$

Thus, the truth value of the left-hand side in Eq. (3.61) is always less than the truth value of the right-hand side, which establishes the formula (3.61) as a tautology.

3)
$$(\sim B) \to (\sim A) \equiv A \to B. \tag{3.64}$$

This proof format is called *contraposition*. In the same "cats-and-dogs" situation it means that the statements "if cats and dogs are not wet, then it is not raining" and "if it is raining, then cats and dogs get wet" are logically equivalent.

To prove the tautology (3.64), we transform the expression for the truth-value of its left-hand side in accordance with formulas (3.19) and (3.9),

$$\begin{aligned} t((\sim B) \to (\sim A)) &= 1 - t(\sim B) + t(\sim B)t(\sim A) \\ &= 1 - (1 - t(B)) + (1 - t(B))(1 - t(A)) \\ &= t(B) + (1 - t(B) - t(A) + t(B)t(A)) \\ &= 1 - t(A) + t(A)t(B) = t(A \to B). \end{aligned} \qquad (3.65)$$

Thus, the truth value of the left-hand side of Eq. (3.64) is always equal to that of the right-hand side, which just means that the statements of the left-hand side and of the right-hand side are really logically equivalent.

4)
$$\sim A \to (\sim B \wedge B) \equiv A. \qquad (3.66)$$

This especially useful proof tool has the very appropriate latin name *reductio ad absurdum*, which mean "reduction to the absurd", the name speaks for itself. This time, the cats and dogs are insufficient to illustrate the statement. We, however, have already had an illustration on usage of this proof format in Chapter 1, when we were proving the uniqueness of the empty set. In that case the statement A we had to prove was "the empty set $\emptyset$ is unique". Correspondingly, the statement $\sim A$ was "the empty set is not unique, that is, there is at least one other empty set, $\emptyset'$". Then, statement B looked like "$\emptyset \neq \emptyset'$", and, correspondingly, statement $\sim B$ was "$\emptyset = \emptyset'$". The assumption $\sim A$ led us to the conclusion that statements B and $\sim B$ must have held at the same time. This contradiction showed that the opposite statement, A, is right.

To prove the tautology (3.66), we again transform the truth value of its left-hand side, making use of the truth-value formulas (3.19), (3.11), and (3.9):

$$\begin{aligned} t((\sim A) \to ((\sim B) \wedge B)) &= 1 - t(\sim A) + t(\sim A)t(\sim B \wedge B) \\ &= 1 - t(\sim A) + t(\sim A)(1 - t(B))t(B) \\ &= 1 - t(\sim A) = 1 - (1 - t(A)) = t(A). \end{aligned} \qquad (3.67)$$

Thus, the truth value of the left-hand side is always equal to that of the right-hand side. So, the formula (3.66) is also a tautology.

In conclusion, we can make some general observations. Note that in establishing tautologies we were interested only in truth values of various logical formulas, not in their atomic-statement content. This was quite natural, because a formula being a tautology does not depend on its particular content *by definition*. The extreme manifestation of this principle is the statement that *all the tautologies are logically equivalent*, because the truth value of each of them equals 1. Thus, we can introduce a *nonspecific tautology*, denoted, say, I. Analogously, we can introduce a *nonspecific contradiction*, denoted O. These new notions of general tautology I and general contradiction O will be of great help in the next section, in constructing mathematical logic as an axiomatic system.

3.9 PROPERTIES OF CONNECTIVES. AN AXIOMATIC APPROACH TO MATHEMATICAL LOGIC

In the previous section we developed, in principle, all the necessary technical apparatus of mathematical logic. Thus, faced with even a very complicated logical formula, we can always determine its truth value through calculations based on our basic formulas for truth values. Sometimes, however, such direct calculations are very tedious. Thus, it is important to establish a complementary technique of attacking complex logical formulas. Before calculating the truth value, we can transform the formula to a more convenient one by making use of combinational properties of logical connectives. We have already discovered some of these properties, while reducing the number of adequate connectives, and while establishing the De Morgan tautologies. Now it is worth listing all these basic and useful properties and representing them in a systematic way.

This goal is more general that it might seem at first sight, because it advances another important aspect of mathematical logic, its architecture as a typical mathematical structure. In this general sense, we consider producing complex logical formulas by means of connectives as *mappings* of the set of logical formulas into itself. More precisely, connective $\sim$ transforms a logical statement into a new (opposite) one. Therefore, it is a mapping from the set $\mathcal{P}$ into itself. Each of the other two basic connectives, $\wedge$ and $\vee$, transforms two "input" logical statements into a new logical formula, thereby performing a mapping from the cross-product set $\mathcal{P} \times \mathcal{P}$ into set $\mathcal{P}$. All the other connectives, and quantifiers, can be expressed through these three connectives, that is, can be represented as superpositions of these three mappings. Thus, mathematical logic is considered as a mathematical structure based on set $\mathcal{P}$ and including the set itself and three specific mappings (or, as they are often called in this regard, operations): one mapping, $\sim$, of type $\mathcal{P} \to \mathcal{P}$, and two mappings, $\wedge$ and $\vee$ of type $\mathcal{P} \times \mathcal{P} \to \mathcal{P}$.

The three mappings are to be specified by describing their properties and interrelations. In what follows, we formulate these properties as axioms, though they are based on (and can be eventually proved by) the previously established rules of truth-value calculations. The following formulas hold for any "incoming" logical statements A, B, and C being elements of the set $\mathcal{P}$:

1) *commutativity*
$$A \wedge B \equiv B \wedge A;$$
$$A \vee B \equiv B \vee A. \qquad (3.68)$$

2) *associativity*
$$A \wedge (B \wedge C) \equiv (A \wedge B) \wedge C;$$
$$A \vee (B \vee C) \equiv (A \vee B) \vee C. \qquad (3.69)$$

3) *absorption*
$$A \wedge (A \vee B) \equiv A;$$
$$A \vee (A \wedge B) \equiv A. \qquad (3.70)$$

These three axioms describes "individual" properties of the operations. Axioms one and two look quite natural and have obvious analogues in arithmetic. The third axiom, however, is quite specific.

4) *distributivity*

$$A \wedge (B \vee C) \equiv (A \wedge B) \vee (A \wedge C);$$
$$A \vee (B \wedge C) \equiv (A \vee B) \wedge (A \vee C). \qquad (3.71)$$

This axiom describes relations between two operations. Note that the two operations here are on equal footing: interchanging $\wedge$ and $\vee$, say, in the first of the formulas (3.71) just transforms it into the second formula, and vise versa. This symmetry is a specific feature of Boolean algebra and has no analogy in the familiar structure of arithmetic.

The last two axioms relate to specific elements of the mathematical structure, the general tautology I and general contradiction O, and to the inversion operation described in terms of these specific elements.

5) *boundedness*

$$A \wedge O \equiv O;$$
$$A \vee I \equiv I. \qquad (3.72)$$

6) *complementation*

$$A \wedge \sim A \equiv O;$$
$$A \vee \sim A \equiv I. \qquad (3.73)$$

Having established these six axioms, we can abstract ourselves from the content of mathematical logic and treat our construction as a formal system of sets and operations satisfying the given axioms. This axiomatically defined mathematical structure, which may lose connections with its "logical" grounds, is called the *Boolean Algebra*.

3.10 SAMPLE PROBLEMS II

In the following problems, we use the properties of logical connectives to simplify complicated logical formulas.

I. Prove the formula

$$(\sim ((\sim Q) \vee (P \rightarrow Q))) \wedge ((Q \leftrightarrow R) \vee R) \qquad (3.74)$$

to be a tautology.

Solution
Before proceeding to truth-value calculations, we try transforming the first part of the

formula:

$$\sim ((\sim Q) \vee (P \to Q)) = (\sim (\sim Q)) \wedge (\sim (P \to Q))$$
$$= Q \wedge (\sim (P \to Q)) = Q \wedge (\sim ((\sim P) \vee Q))$$
$$= Q \wedge ((\sim (\sim P)) \wedge (\sim Q)) = Q \wedge (P \wedge (\sim Q))$$
$$= (Q \wedge (\sim Q)) \wedge P = O \wedge P = O. \tag{3.75}$$

Thus, the whole formula under consideration can be rewritten as:

$$O \wedge ((Q \leftrightarrow R) \vee R) \equiv O. \tag{3.76}$$

Thus, this is a *contradiction*. Note, that we come to this conclusion even without the truth-value calculations.

II. Prove the following formula with quantifiers,

$$(\forall x_i (P(x_i) \wedge Q(x_i))) \to ((\forall x_i P(x_i)) \wedge (\forall x_i Q(x_i))), \tag{3.77}$$

to be a tautology.

Solution
Since the quantifier $\forall$ can be reduced to multiple use of the connective $\wedge$, it is sufficient to check the following statement for a set of only two logical objects, x_1 and x_2:

$$(P(x_1) \wedge Q(x_1)) \wedge (P(x_2) \wedge Q(x_2)) \to$$
$$(P(x_1) \wedge P(x_2)) \wedge (Q(x_1) \wedge Q(x_2)). \tag{3.78}$$

According to commutativity of the connective $\wedge$, the left-hand side and the right-hand side in this formula are just the same. So, this is a tautology (and it will be tautology even if we write $\leftrightarrow$ instead of $\to$).

3.11 BOOLEAN ALGEBRA. THE ANALOGY BETWEEN MATHEMATICAL LOGIC AND SET THEORY

Now we can make use of the generalization we undertook in the preceding section, and substantiate the analogy between the logical connectives, $\vee$ and $\wedge$, and the operations of union and intersection of sets which we have already mentioned and even used in Section 3.4 to illustrate the relations between connectives $\vee$ and $\to$. Now, we will construct a system in the realm of sets, which would literally repeat the formal structure of mathematical logic.

In this new system, the role of its elements, analogous to logical statements in mathematical logic, should be taken by *sets*. Then, the operations of union and intersection of sets, $\cup$ and $\cap$, are really analogous to connectives $\vee$ and $\wedge$, respectively, including obeying the axioms of associativity, commutativity, and distributivity. This just follows from the fact that we defined operations ∪ and ∩ (in Section 1.8 of Chapter 1) by means of connectives ∨ and ∧.

To go further, however, we should find in the realm of sets the analogy of logical negation, $\sim$. This is not so straightforward, but if we restrict our consideration to sets being subsets of some given set A, we can define, as analog of logical negation, the so-called operation of *complementation*. For a given set $B \subset A$ its *complement* $\bar{B} \subset A$ is defined as the set containing all the elements a of the set A which are not included in B,

$$\forall a \, (a \in \bar{B} \leftrightarrow ((a \in A) \wedge (a \notin B))). \tag{3.79}$$

This definition clearly implies the following properties of set complementation:

$$\begin{aligned} \bar{B} \cup B &= A; \\ \bar{B} \cap B &= \emptyset. \end{aligned} \tag{3.80}$$

Example on complementation

Consider again our sample set of all natural numbers less than 5, that of Eq. (1.5), $D = \{1, 2, 3, 4\}$. Let $B \subset D$ be the set of all odd natural numbers less than 5, $B = \{1, 3\}$. Its complement is evidently the set of all even natural numbers less than 5, $\bar{B} = \{2, 4\}$. Then,

$$\begin{aligned} B \cup \bar{B} &= \{1, 3\} \cup \{2, 4\} = \{1, 3, 2, 4\} = \{1, 2, 3, 4\} = D; \\ B \cap \bar{B} &= \{1, 3\} \cap \{2, 4\} = \emptyset. \end{aligned} \tag{3.81}$$

Comparing these properties with the axioms of complementation and boundedness, Eqs. (3.73) and (3.72), we see that those axioms also hold for operations on sets, if the empty set takes the role of element O, and the set A itself takes the role of element I. Noticing, finally, that all the subsets of given set A just constitute the power set $\mathcal{P}(A)$ (we cannot help mentioning the ironic coincidence in the notations of the power set and of the set of logical formulas), we obtain our new mathematical structure as follows.

Given set A, the basic set of our structure is $\mathcal{P}(A)$, and its elements are subsets of A. This structure includes one mapping of type $\mathcal{P}(A) \to \mathcal{P}(A)$, corresponding to operation of complementation, $B \to \bar{B}$, and two different mappings of type $\mathcal{P}(A) \times \mathcal{P}(A) \to \mathcal{P}(A)$, corresponding to operations of union, $(B, C) \to B \cup C$, and intersection, $(B, C) \to B \cap C$, where B and C are subsets of the set A. As distinguished elements of the set $\mathcal{P}(A)$ playing a special role with respect to the introduced three operations, we note the empty set, $\emptyset$, and the set A itself. Given these distinguished elements, the three mentioned mappings satisfy all the six axioms of Boolean Algebra, Eqs. (3.68) through (3.73).

Thus, from the formal point of view, the system described in the realm of sets is a proper Boolean Algebra. It is not only analogous to Mathematical Logic, but formally indistinguishable from it!

3.12 SAMPLE PROBLEMS III

The following rather complicated problems might be used as a comprehension test. They invoke all the variety of methods and tricks we have learned in this chapter.

Classify the following logical formulas, that is, determine whether a given formula is a tautology, a contradiction, or neither.

I. $\exists x(P(x) \wedge Q(x)) \to ((\exists x P(x)) \wedge (\exists x Q(x)))$.

Solution
Similarly to what we did with the quantifier $\forall$ we mention that the quantifier $\exists$ can be reduced to multiple use of the connective $\vee$. Therefore, it is sufficient to check the formula for two objects, x_1 and x_2. In this case, the formula turns to be

$$(P(x_1) \wedge Q(x_1)) \vee (P(x_2) \wedge Q(x_2)) \to$$
$$(P(x_1) \vee P(x_2)) \wedge (Q(x_1) \vee Q(x_2)). \quad (3.82)$$

We formally transform the left-hand side as follows:

$$(P_1 \wedge Q_1) \vee (P_2 \wedge Q_2)$$
$$= (P_1 \vee (P_2 \wedge Q_2)) \wedge (Q_1 \vee (P_2 \wedge Q_2))$$
$$= ((P_1 \vee P_2) \wedge (P_1 \vee Q_2)) \wedge ((Q_1 \vee P_2) \wedge (Q_1 \vee Q_2))$$
$$= (P_1 \vee P_2) \wedge (Q_1 \vee Q_2) \wedge (P_1 \vee Q_2) \wedge (Q_1 \vee P_2);$$

$$t(\text{all this}) \leq t((P_1 \vee P_2) \wedge (Q_1 \vee Q_2)), \quad (3.83)$$

that is, the truth value of the left-hand side of the formula is always less or equal to the truth value of its right-hand side. Thus, this formula is a tautology.

II. $(R \wedge (Q \vee P) \wedge (Q \leftrightarrow P) \wedge (\sim (Q \wedge P))) \to (((\sim Q) \vee R) \leftrightarrow (P \wedge R))$

Solution
We transform the left-hand side as:

$$R \wedge (Q \vee P) \wedge (Q \leftrightarrow P) \wedge (\sim (Q \wedge P))$$
$$= R \wedge (Q \leftrightarrow P) \wedge ((Q \vee P) \wedge ((\sim Q) \vee (\sim P))). \quad (3.84)$$

Then,

$$(Q \vee P) \wedge ((\sim Q) \vee (\sim P))$$
$$= (Q \wedge ((\sim Q) \vee (\sim P))) \vee (P \wedge ((\sim Q) \vee (\sim P)))$$
$$= ((Q \wedge (\sim Q)) \vee (Q \wedge (\sim P))) \vee ((P \wedge (\sim Q)) \vee (P \wedge (\sim P)))$$
$$= (O \vee (Q \wedge (\sim P))) \vee ((P \wedge (\sim Q)) \vee O)$$
$$= (Q \wedge (\sim P)) \vee (P \wedge (\sim Q))$$

$$=\sim((\sim(Q \land (\sim P))) \land (\sim (P \land (\sim Q)))$$
$$=\sim(((\sim Q) \lor P) \land ((\sim P) \lor Q)) =\sim ((Q \to P) \land (P \to Q)). \quad (3.85)$$

Thus, we can rewrite the left-hand side as:

$$R \land (Q \leftrightarrow P) \land (\sim ((Q \to P) \land (P \to Q)))$$
$$= R \land (Q \leftrightarrow P) \land (\sim (Q \leftrightarrow P)) = R \land O = O. \quad (3.86)$$

Thus,

$$t(\text{left-hand side}) = 0, \quad (3.87)$$

which is always less or at least equal to the truth value of the right-hand side, whatever the latter is. Thus, the whole formula is a tautology.

III. $\forall v_i (P \to Q) \to \sim \exists v_j (P \land ((\sim Q) \lor (\sim P)))$.

Solution

We first transform the logical formula of the right-hand-side quantifier. Using the distributivity of the connectives, we get

$$(P \land ((\sim Q) \lor (\sim P))) = (P \land (\sim Q)) \lor (P \land (\sim P))$$
$$= (P \land (\sim Q)) \lor O = P \land (\sim Q)$$
$$=\sim ((\sim P) \lor (\sim (\sim Q))) =\sim ((\sim P) \lor Q) =\sim (P \to Q). \quad (3.88)$$

The original formula then turns out to be:

$$\forall v_i (P \to Q) \to \sim \exists v_j (\sim (P \to Q)). \quad (3.89)$$

We can denote $A = P \to Q$, and obtain the formula in the form:

$$\forall v_i (A) \to \sim \exists v_j (\sim A). \quad (3.90)$$

For a trial set of *two* logical objects this reads:

$$(A(v_1) \land A(v_2)) \to \sim ((\sim A(v_1)) \lor (\sim A(v_2))). \quad (3.91)$$

The right-hand side of this formula can be transformed as:

$$\sim ((\sim A_1) \lor (\sim A_2)) = (\sim (\sim A_1)) \land (\sim (\sim A_2))$$
$$= A_1 \cap A_2 = A(v_1) \land A(v_2), \quad (3.92)$$

which coincides with its left-hand side. Thus, this formula, as well as the original one, is a tautology.

IV. $((A \land (\sim B) \land (A \to B)) \lor ((\sim A) \land (\sim B) \land (A \to B))) \to$
$$(((\sim A) \land B) \lor (\sim (A \lor B))).$$

Solution

Both sides of this formula just seem made for using the distributivity properties of Eq. (3.71):

$$\text{left-hand side} = ((A \wedge (\sim B)) \vee ((\sim A) \wedge (\sim B))) \wedge (A \to B)$$
$$= ((A \vee (\sim A)) \wedge (\sim B)) \wedge (A \to B) = (I \cap (\sim B)) \wedge (A \to B)$$
$$= (\sim B) \wedge (A \to B) = (\sim B) \wedge ((\sim A) \vee B)$$
$$= ((\sim B) \wedge (\sim A)) \vee ((\sim B) \wedge B) = ((\sim B) \wedge (\sim A)) \vee O$$
$$= (\sim B) \wedge (\sim A). \tag{3.93}$$

On the other hand,

$$\text{right-hand side} = ((\sim A) \wedge B) \vee ((\sim A) \wedge (\sim B))$$
$$= (\sim A) \wedge (B \vee (\sim B)) = (\sim A) \wedge I = \sim A. \tag{3.94}$$

Then,

$$t((\sim B) \wedge (\sim A)) = t(\sim B) \cdot t(\sim A) \leq t(\sim A). \tag{3.95}$$

Thus, the truth value of the left-hand side is always less or equal to that of the right-hand side. So, the formula is a tautology.

3.13 SUMMARY OF CHAPTER 3

1) Mathematical Logic operates with *logical statements*, assertions having truth values and being made with respect to *logical objects*. The set of all logical objects is denoted $\mathcal{U}$, the set of all logical statements is denoted $\mathcal{P}$.

2) Given a relation R in the cross-product set $V_1 \times V_2 \times \ldots V_n$, there is a special mapping that transforms the n-tuples of the cross-product set into elementary (*atomic*) logical statements: $(v_1, \ldots v_n) \to ((v_1, \ldots v_n) \in R)$. This mapping $\mathcal{U} \to \mathcal{P}$ is called the *n-ary predicate function*.

3) The atomic statements and all the logical statements constructed from them by means of *logical connectives* are *logical formulas*.

4) Any logical connective can be reduced to a combination of the elements of the *adequate set of connectives*: either $\{\vee, \sim\}$ or $\{\wedge, \sim\}$.

5) Each logical formula has its *truth value*, either 0 (false), or 1 (true). The mapping of the set of logical formulas to the set of truth values, $\mathcal{B} = \{0, 1\}$, $\mathcal{P} \to \mathcal{B}$ is called the truth-value mapping or the *truth-value function*. The truth value of a logical formula A is denoted as $t(A)$.

6) Atomic statements have ascribed truth values. For any complex logical formula A its truth value, $t(A)$, can be calculated on the basis of the *Table of Truth-Value Formulas*, Table 3.1.

7) There exist nonspecific logical formulas, that is, formulas whose truth values do not depend on the truth-values of the included atomic statements. Such formulas

are called *tautologies* when their truth value is 1 and *contradictions* when their truth value is 0. The most important and useful among them are *De Morgan laws* and *proof formats*.

8) Mathematical logic can be represented as a formal mathematical structure based on the set of logical statements and on the basic connectives, $\sim$, $\wedge$, and $\vee$, and defined by the system of six axioms, which describe the properties of connectives. This formal structure is called the *Boolean Algebra*.

9) The mathematical structure of Set Theory developed in Chapter 1, when applied to a given set A, is the *Boolean Algebra* based on the power set $\mathcal{P}(A)$ and operations of complementation, intersection, and union. In this sense, it is equivalent to Mathematical Logic.

3.14 PROBLEMS

In each of the following problems classify the given logical formulas, that is, determine for each of them whether the formula is a tautology, a contradiction, or neither.

I.

$$\bigl((\sim(A \vee B)) \to ((\sim A) \wedge (\sim B))\bigr) \wedge \bigl(((\sim A) \wedge (\sim B)) \vee (A \vee B)\bigr).$$

II.

$$\bigl((\sim A) \to \bigl(\sim \bigl(((A \to B) \wedge (\sim B)) \to (\sim A)\bigr)\bigr)\bigr) \leftrightarrow A.$$

III.

$$\bigl(((\sim B) \to (\sim A)) \wedge (\sim (A \to B))\bigr) \vee \\ \bigl(\sim ((A \to B) \to ((\sim B) \to (\sim A)))\bigr).$$

IV.

$$\bigl(((A \to B) \wedge A) \to B\bigr) \wedge \bigl((A \to (A \vee B)) \to B\bigr).$$

3.15 FURTHER READING

The basic ideas of Mathematical Logic which are presented in this chapter can be found exposed in more detail in the following two general introductory books for senior level students:

E. Mendelson, *Introduction to Mathematical Logic*, Van Nostrand, New York, 1979;

R. L. Wilder, *Introduction to the Foundations of Mathematics* (2nd ed.), Wiley, New York, 1965.

A more rigorous consideration of the fundamental concepts of Mathematical Logic can be found in

H. B. Carry, *Foundations of Mathematical Logic*, Dover, New York, 1977.

A reader interested in particular topics can find helpful more specific texts such as:

J. C. Abbott, *Sets, Lattices, and Boolean Algebras*, Allyn and Bacon, Boston, 1969.

We have already mentioned this classical text in Chapter 2. It contains a precise and readable treatment of the formal structure of Mathematical Logic and its connections with Set Theory.

J. R. Shoenfield, *Mathematical Logic*, Addison-Wesley, Reading, MA, 1967.

This introductory text includes a treatment of formal systems and mathematical theory of models.

H. De Long, *A Profile of Mathematical Logic*, Addison-Wesley, Reading, MA, 1970.

This book mainly deals with the historical development of Mathematical Logic and with its general implications in the whole system of human knowledge.

The following two books are mainly devoted to computer implementations of Mathematical Logic:

M. Minsky, *Computation: Finite and Infinite Machines*, Prentice-Hall, Englewood Cliffs, NJ, 1967;

Z. Manna, *Mathematical Theory of Computation*, McGraw-Hill, New York, 1974.

4
Algebraic Structures: Group through Linear Space

4.1 CLASSIFICATION OF ALGEBRAIC STRUCTURES

In the end of Chapter 3, Mathematical Logic was introduced as a formal mathematical structure determined by properties of connectives and their interrelations expressed by axioms. This formal mathematical structure is rather complicated. The same principles, however, can be used to construct and classify much simpler mathematical structures. In the first part of this chapter we consider these principles in more detail, and present an hierarchy of algebraic structures, from very simple ones, being mainly of academic interest, to ones that are sufficient to adequately represent complicated real-world situations.

The primary classifying feature of a mathematical structure is the number and character of *operations* involved. The term *operation* in this context is used to denote a mapping from the cross-product set

$$V^m = \underbrace{V \times V \times V \times \cdots V}_{m \text{ times}} \qquad (4.1)$$

into the set V itself. The number m in Eq. (4.1) is called *arity* of the operation. Note, there is some terminological ambiguity here, since the *arity of operation* is something different from the *arity of a relation* introduced previously, in Chapter 2. This notational discord is due to historical reasons, and we can only be aware of the fact that an m-ary *operation* corresponds to (and may be treated as) an $(m+1)$-ary *relation*.

Examples

It is easy to indicate the arities of customary operations for which the set V of Eq. (4.1) is the set of real numbers, $\mathcal{R}$. Thus, the operation of negation, $x \to -x$, and the operation of absolute value, $x \to |x|$, are unary operations; the operation of addition $(x, y) \to x + y$ is a binary operation; the logic operation "if $x = 0$ then y else z" is a ternary operation.

With the notion of an n-ary operation, we are ready to consider various mathematical structures based on a given set V. Such mathematical structures are generally called *algebraic structures* or just *algebras*. The basic classifying elements of an algebraic structure are the following:
a) the set V itself, which is called the *carrier set* of the algebra, and which can be of any nature;
b) several *operations* of various "arities";
c) the identified special elements of the carrier set, which manifest special behavior with respect to the operations, and which are called *constants* of the algebra.

In the formal system of notations, incorporating these three basic elements, *algebra* is denoted as an n-tuple, whose elements are the carrier set, the operations (in the order of descending arities) and the constants. This n-tuple is (quite naturally) called the *signature* of the algebra. Using these notations, the signature of Mathematical Logic is

$$(\mathcal{P}, \wedge, \vee, \sim, I, O), \tag{4.2}$$

which incorporates the set of logical statements, $\mathcal{P}$, as the carrier set, the three logical operations (connectives), $\wedge$, $\vee$, and $\sim$, and two constants, the general tautology, I, and the general contradiction, O.

However, an algebra is not completely defined by its signature. What really matters, and what distinguishes one structure from another of the same signature, is the properties of the operations. These properties are expressed in the form of *axioms* of the algebra. Let two algebras have the same signature and let their operations satisfy the same set of axioms. Such algebras are indistinguishable as mathematical structures, though they might be based on quite different carrier sets and the particular meaning of their operations might have nothing in common. We refer to such algebras as to algebras of the same *variety*. Thus, the above-mentioned signature of Mathematical Logic and the set of its axioms listed in the end of the preceding chapter, define its variety. The seemingly different structure of operations on a set V has the signature $(\mathcal{P}(V), ^-, \cup, \cap, V, \emptyset)$. This signature is the same as that of Mathematical Logic, in the sense that it contains the same number of operations of given arities (one unary operation, the complementation $^-$, and two binary operations, the union $\cup$ and the intersection $\cap$), and the same number of constants (the set V and the empty set $\emptyset$). Having seen that the operations of these two structures satisfy the same axioms, we come to the conclusion that these are algebras of the same variety, called *Boolean Algebra*.

Quite often, it is possible to detach from the carrier set some subset which is closed with respect to all the operations of the given algebra, that is, the result of any operation on elements of the subset is an element of the same subset. In this case, the algebraic structure having this subset as its carrier set and containing all the other elements of the algebra is called a *subalgebra of the given algebra*. Note, that all the possible subalgebras of a given algebra must have at least the constants of this algebra as their common elements, because these constants are used in the system of axioms defining the structure (e.g., in axioms of boundedness and complementation of Boolean Algebra, Section 3.9).

Some varieties of algebras

To substantiate the general scheme, we consider here some simple algebraic structures.

A) *Semigroup* is the algebraic structure having signature $(V, \cdot)$, where $\cdot$ stands for a binary *associative* operation, that is, an operation satisfying the axiom

$$v_1 \cdot (v_2 \cdot v_3) = (v_1 \cdot v_2) \cdot \cdot v_3 \tag{4.3}$$

Note that no constants should be indicated in this structure. Note also that if a subset of the carrier set V is closed with respect to the operation $\cdot$, this operation is apparently associative on the subset. Thus, any such subset can be used as the carrier set of a subalgebra, *subsemigroup*. Since there are no required constants, various subsemigroups of a given semigroup may have no common elements.

Examples of semigroups

a) The set of natural numbers with operation of usual addition, $(\mathcal{N}, +)$;
b) The structure $(S_k, +)$ where set S_k is defined as $S_k = \{s \mid s \in \mathcal{I}, s \geq k\}$, that is, all the integer numbers greater than given number k, and the operation $+$ is again the usual addition. Note that this structure makes sense only for $k > 0$, and in this case this is a subalgebra of the semigroup defined in a). If $k < 0$, addition of two elements of the carrier set can give a number which lies outside the carrier set, say, $k + k$ is a negative number less than k. Therefore, in this case the carrier set is nonclosed with respect to addition, and the structure is not an algebra at all;
c) Another example of a structure which might pretend to be a semigroup, is $(\mathcal{I}, -)$, where $-$ means the usual subtraction. Careful inspection reveals, however, that this structure is not a semigroup, because its operation is not associative (say, $(5-2)-1 = 2$ while $5 - (2 - 1) = 4$).

B) *Monoid*, with the signature $(V, \cdot, 1)$, is a semigroup with identity element, which we symbolically denote 1, and whose properties are determined by the axiom

$$1 \cdot v = v \cdot 1 = v. \tag{4.4}$$

An important property of the identity element is that it is unique. We can easily show this by repeating, almost verbatim, our *reductio ad absurdum* consideration of the empty set in Chapter 1 (Section 1.8). Suppose there exist two *different* identity elements, 1 and $1'$, $1 \neq 1'$. According to Eq. (4.4),

$$1 \cdot 1' = 1'. \tag{4.5}$$

On the other hand, according to the same Eq. (4.4),

$$1 \cdot 1' = 1. \qquad (4.6)$$

Thus, $1 = 1'$, in contradiction to our assumption. Thus, the identity element is unique.

Examples of monoids
a) The set of natural numbers with the usual multiplication as the operation and with the number 1 as the identity element $(\mathcal{N}, \cdot, 1)$.
b) The structure $(S_0, +, 0)$, where S_0 is a set S_k of the example b) on semigroups (see the previous page), for $k = 0$, the operation $+$ is the usual addition, and the identity element is the number 0.

C) *Group*, the most celebrated of simple algebraic structures. This is a monoid with the operation of inversion, $(V, \cdot, ^-, 1)$, where the properties of this additional operation, inversion¯, are defined by the axiom

$$v \cdot \bar{v} = \bar{v} \cdot v = 1. \qquad (4.7)$$

Since this variety is of special importance, we summarize the three basic properties of the group as follows:
i) *associativity*

$$v_1 \cdot (v_2 \cdot v_3) = (v_1 \cdot v_2) \cdot v_3; \qquad (4.8)$$

ii) *identity*

$$v \cdot 1 = 1 \cdot v = v; \qquad (4.9)$$

iii) *inversion*

$$v \cdot \bar{v} = \bar{v} \cdot v = 1. \qquad (4.10)$$

On the basis of these three axioms, the following important statements regarding groups are established.
1) For a given element of a group, x, its inverse, $\bar{x}$, is unique. Employing *reductio ad absurdum*, suppose that for a given x there are two different inverse elements, $\bar{x}$ and $\bar{x}'$. Then, according to the properties of Eqs. (4.8) and (4.10),

$$\bar{x}' \cdot x \cdot \bar{x} = \bar{x}' \cdot (x \cdot \bar{x}) = \bar{x}' \cdot 1 = \bar{x}'. \qquad (4.11)$$

On the other hand, the same product can be transformed as

$$\bar{x}' \cdot x \cdot \bar{x} = (\bar{x}' \cdot x) \cdot \bar{x} = 1 \cdot \bar{x} = \bar{x}. \qquad (4.12)$$

Thus, $\bar{x}'$ must equal $\bar{x}$, which contradicts our assumption and thereby establishes uniqueness of the inverse element.

CLASSIFICATION OF ALGEBRAIC STRUCTURES 71

2) For a, b, and c being elements of a group,

$$\text{if } a \cdot c = b \cdot c, \text{ then } a = b; \tag{4.13}$$

$$\text{if } c \cdot a = b \cdot c \text{ then also } a = b. \tag{4.14}$$

These statements are known as right and left *cancellation laws*, respectively. To prove, for instance, the statement of Eq. (4.13), we just multiply both sides of this equation by the inverse element $\bar{c}$ on the right,

$$(a \cdot c) \cdot \bar{c} = (b \cdot c) \cdot \bar{c}, \tag{4.15}$$

and use the property of associativity to transform this to

$$a \cdot (c \cdot \bar{c}) = b \cdot (c \cdot \bar{c}). \tag{4.16}$$

Now, $c \cdot \bar{c} = 1$ in both sides of Eq. (4.16), and statement (4.13) holds.
3) Equation

$$a \cdot x = b \tag{4.17}$$

always has a unique solution on a group. This is the most important property of a group and the reason for the extensive popularity of this algebraic structure. To establish this property, we explicitly find the solution by multiplying both sides of Eq. (4.17) on the left by the *unique* inverse element $\bar{a}$,

$$\bar{a} \cdot (a \cdot x) = \bar{a} \cdot b \Rightarrow x = \bar{a} \cdot b. \tag{4.18}$$

In other words, on a group V the mapping $V \to V$ transforming an element x into the element $a \cdot x$, where $a \in V$ is a bijection.

For many important groups, in addition to the three basic properties just mentioned, the following axiom holds:
iv) *commutativity*

$$x \cdot y = y \cdot x. \tag{4.19}$$

Such four-axiom groups are called *commutative*, or *Abelian* groups. In fact, most of the groups we encounter in solving engineering problems, are Abelian groups. We will, however, meet an important non-Abelian group in the next chapter.

It is important to note that in describing groups (especially Abelian groups) an equivalent system of notation is also in use, with the symbol $+$ for the group operation. In this way, the inverse element to an element v is naturally denoted $(-v)$, and the identity element is "generalized zero" denoted as θ. In these symbols, the group signature becomes

$$(V, +, -, \theta), \tag{4.20}$$

while the Abelian group axioms (4.8)–(4.11) look like

a) $v_1 + (v_2 + v_3) = (v_1 + v_2) + v_3$, the axiom of associativity;
b) $v_i + \theta = \theta + v_i = v_i$, the axiom of identity;
c) $\bar{v}_i : v_i + (-v_i) = \theta$, the axiom of inversion;
d) $v_1 + v_2 = v_2 + v_1$, the axiom of commutativity.

In particular, we will use this equivalent system of notation later in this chapter, when dealing with linear spaces.

Examples of groups
a) The best known example of a group is the *group of integer numbers* with respect to the operation of addition, $(\mathcal{I}, +, -, 0)$. In this group, the operation of inversion is usual negation, $v \to -v$, and the identity element is the number 0.
b) Another common example of a group is the group of real numbers with respect to operation of multiplication. Strictly speaking, the carrier set of this group is the set of real numbers *except 0*. The group operation is the usual multiplication, the identity element is the number 1, and the inversion proceeds as $v \to 1/v$.
c) A less trivial example of a group is a group of symmetries of a plane curve (of a geometric figure). The elements of such a group are all the possible rigid motions of the plane that transform the curve into itself. The group operation is a combination, or successive application, of two such motions. Then, the identity element is "no-motion", the absence of motion. The inverse element is the inverse motion. In a simple case of the parabola of Fig. 2.9, there is only one non-identical motion of the plane which transforms the curve into itself: the reflection in the y-axis. This motion is its own inverse. Thus, the carrier set consists of only two elements.

Note that all these examples dealt with Abelian groups.

D) The most useful of simple algebraic structures with two operations is a *lattice*. It has the signature $(S, \cdot, +)$. Here $\cdot$ and $+$ stand for two different binary operations (of course, these operations are not necessarily the usual addition and multiplication!), which satisfy the following axioms:
i) associativity (for both operations);
ii) commutativity (for both operations);
iii) distributivity, that is, property of interrelations between the two operations,

$$(s_1 + s_2) \cdot s_3 = s_1 \cdot s_3 + s_2 \cdot s_3 \tag{4.21}$$

and

$$(s_1 \cdot s_2) + s_3 = (s_1 + s_3) \cdot (s_2 + s_3). \tag{4.22}$$

E) In the *Boolean Algebra*, with which we started our journey in algebraic structures, the two operations of the *lattice*, $\cdot$ and $+$, acquire their constants and inverses. The two constants, I and O, the identity elements for the operations $\cdot$ and $+$, respectively, are specified in the carrier set, so that for any element s of this set

$$s \cdot I = s \quad \text{and} \quad s + O = s. \tag{4.23}$$

The peculiar thing with the Boolean Algebra is that the operation of inversion is *the same* for both $\cdot$ and $+$ operations. This frugal inversion is called *complementation*, $\bar{s}$

and its properties are defined by the axioms,

$$s \cdot \bar{s} = O \quad (4.24)$$

and

$$s + \bar{s} = I. \quad (4.25)$$

Note the mixup of operations and constants here: if $\bar{s}$ were a regular *group* inversion, it would be I in Eq. (4.24) and O in Eq. (4.25)! Thus, in our classification, Boolean Algebra is lattice with unary operation of complementation and two constants, $(S, \cdot, +, ^-, O, I)$. The whole list of its axioms has been already given in the end of Chapter 3 (Section 3.9).

Our goal is not to exhaustively describe all the possible algebraic structures. Thus, we will not discuss the *algebra of real numbers*. We assume the reader to be well acquainted with real numbers and the properties of the operations of usual addition and multiplication, which have been repeatedly dealt with in various mathematical courses.

4.2 ALGEBRA OF COMPLEX NUMBERS

Besides the algebra of real numbers, there is another mathematical structure which is very important in applications, so important that we consider it in this separate section. This is the *algebra of complex numbers*. In this case the carrier set is $\mathcal{C} = \mathcal{R} \times \mathcal{R}$; its elements are two-tuples (ordered pairs) with their components being real numbers, $z = (x, y)$. The first component in such a pair, x, is called the *real part* of the complex number, the second component, y, the *imaginary part* of the complex number.

Two binary operations of this structure, addition and multiplication, are defined in terms of the addition and multiplication of real numbers in the following way.

For two complex numbers, $z_1 = (x_1, y_1)$ and $z_2 = (x_2, y_2)$, the result of their addition is defined (quite naturally) as

$$z_1 + z_2 = (x_1 + x_2, y_1 + y_2). \quad (4.26)$$

Correspondingly, the unary operation of inversion with respect to addition of Eq. (4.26) is defined as

$$-z = (-x, -y), \quad (4.27)$$

and the related constant, the identity element with respect to addition, comes out as

$$O = (0, 0). \quad (4.28)$$

In contrast, the operation of multiplication is defined in a seemingly weird way,

$$z_1 \cdot z_2 = (x_1 \cdot x_2 - y_1 \cdot y_2, \ x_1 \cdot y_2 + x_2 \cdot y_1), \quad (4.29)$$

with the related identity element being

$$1 = (1, 0) \tag{4.30}$$

and unary operation of inversion defined as

$$\frac{1}{z} = \left(\frac{x}{x^2 + y^2}, -\frac{y}{x^2 + y^2} \right). \tag{4.31}$$

We leave it to the reader to check that each of the defined operations, that is, both addition and multiplication, along with their constants and inverses, satisfy all the axioms of an Abelian group. The interrelation between these two operations (or between these two Abelian groups based on the same carrier set) is established by the axiom of distributivity,

$$z_1 \cdot (z_2 + z_3) = z_1 \cdot z_2 + z_1 \cdot z_3. \tag{4.32}$$

(Note the difference between this property and the "symmetrical" distributivities of the Boolean Algebra, Eqs. (4.21, 4.22).)

Overall, the operations of our structure satisfy all the axioms possessed by the algebra of real numbers. Represented as a signature, this new construction is

$$\left(C, \cdot, +, -, \frac{1}{z}, (1,0), (0,0) \right). \tag{4.33}$$

Looking at the constants $(1, 0)$ and $(0, 0)$, we are tempted to ask: "what about the complex number $(0, 1)$?" Does this number also play some role as yet another special element of the carrier set? Actually, it does. It is the so-called *imaginary unity*, which is usually denoted i in mathematical texts and j in electrical engineering texts. The main property of this complex number is

$$i^2 = -(1, 0) = (-1, 0). \tag{4.34}$$

This property allows us to modify the system of notation and to make it more handy. Based on Eq. (4.26), we can decompose any complex number into two numbers, one having zero imaginary part and the other having zero real part:

$$z = (x, y) = (x, 0) + (0, y) = (x, 0) + (y, 0)(0, 1). \tag{4.35}$$

Making use of Eq. (4.29), we rewrite this as

$$z = (x, 0) + (y, 0)(0, 1). \tag{4.36}$$

Then, we set a convention to denote the zero-imaginary-part numbers just as real numbers,

$$(x, 0) \equiv x; \quad (y, 0) \equiv y; \quad \text{etc.} \tag{4.37}$$

Thus, we arrive to the new notation,

$$z = x + i \cdot y, \qquad (4.38)$$

which allows one to operate with complex numbers by directly applying the usual operations with real numbers taken along with the property $i^2 = -1$ of Eq. (4.34). For instance,

$$\begin{aligned} z_1 \cdot z_2 &= (x_1 + i \cdot y_1) \cdot (x_2 + i \cdot y_2) \\ &= x_1 \cdot x_2 + i \cdot y_1 \cdot x_2 + i \cdot x_1 \cdot y_2 + i^2 \cdot y_1 \cdot y_2 \\ &= (x_1 \cdot x_2 - y_1 \cdot y_2) + i \cdot (y_1 \cdot x_2 + x_1 \cdot y_2), \end{aligned} \qquad (4.39)$$

cf. Eq. (4.30). The reader can verify that this new system of notation is completely equivalent to the original one.

The backlash of the additional constant, i, having appeared in the algebra is that an additional operation emerges. This is the unary operation of *complex conjugation*,

$$z^* = x - i \cdot y, \qquad (4.40)$$

that is, changing the sign of the imaginary part. The following relations between the operation of complex conjugation and the other operations of the algebra can be verified by direct calculations:

$$(z_1 \cdot z_2)^* = z_1^* \cdot z_2^*; \qquad (4.41)$$

$$(z_1 + z_2)^* = z_1^* + z_2^*. \qquad (4.42)$$

Then, in the new system of notation we write for constants $(1, 0)$ and $(0, 0)$ just 1 and 0, respectively. The completed signature of our structure thus takes the form

$$\left(\mathcal{C}, +, -, \cdot, \frac{1}{z}, z^*, 1, 0, i \right). \qquad (4.43)$$

Taken along with all the above-mentioned axioms, this is the complete specification of the *algebra of complex numbers*, the mathematical structure to which we will often return during this course.

4.3 GEOMETRIC REPRESENTATION OF COMPLEX NUMBERS

The sole purpose of this section is to introduce a useful representation of complex numbers and a corresponding system of notation which we are going to use in the following chapters. Being a two-tuple with real-number components, a complex number, $z = (x, y)$, can be represented as a point in the Cartesian plane (see Section 2.4), as shown in Fig. 4.1. In this case the Cartesian plane is called the *complex plane z*; the x and y axes are called the *real axis* and the *imaginary axis*, respectively.

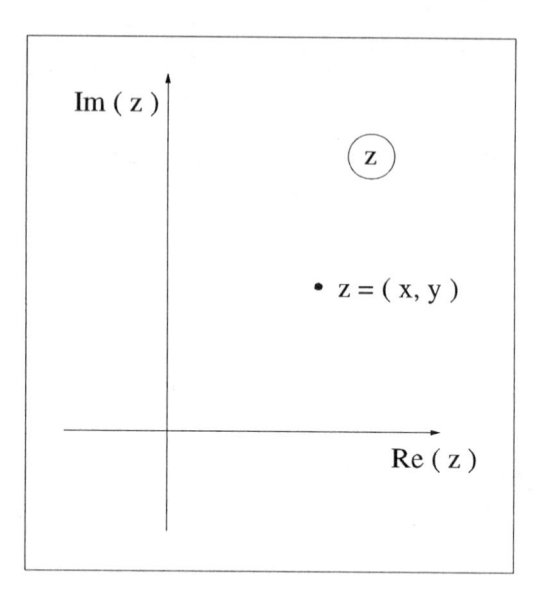

Fig. 4.1 The complex plane z.

Once the complex numbers are represented as points in the plane, their description is not restricted by the cartesian coordinate system. For instance, the points of the z-plane can be described by *polar coordinates*. In this coordinate system any complex number is specified by its magnitude (or *amplitude*), denoted $|z|$, and by angle (or *argument*), φ, with respect to the real axis, see Fig. 4.2. Combining the known expressions for the polar coordinates with the expressions of Eqs. (4.38) and (4.39), we obtain the following formulas for $|z|$ and φ:

$$|z| = \sqrt{x^2 + y^2} = \sqrt{z \cdot z^*}; \qquad (4.44)$$

$$\varphi = \tan^{-1}\left(\frac{y}{x}\right) = \tan^{-1}\left(\frac{z - z^*}{i \cdot (z + z^*)}\right). \qquad (4.45)$$

Inversely, the real and imaginary parts of the number z can be expressed in terms of its amplitude and argument as

$$x = |z| \cdot \cos\varphi, \quad y = |z| \cdot \sin\varphi, \qquad (4.46)$$

so that,

$$z = |z| \cdot (\cos\varphi + i \cdot \sin\varphi). \qquad (4.47)$$

In the case where $|z| = 1$ this relates to *Euler's formula*,

$$e^{i \cdot \varphi} = \cos\varphi + i \cdot \sin\varphi. \qquad (4.48)$$

So, the complex number z is expressed as

$$z = |z| e^{i \cdot \varphi}, \tag{4.49}$$

while its conjugate is

$$z^* = |z| e^{-i \cdot \varphi}. \tag{4.50}$$

From formulas (4.49) and (4.50) we can also write for the argument φ,

$$\varphi = \frac{1}{2 \cdot i} \cdot \ln\left(\frac{z}{z^*}\right), \tag{4.51}$$

to be used instead of (4.45).

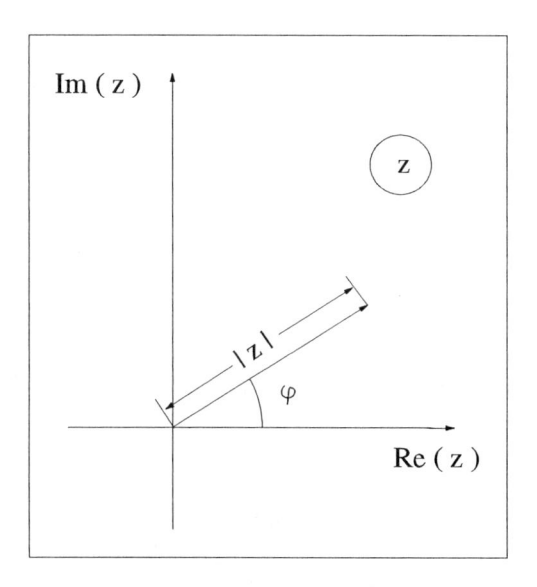

Fig. 4.2 The amplitude $|z|$ and the argument φ of a complex number z.

4.4 LINEAR SPACE: AXIOMS AND EXAMPLES

Following the discourse on mathematical structures in the first section of this chapter, the reader could ask, why we committed ourselves to structures based on only one carrier set? In reality, this is quite artificial, and now we will consider structures dealing with more than one set. However, this can lead to an extremely wide variety of structures, so we will consider only that one structure which is of most importance in various applications and, consequently, that for which the theory is most developed. This structure is called a *linear space*.

A linear space is constructed in the following way:

A) First, an Abelian group of Eq. (4.20) is taken, with its carrier set, V, binary operation + (symbolically called addition), inversion −, the identity element θ, and four axioms.
B) Then, the algebra of complex numbers is added, with all its numerous properties and axioms.
C) Finally, an operation is introduced to connect these two realms. This additional operation is the mapping $V \times C \to V$, which transforms a two-tuple consisting of an arbitrary element of the carrier set, v, and an arbitrary complex number, α, into a certain element of the carrier set, $v' = \alpha \cdot v$. (Now we see why it was wise to take the symbol + for the group operation.)

The basic properties of this new operation are determined by the following four axioms):

1) *associativity*

$$\alpha_1 \cdot (\alpha_2 \cdot v) = (\alpha_1 \cdot \alpha_2) \cdot v; \qquad (4.52)$$

2) *first distributivity*

$$\alpha_1 \cdot (v_1 + v_2) = \alpha_1 \cdot v_1 + \alpha_1 \cdot v_2; \qquad (4.53)$$

3) *second distributivity*

$$(\alpha_1 + \alpha_2) \cdot v_1 = \alpha_1 \cdot v_1 + \alpha_2 \cdot v_1; \qquad (4.54)$$

4) *identity*

$$0 \cdot v = \theta \quad \text{and} \quad 1 \cdot v = v, \qquad (4.55)$$

for any element v of the carrier set. Here, θ is the identity element with respect to the group operation (since this group operation is thought of as addition, this identity element is most often referred to as the *zero element*).

Notes on notations

1) While considering the linear space, it is convenient and generally accepted to denote complex numbers with lower-case Greek letters, such as α, β, and so on, to distinguish them from the elements of the carrier set which are denoted by lower-case Roman letters. This contradicts to the notation of complex numbers as z which is used when these numbers are considered separately, but this is not the only terminological discord between different fields of mathematics.

2) Sometimes, in the case of a linear space the elements of the carrier set are called *vectors*, and the complex numbers are called *scalars*. Correspondingly, the linear space in this case is called the *vector space*, the group operation is called vector addition, and the additional operation of the linear space is called multiplication of a vector by a scalar.

3) The algebra of complex numbers has been taken as the second constructive unit for the linear space. We could make another choice, and take the algebra of real numbers to provide the scalars. In order to distinguish these two cases, the resulting

LINEAR SPACE: AXIOMS AND EXAMPLES 79

constructions are called a *complex linear space* and a *real linear space*.

Examples of linear spaces

The simplest example of a complex linear space is the algebra of complex numbers itself. In this degenerate case, the carrier set of the space coincides with the set of complex numbers, $\mathcal{C}$, and the operation of multiplication of a vector by a scalar coincides with the usual multiplication of complex numbers. We invite the reader to check that all properties of a linear space hold in this case.

Another example is a *real* vector space $\mathcal{R}^3$, whose elements are the usual three-dimensional vectors expressed as columns of their cartesian coordinates. Then the addition and multiplication by a scalar are defined as:

$$\begin{pmatrix} a \\ b \\ c \end{pmatrix} + \begin{pmatrix} a_1 \\ b_1 \\ c_1 \end{pmatrix} = \begin{pmatrix} a + a_1 \\ b + b_1 \\ c + c_1 \end{pmatrix}; \quad (4.56)$$

$$\alpha \cdot \begin{pmatrix} a \\ b \\ c \end{pmatrix} = \begin{pmatrix} \alpha \cdot a \\ \alpha \cdot b \\ \alpha \cdot c \end{pmatrix}. \quad (4.57)$$

Again, the curious reader can verify that in this case all the axioms of a linear space are satisfied.

The third, less trivial, example is the set of all *real-valued functions*, $f(x)$, of the real variable x on a given interval, say, $x \in [0, 1]$. By the sum of two functions, $f_1(x)$ and $f_2(x)$, we understand the function $f_1(x) + f_2(x)$, whose value at any point of the interval equals the sum of the values of those two functions, as we are accustomed to in the usual calculus. Then, also in accordance with the usual calculus, by multiplication of a function $f(x)$ by a real number, α, we understand the function $\alpha \cdot f(x)$, whose value at any point x is α times the value of the function $f(x)$. Thus, in the case of functions the operations of addition and multiplication by a scalar are defined though addition and multiplication of numbers, the values of the functions at the points of the interval. Therefore, these operations with functions satisfy all the axioms of the operations with numbers. This, in turn, means that all the axioms of a linear space are satisfied. So, the real-valued functions on the interval $[0, 1]$ constitute a *real* linear space. Note, that the identity (zero) element in this space is the function $f(x) = 0$ that equals zero on the whole interval. The elements of this linear space (and operations on them) are easy to visualize as various curves in the cartesian plane with the argument x as the abscissa and a function $f(x)$ as the ordinate.

In the same vein, we can consider *complex-valued* functions of a real variable on a given interval, say $[0, 1]$. Using complex numbers as scalars, we construct a *complex* linear space of such functions, though such a space is much harder to visualize than its *real* sibling.

4.5 LINEAR COMBINATION AND SPAN. LINEAR MAPPINGS

Let $v_1, v_2, \ldots v_n$ be elements of the linear space V so that they constitute some subset

$$\{v_i\} = \{v_1, v_2, \ldots, v_n\} \tag{4.58}$$

of the carrier set V. A sum of the products of these elements with some scalar coefficients, $\sum_{i=1}^{n} \alpha_i \cdot v_i$, is called a *linear combination* of the elements $v_1, v_2, \ldots v_n$. Note, that the subset $\{v_i\}$ may contain an infinite number of elements; correspondingly, the linear combination may contain an infinite number of terms.

If a linear combination produces the zero element of the space V,

$$\sum_{i=1}^{n} \alpha_i \cdot v_i = \theta, \tag{4.59}$$

it is called a *trivial* linear combination; otherwise a linear combination is called *non-trivial*. It is evident that one can always obtain a trivial linear combination by taking all the coefficients α_i to be zero.

Let us now construct all possible linear combinations of the elements of set $\{v_i\}$ that is, the combinations with all the possible choices of coefficients α_i. The infinite set of all these combinations is called the *span* of set $\{v_i\}$ and denoted $[\{v_i\}]$. Sometimes the set $\{v_i\}$ is said to span (or to generate) set $[\{v_i\}]$.

The notion of linear combination is so fundamental that a natural question arises, what happens with a linear combination when a linear space is being mapped into another linear space. For a mapping in general, nothing can be concluded, since there are too many various kinds of mappings. However, there exists a vast class of mappings that preserve linear combination. A mapping F from a linear space V into another linear space, W is called a *linear mapping*, if

$$f\left(\sum_{i=1}^{n} \alpha_i \cdot v_i\right) = \sum_{i=1}^{n} \alpha_i \cdot f(v_i), \tag{4.60}$$

that is, a linear combination of elements of the space V is mapped into a linear combination of their values with *the same coefficients*. We will meet linear mappings many times and in many fields throughout this book.

4.6 LINEAR SUBSPACES

If some subset $V_1 \subseteq V$ happens to coincide with its own span, $V_1 = [V_1]$, such a set V_1 is called the *linear subspace of space* V. More formally: a subset $V_1 \subseteq V$ is called a linear subspace of the space V if it contains all the linear combinations of its elements, that is, for any elements v_1 and v_2 of the subset V_1 and for any complex numbers α_1 and α_2, the combination $\alpha_1 \cdot v_1 + \alpha_2 \cdot v_2$ is also an element of V_1:

$$(v_1, v_2 \in V_1) \wedge (\alpha_1, \alpha_2 \in \mathcal{C}) \rightarrow (\alpha_1 \cdot v_1 + \alpha_2 \cdot v_2 \in V_1). \tag{4.61}$$

Thus, a linear subspace is a nonempty subset which is closed under addition and scalar multiplication.

In particular, any subspace must contain the trivial linear combination of its elements, thus the zero element θ is necessarily an element of any linear subspace. Thus, the smallest possible subspace (the trivial subspace) contains only the zero element, θ. Obviously, if the subspace is not the trivial one, it must contain an infinite number of elements. The largest possible subspace of the space V is the space V itself. The concept of linear subspace, as we will shortly see, plays a crucial role in the construction and application of linear spaces.

Examples of linear subspaces

1) Let us return to the example of the real linear space $\mathcal{R}^3$ of Section 4.5. Consider now a very simple subset in this space, the subset consisting of only one element, the vector:

$$\begin{pmatrix} 1 \\ 0 \\ 0 \end{pmatrix}. \tag{4.62}$$

The span of this subset contains all the vectors of the form:

$$\begin{pmatrix} \alpha \\ 0 \\ 0 \end{pmatrix}, \tag{4.63}$$

where α is an arbitrary real number. Clearly, any linear combination of such vectors produces a vector of the same form:

$$\alpha_1 \cdot \begin{pmatrix} \alpha \\ 0 \\ 0 \end{pmatrix} + \alpha_2 \cdot \begin{pmatrix} \beta \\ 0 \\ 0 \end{pmatrix} = \begin{pmatrix} \alpha_1 \cdot \alpha + \alpha_2 \cdot \beta \\ 0 \\ 0 \end{pmatrix}. \tag{4.64}$$

Thus, this span is a subspace. (Actually, this is a *line* in a three-dimensional space. Note, that a half-line, say, all the vectors of the form of Eq. (4.63) with *positive* α, is not a subspace: the linear combination of Eq. (4.64) may well produce a vector with *negative* α!)

2) Having turned to the third example of linear space of Section 4.5, that of the real-valued functions on the interval $[0, 1]$, we consider a subset of such functions, the set of polynomials of x. Indeed, any linear combination of polynomials of x is a polynomial of x, so this subset coincides with its span. Another example of a subspace in the space of real-valued functions is the set of functions all of which are zero at the point $x = 1/2$. The reader is invited to show that this subset is also a linear subspace.

Note that if V_1 and V_2 are two subspaces of a linear space V, then their union and intersection are also subspaces of V. Let us show this for the case of intersection, $V_1 \cap V_2$. It can be seen that if two elements of the space V, v_1 and v_2, both belong to $V_1 \cap V_2$, then both of them belong to V_1 *and* both of them belong to V_2, $v_1, v_2 \in$

$V_1 \wedge v_1, v_2 \in V_2$. But this implies that for arbitrary constants α_1 and α_2, $(\alpha_1 v_1 + \alpha_2 v_2 \in V_1) \wedge (\alpha_1 v_1 + \alpha_2 v_2 \in V_2)$. Thus, this linear combination, $\alpha_1 v_1 + \alpha_2 v_2$ belongs to the set $V_1 \cap V_2$, which means that this intersection is a linear subspace, according to Eq. (4.61).

By its very definition, the span of a set contains all the possible linear combinations of its elements, that is, it is closed with respect to the operations of the linear space. In other words, for any subset $V_1 \in V$, its span $[V_1]$ constitutes a subspace. Actually, it is the smallest subspace containing the set V_1. Moreover, if some subspace V_2 contains the whole set V_1 then this subspace V_2 necessarily contains all the span $[V_1]$.

4.7 LINEAR INDEPENDENCE

A set $\{v_i\} = \{v_1, v_2, \ldots v_n\}$ is called *linearly independent* if the only trivial (in the sense of Eq. (4.59)) linear combination of the vectors v_i is that with all zero coefficients,

$$\sum_{i=1}^{n} \alpha_i \cdot v_i = \theta \Rightarrow \forall \alpha_i = 0. \tag{4.65}$$

If there is a trivial linear combination of the vectors $\{v_i\}$ with at least one non-zero coefficient, then the vectors $v_1, \ldots, v_n$ (or the subset $\{v_i\}$) are called *linearly dependent*.

Examples of linearly dependent sets
1) If even one of the vectors $\{v_i\}$ is θ, then $\{v_1, \ldots v_n\}$ are linearly dependent, since we always can construct a trivial linear combination of these vectors with just one non-zero coefficient that stays at the zero vector.
2) Any set of *collinear* vectors in the space $\mathcal{R}^3$ is linearly dependent, for example, the two vectors,

$$\begin{pmatrix} a \\ 0 \\ 0 \end{pmatrix} \text{ and } \begin{pmatrix} b \\ 0 \\ 0 \end{pmatrix}, \tag{4.66}$$

are clearly linearly dependent: multiplying the first vector by b/a an subtracting the result from the second vector, we obtain the zero vector.

4.8 BASIS AND DIMENSION OF A LINEAR SPACE

Definition: A finite set $\{v_1, \ldots v_n\}$ of *linearly independent* vectors is said to be a *basis* for the linear space V, if $\{v_1, \ldots v_n\}$ *generates* the whole space V that is, $V = [\{v_1, \ldots v_n\}]$.

A vector space having a finite basis is called *finite-dimensional*. An example of finite-dimensional linear space is the space $\mathcal{R}^3$ (later in this section we will construct a basis in this space). We will also consider as finite-dimensional the special one-element linear space, the so-called *trivial* space, $V = \{\theta\}$. (The complication here is that the sole element of this space, $v = \theta$, cannot be a basis, because linearly independent vectors must be non-zero vectors.) A vector space, which cannot be represented as the span of a finite basis, is called *infinite-dimensional*. An example of an infinite-dimensional linear space is the space of functions on the interval $[0, 1]$. In fact, the usefulness of the concept of basis, which will be demonstrated in this chapter, makes it highly desirable to introduce some analog of this concept in the infinite-dimensional case too. However, the ensuing problems occur to be quite complicated. They require a whole bunch of new concepts and notions, so that we will only be able to properly handle the infinite-dimensional case in Chapter 8.

The definition of basis implies that any vector $\tilde{v} \in V$ can be expressed as a linear combination of the vectors of a basis,

$$\tilde{v} = \sum_{i=1}^{n} \alpha_i \cdot v_i \quad (v_i \in \{v_1, \ldots v_n\}). \tag{4.67}$$

The important thing here is that this representation (also called the decomposition) of the vector $\tilde{v}$ is unique. We can prove this important statement by our favorite method, *reductio ad absurdum*. Suppose we have another decomposition of the same vector $\tilde{v}$, with different coefficients,

$$\tilde{v} = \sum_{i=1}^{n} \beta_i \cdot v_i. \tag{4.68}$$

Then, as the expressions of Eq. (4.67) and Eq. (4.68) represent the same vector,

$$\sum_{i=1}^{n} \alpha_i \cdot v_i = \sum_{i=1}^{n} \beta_i \cdot v_i. \tag{4.69}$$

This means,

$$\sum_{i=1}^{n} \alpha_i \cdot v_i - \sum_{i=1}^{n} \beta_i \cdot v_i = \sum_{i=1}^{n} (\alpha_i - \beta_i) \cdot v_i = \theta. \tag{4.70}$$

The vectors $\{v_i\}$ are linearly independent, hence the last equation implies

$$\alpha_i = \beta_i \ \forall i, \tag{4.71}$$

which contradicts our assumption and thereby proves the uniqueness of the representation.

The other important thing is that although the choice of a basis is not unique, all the bases of a given linear space contain *the same number* of elements. Again, suppose $\{x_1, \ldots x_n\}$ and $\{y_1, \ldots y_m\}$ are two bases of a linear space V; and let $m \neq n$. To be specific, let $m > n$. We can represent the elements of the "y-basis" through the elements of the "x-basis",

$$y_1 = \sum_{i=1}^{n} \alpha_i^{(1)} \cdot x_i$$

$$y_2 = \sum_{i=1}^{n} \alpha_i^{(2)} \cdot x_i$$

$$y_3 = \sum_{i=1}^{n} \alpha_i^{(3)} \cdot x_i$$

$$y_4 = \sum_{i=1}^{n} \alpha_i^{(4)} \cdot x_i$$

$$\vdots$$

$$y_m = \sum_{i=1}^{n} \alpha_i^{(m)} \cdot x_i. \qquad (4.72)$$

Now, we can treat these expressions of y_j through x_i as a system of linear equations for x_i. If y_j are linearly independent and $m > n$, this system is inconsistent (overspecified). This means that we *cannot* express the vectors $\{x_i\}$ through the vectors $\{y_j\}$. But this must not be true, because the latter constitute a basis. Therefore, the assumption $m \neq n$ was wrong. Thus, $m = n$.

The number of elements in a basis (which we just have shown to be the same for any basis) is called the *dimension* of the linear space V. It is a fundamental characteristic of a linear space. Of course, the notion of dimension makes sense only for finite-dimensional spaces.

Example of a basis
In the space $\mathcal{R}^3$ we can choose as a basis the following set:

$$v_1 = \begin{pmatrix} 1 \\ 0 \\ 0 \end{pmatrix}, \quad v_2 = \begin{pmatrix} 0 \\ 1 \\ 0 \end{pmatrix}, \quad v_3 = \begin{pmatrix} 0 \\ 0 \\ 1 \end{pmatrix}. \qquad (4.73)$$

Both of the requirements for a basis are satisfied with this choice. The linear combination

$$\alpha_1 \cdot v_1 + \alpha_2 \cdot v_2 + \alpha_3 \cdot v_3 = \begin{pmatrix} \alpha_1 \\ \alpha_2 \\ \alpha_3 \end{pmatrix}, \qquad (4.74)$$

can be the zero vector only in the case when $\alpha_1 = \alpha_2 = \alpha_3 = 0$, that is, the vectors are linearly independent. On the other hand, looking at the same linear combination, we see that it represents an *arbitrary* three-component vector. Therefore, $\{v_1, v_2, v_3\}$

is a basis. Another choice of a basis can be realized, for instance, by the vectors:

$$v'_1 = \begin{pmatrix} 1 \\ 1 \\ 0 \end{pmatrix}, \quad v'_2 = \begin{pmatrix} 1 \\ -1 \\ 0 \end{pmatrix}, \quad v'_3 = \begin{pmatrix} 0 \\ 0 \\ 1 \end{pmatrix}. \qquad (4.75)$$

(Show that this is also a basis!)

4.9 DECOMPOSITION AS MAPPING. ISOMORPHISM OF FINITE-DIMENSIONAL LINEAR SPACES

The concept of basis allows us to make a very important generalization. The decomposition of an element of the linear space V over a given basis being unique means that we obtain a unique representation of all the elements of the linear space through the coefficients of their decompositions. There is, however, one delicate point here. If for representing an element $v \in V$ we intend to keep only the coefficients of its decomposition in Eq. (4.67), we need to know which coefficient corresponds to which vector of the basis. The easiest way to establish this correspondence is to fix the order of the elements of the basis, that is, to consider the basis not as a *set* $\{v_1, v_2, v_3, \ldots, v_n\}$ but as an *n-tuple*, $(v_1, v_2, v_3, \ldots, v_n)$. Then, the coefficients of the decomposition of Eq. (4.67) are also arranged into an n-tuple. The element v can be always reconstructed from this n-tuple of coefficients by using these coefficients in the mentioned equation. In other words, we have just established a *one-to-one* mapping from the set V into the set of n-tuples, $\mathcal{C}^n$ (or $\mathcal{R}^n$, in the case of a real linear space),

$$v \to (\alpha_1, \ldots \alpha_n), \qquad (4.76)$$

which can be called the n-tuple transform of an element v. Moreover, the fact that *any* choice of the coefficients in Eq. (4.67) must give some element of the linear space V shows that the mapping (4.76) is also a surjection.

The n-tuple representation allows us to reduce any linear space of given dimension n to some standard linear space. Indeed, when dealing with any particular linear space, we are now able to substitute the carrier set V of arbitrary nature by the "standard" carrier set of n-tuples (or n-component column vectors). Then, the results can always be reformulated in terms of the original linear space V, by reconstructing its elements from the representing n-tuples through the formula of Eq. (4.67). In particular, this means that *all* linear spaces of the same dimension n are represented by the same "standard" space; in this sense, they are indistinguishable.

4.10 SAMPLE PROBLEMS

I. Consider the following subset of the space of functions $x(t)$ of real variable t on the interval $[0, 1]$:

$$\{x(t) | x(t) \text{ is real-valued} \wedge \frac{d^3 x}{dt^3} + 2x = 0, \wedge$$
$$\left.\frac{dx}{dt}\right|_{t=0} = 0, \wedge x(0) = 0\}. \quad (4.77)$$

Is this subset a real linear subspace? If yes, determine the dimension and find a basis of this subspace.

Solution
1) We first check if the subset is a linear subspace, that is, we check whether the linear combination $\alpha_1 x_1(t) + \alpha_2 x_2(t)$ satisfies all three conditions of Eq. (4.77), provided both $x_1(t)$ and $x_2(t)$ do:
I-st condition:

$$\frac{d^3}{dt^3}(\alpha_1 x_1 + \alpha_2 x_2) + 2(\alpha_1 x_1 + \alpha_2 x_2)$$
$$= \alpha_1 \left(\frac{d^3 x_1}{dt^3} + 2x_1\right) + \alpha_2 \left(\frac{d^3 x_2}{dt^3} + 2x_2\right) = 0. \quad (4.78)$$

II-nd condition:

$$\left.\frac{d}{dt}(\alpha_1 x_1 + \alpha_2 x_2)\right|_{t=0} = \alpha_1 \left.\frac{dx_1}{dt}\right|_{t=0} + \alpha_2 \left.\frac{dx_2}{dt}\right|_{t=0} = 0. \quad (4.79)$$

III-rd condition:

$$(\alpha_1 x_1(t) + \alpha_2 x_2(t))|_{t=0} = \alpha_1 x_1(0) + \alpha_2 x_2(0) = 0. \quad (4.80)$$

Thus, the subset is really a linear subspace.
2) To find the dimension and the basis of this subspace, we solve Eq. (4.77) explicitly. The equation being a linear ordinary differential equation, we look for a solution in the form:

$$x = e^{\lambda t}. \quad (4.81)$$

Then, the characteristic equation for λ has the form:

$$\lambda^3 + 2 = (\lambda + a)(\lambda^2 - a\lambda + a^2) = 0, \quad (4.82)$$

where $a = 2^{1/3}$.
The roots of the characteristic equation are obtained as:

$$\lambda_1 = -a; \quad \lambda_{2,3} = \frac{a}{2} \pm \sqrt{\frac{a^2}{4} - a^2} = \frac{a}{2}(1 \pm i\sqrt{3}), \quad (4.83)$$

$$\lambda_1 = -a; \quad \lambda_2 = \frac{a}{2}(1 + i\sqrt{3}); \quad \lambda_3 = \frac{a}{2}(1 - i\sqrt{3}). \tag{4.84}$$

Thus, three independent *complex-valued* solutions of the differential equation are:

$$x_1(t) = e^{-at}; \quad x_2(t) = e^{\frac{at}{2} + i\frac{\alpha}{2}\sqrt{3}t}; \quad x_3(t) = e^{\frac{at}{2} - i\frac{\alpha}{2}\sqrt{3}t} \tag{4.85}$$

To obtain three *real* independent solutions, we take standard real-valued combinations of the complex exponentials of Eq. (4.85) as:

$$\tilde{x}_1 = e^{-at}; \quad \tilde{x}_2 = e^{\frac{at}{2}} \cos\left(\frac{\alpha}{2}\sqrt{3}t\right); \quad \tilde{x}_3 = e^{\frac{at}{2}} \sin\left(\frac{\alpha}{2}\sqrt{3}t\right). \tag{4.86}$$

The general real solution has the form:

$$x(t) = A\tilde{x}_1(t) + B\tilde{x}_2(t) + C\tilde{x}_3(t), \tag{4.87}$$

where A, B, and C are arbitrary real numbers.

The initial conditions of Eq. (4.77) impose the following constraints on the constants A, B, and C. First,

$$x(0) = A + B = 0, \tag{4.88}$$

which gives:

$$B = -A. \tag{4.89}$$

Second,

$$\left.\frac{dx}{dt}\right|_{t=0} = -aA + \frac{a}{2}B + \frac{a}{2}\sqrt{3}C = 0, \tag{4.90}$$

which gives:

$$C = \sqrt{3}A. \tag{4.91}$$

Thus, the general form of the function satisfying all the three conditions of the problem is:

$$x(t) = A(\tilde{x}_1(t) - \tilde{x}_2(t) + \sqrt{3}\tilde{x}_3(t)). \tag{4.92}$$

All such functions are just proportional to

$$\bar{x}(t) = e^{-at} - e^{\frac{at}{2}} \cos\left(\frac{a}{2}\sqrt{3}t\right) + \sqrt{3}\, e^{\frac{at}{2}} \sin\left(\frac{a}{2}\sqrt{3}t\right). \tag{4.93}$$

In other words, all the elements of the subspace are linear combinations of just one element, $\bar{x}(t)$. Thus, this is a *one-dimensional* subspace, and its one-element basis is

$\bar{x}(t)$ given by Eq. (4.93).

II. Consider the following subset in the *real* linear space of real-valued functions $f(t)$ of the real variable t on the interval $[0, 1]$:

$$\left\{ \phi(t) \mid \phi(1-t) = \frac{d\phi(t)}{dt} \right\}. \tag{4.94}$$

Is this subset a real linear subspace? If yes, determine the dimension and find a basis of this subspace. *Hint.* Try to reduce the problem to an ordinary differential equation.

Solution
1) We first check whether the subset in question is a linear subspace. Let two functions, $y_1(t)$ and $y_2(t)$, satisfy the given condition, that is,

$$\frac{dy_1}{dt} = y_1(1-t),$$

$$\frac{dy_2}{dt} = y_2(1-t). \tag{4.95}$$

We check the condition of Eq. (4.94) regarding their linear combination,

$$\tilde{y}(t) = \alpha_1 \cdot y_1(t) + \alpha_2 \cdot y_2(t), \tag{4.96}$$

and get

$$\frac{d\tilde{y}}{dt} = \alpha_1 \cdot \frac{dy_1}{dt} + \alpha_2 \cdot \frac{dy_2}{dt}$$
$$= \alpha_1 \cdot y_1(1-t) + \alpha_2 \cdot y_2(1-t) = \tilde{y}(1-t). \tag{4.97}$$

Thus, the linear combination $\tilde{y}$ satisfies the same condition, that is, it belongs to the same subset. Therefore, this subset is a *linear subspace*.

2) To determine the *dimension* and to find a *basis* of this subspace, we need to solve the given equation (4.94) explicitly. To do so, we reduce it to an ordinary differential equation by the following trick. Let us differentiate both sides of Eq. (4.94) with respect to t:

$$\frac{d^2 y}{dt^2} = \frac{d}{dt}(y(1-t)). \tag{4.98}$$

Now, we transform the right-hand side of Eq. (4.98) as

$$\frac{d}{dt}(y(1-t)) = -\frac{d}{d(1-t)}(y(1-t)). \tag{4.99}$$

Then, by virtue of the original Eq. (4.94),

$$-\frac{d}{d(1-t)}(y(1-t)) = -y(1-(1-t)) = -y(t). \tag{4.100}$$

Thus, the function $y(t)$ satisfies an ordinary differential equation of second order,

$$\frac{d^2y}{dt^2} + y(t) = 0. \tag{4.101}$$

Using the standard method, we look for a solution of Eq. (4.101) in the form,

$$y(t) = e^{\lambda t}, \tag{4.102}$$

and obtain the characteristic equation for the coefficient λ,

$$\lambda^2 + 1 = 0. \tag{4.103}$$

The solution of Eq. (4.103),

$$\lambda_{1,2} = \pm i, \tag{4.104}$$

provides the general solution of Eq. (4.101) in the form,

$$y(t) = \tilde{A}e^{it} + \tilde{B}e^{-it}, \tag{4.105}$$

with arbitrary complex constants $\tilde{A}$ and $\tilde{B}$.

Since we are required to find the *real* solution, we should operate with real combinations of the exponentials of Eq. (4.105). Thus, we obtain the general solution in the form,

$$y(t) = A\cos t + B\sin t, \tag{4.106}$$

where A and B are arbitrary *real* constants.

However, by raising the order of the initial differential equation from 1 to 2, we might add some extra solutions. So, we need to check the general solution of Eq. (4.106) using the original Eq. (4.94):

$$\frac{dy}{dt} = -A\sin t + B\cos t; \tag{4.107}$$

$$\begin{aligned} y(1-t) &= A\cos(1-t) + B\sin(1-t) \\ &= A(\cos 1 \cos t + \sin 1 \sin t) + B\cdot(\sin 1 \cos t - \cos 1 \sin t) \\ &= \cos t(A\cos 1 + B\sin 1) + \sin t(A\sin 1 - B\cos 1). \end{aligned} \tag{4.108}$$

We substitute the expressions of Eq. (4.107) and (4.108) in the left-hand side and right-hand side of the original Eq. (4.94), respectively, and obtain:

$$\begin{aligned} -A\sin t + B\cos t \\ = (A\cos 1 + B\sin 1)\cos t + (A\sin 1 - B\cos 1)\sin t. \end{aligned} \tag{4.109}$$

The equality of Eq. (4.109) must hold for any t. We, therefore, equate the coefficients of $\sin t$ and $\cos t$ separately,

$$-A = A\sin 1 - B\cos 1; \tag{4.110}$$

$$B = A\cos 1 + B\sin 1, \tag{4.111}$$

and thus obtain a system of linear algebraic equations for the constants A and B:

$$\begin{cases} A(1 + \sin 1) - B\cos 1 = 0 \\ A\cos 1 + B(\sin 1 - 1) = 0 \end{cases}. \tag{4.112}$$

The determinant of this system is

$$\begin{aligned} -(1 + \sin 1)(1 - \sin 1) + \cos^2 1 \\ = -1 + \sin^2 1 + \cos^2 1 = -1 + 1 = 0. \end{aligned} \tag{4.113}$$

This means that the system of Eq. (4.112) is consistent, that is it has a non-trivial (non-zero) solution,

$$B = A\frac{\cos 1}{1 - \sin 1}. \tag{4.114}$$

Thus, the general solution of Eq. (4.94) is determined by only *one* arbitrary constant, and has the form,

$$\begin{aligned} y(t) &= A\left(\cos t + \frac{\cos 1}{1 - \sin 1}\sin t\right) \\ &= \frac{A}{1 - \sin 1}(\cos t + \cos 1 \sin t - \sin 1 \cos t) \\ &= A'(\cos t - \sin(1 - t)), \end{aligned} \tag{4.115}$$

where the constant

$$A' = \frac{A}{1 - \sin 1}. \tag{4.116}$$

Therefore, any function which is a solution of the original Eq. (4.94) is proportional to

$$\bar{y}(t) = \cos t - \sin(1 - t). \tag{4.117}$$

Thus, the linear subspace under consideration is *one-dimensional*, and its one-element basis is the function $\bar{y}(t)$ of Eq. (4.117).

4.11 SUMMARY OF CHAPTER 4

1) *Algebraic Structure* is determined by its *carrier set, operations, and constants*, and specified by a *signature* and *axioms*.
2) *Group* is an algebraic structure with signature $(V, +, -, \theta)$. It has one binary operation, say, $+$, and one unary operation, inversion $-$, which satisfy three axioms:

a) $v_1 + (v_2 + v_3) = (v_1 + v_2) + v_3$, the axiom of associativity;
b) $v_i + \theta = \theta + v_i = v_i$, the axiom of identity;
c) $\bar{v}_i : v_i + (-v_i) = \theta$, the axiom of inversion.

An *Abelian Group* satisfies one additional axiom,

d) $v_1 + v_2 = v_2 + v_1$ — commutativity.

3) The *algebra of complex numbers* is an algebraic structure with signature $(\mathcal{C}, \cdot, +, -, \cdot, \frac{1}{z}, z^*, 1, 0, i)$.
It is an Abelian group with respect to two operations, addition, $+$, and multiplication, $\cdot$, and it has an additional unary operation, that of complex conjugation, z^*. Complex numbers correspond to points in the *complex plane*, with their cartesian coordinates being the *real* and *imaginary* parts of the number.

4) A *linear space* is an algebraic structure based on two sets; one is a carrier set of arbitrary nature, V, the other is the set of complex numbers, $\mathcal{C}$ (or the set of real numbers, $\mathcal{R}$). The structure is an Abelian group with respect to the set V and to the "addition" operation. The set $\mathcal{C}$ is taken with all the operations and all the properties of the algebra of complex numbers. Besides, there is a specific operation of vector-by-number multiplication, the mapping $\mathcal{C} \times V \to V$, which for any pair $x \in V$, $\alpha \in \mathcal{C}$ produces an element $x' = \alpha \cdot x \in V$. This additional operation satisfies four additional axioms:

e) $\alpha \cdot (\beta \cdot x) = (\alpha \cdot \beta) \cdot x$, associativity;
f) $(\alpha + \beta) \cdot x = \alpha \cdot x + \beta \cdot x$, first distributivity;
g) $\alpha \cdot (x_1 + x_2) = \alpha \cdot x_1 + \beta \cdot x$, second distributivity;
h) $0 \cdot x = \theta$, $1 \cdot x = x$, identity.

5) *Linear independence and basis*: n vectors, $\{v_1, v_2, \ldots, v_n\}$, are linearly independent if their only trivial linear combination is that where all coefficients are zero. A finite set of n vectors is a basis of a given linear space, if:

a) they are linearly independent;
b) the set generates (spans) the entire space V.

A space is called finite-dimensional, if it has a finite basis or if it is a trivial space $V = \{\theta\}$; otherwise the space is called infinite-dimensional. All the bases of a given finite-dimensional linear space have the same number of elements. This number is called the *dimension* of the space.

6) Any element v of a finite-dimensional linear space V can be uniquely represented as a linear combination of the elements of a basis, $\{v_1, \cdots v_n\} : v = \sum_{i=1}^{n} \alpha_i \cdot v_i$. This *decomposition* is a bijection mapping, $V \to \mathcal{C}^n$, which transforms an element v of the space V into n-tuple $(\alpha_1, \alpha_2, \alpha_3, \ldots \alpha_n)$ of the coefficients of its decomposition. Thus, *any* linear space of a given dimension, n, can be transformed into the standard n-dimensional space of n-tuples (or n-component column vectors) of the coefficients.

4.12 PROBLEMS

I. Show that if $(A, \cdot, ^-, 1)$ is a group, then for any $a \in A$ the inverse of the inverse of a is a itself, $\bar{\bar{a}} = a$.

II. Consider the same group. Show that if B is a nonempty subset of the set A such that for any two of its elements, $b_1 \in B$ and $b_2 \in B$, $b_1 \cdot b_2 \in B$, then $(B, \cdot, ^-, 1)$ is a subgroup of the group under consideration.

III. Show, that for any set A the mathematical structure $(\{A, \emptyset\}, \cup, \cap, ^-, A, \emptyset)$ is a Boolean subalgebra of the Boolean Algebra $(\mathcal{P}(A), \cup, \cap, ^-, A, \emptyset)$.

IV. A computer operates only with nonnegative integers and has k bits to represent them in binary notation. Consider addition on such a computer, if the overflow causes the result to be set to the largest representable number. Which of the algebraic structures considered in Section 4.1 models the addition? How many elements does the carrier set contain? Will the answer differ if the overflow just causes a loss of the high order bits?

V. Consider the following subset of continuous functions of the real variable t on the interval $[0, 1]$:

$$\left\{ x(t) \mid \frac{d^2 x}{dt^2} - 2x = 0, \quad x(0) = 0 \right\}.$$

Is this subset a linear subspace? If yes, determine the dimension and find a basis of this subspace.

VI. Answer the same questions for the subset

$$\left\{ x_1(t) \mid 2\frac{d^3 x_1}{dt^3} - 6x_1 = 0, \quad \left.\frac{dx_1}{dt}\right|_{t=0} = 0 \right\}.$$

4.13 FURTHER READING

S. H. Friedberg, A. J. Insel, L. E. Spence, *Linear Algebra*, Prentice-Hall, Upper Saddle River, NJ, 1997.
This book is designed as a second graduate course in Linear Algebra. Its primary purpose is to present a careful treatment of the principal topics of Linear Algebra and to illustrate the power of the subject through a variety of applications.

T. Banchoff, J. Wermer, *Linear Algebra Through Geometry*, Springer-Verlag, New York, 1983.

This is a textbook for a senior-level course. Its efforts to implement intuitive geometric images and ideas for understanding basic concepts of Linear Algebra are noteworthy.

S. Axler, *Linear Algebra Done Right*, Springer-Verlag, New York, 1996.

This is a textbook for an advanced graduate course in Linear Algebra. It emphasizes understanding the structure of linear mappings from a vector space into itself.

5
Linear Mappings and Matrices

5.1 LINEAR MAPPINGS AND COLUMN VECTORS

Toward the end of Chapter 4, we have shown that the finite-dimensional vector space V of dimension n is in one-to-one correspondence to the vector space of n-tuples of complex numbers $\mathcal{C}^n$ (or n-tuples of real numbers $\mathcal{R}^n$, if V is a real vector space). The components of the n-tuples are just the coefficients of the decomposition of the vectors in V over some chosen basis. Consider now a mapping from this space V into an m-dimensional vector space W, $V \to W$. Of course, vectors in W also correspond to m-tuples of complex numbers, $\mathcal{C}^m$. This means that the mapping $V \to W$ itself can be uniquely represented by the mapping $\mathcal{C}^n \to \mathcal{C}^m$. Moreover, when the original mapping $V \to W$ is a *linear mapping*, the mapping $\mathcal{C}^n \to \mathcal{C}^m$ can be written in a standard form, completely specified by $n \times m$ coefficients. In this chapter we will consider such mappings between column vectors of complex numbers, develop apparatus for describing these mappings, and learn some important techniques useful in applications.

5.2 MATRICES

Let the basis in the vector space V be $\{\tilde{v}_1, \tilde{v}_2, \tilde{v}_3 ..... \tilde{v}_n\}$. This basis, which we will call the "$\tilde{v} - basis$", determines the particular correspondence $V \leftrightarrow \mathcal{C}^n$. Also, let the basis determining the particular correspondence $W \leftrightarrow \mathcal{C}^m$, the "$\tilde{w} - basis$", be

$\{\tilde{w}_1, \tilde{w}_2, \tilde{w}_3, \ldots\ldots\tilde{w}_m\}$. Then, the linear mapping $V \to W$ transforms an element $v \in V$ into an element $w = f(v) \in W$.

First, consider the mapping of the elements of the "$\tilde{v} - basis$". They are being transformed into elements of the W space,

$$\{w_1 = f(\tilde{v}_1), w_2 = f(\tilde{v}_2), w_3 = f(\tilde{v}_3), \ldots, w_n = f(\tilde{v}_n)\}. \tag{5.1}$$

Each of these values can be then decomposed over the "$\tilde{w} - basis$",

$$w_i = f(\tilde{v}_i) = \sum_{j=1}^{m} f_{ji}\tilde{w}_j, \tag{5.2}$$

where f_{ji} are the coefficients of this decomposition, necessarily bearing two indices, one for the "$\tilde{w} - basis$" and the other for the "$\tilde{v} - basis$".

Then, an arbitrary element v of space V is represented through its decomposition over the "$\tilde{v} - basis$",

$$v = \sum_{i=1}^{n} \alpha_i \tilde{v}_i. \tag{5.3}$$

The value of this element under the linear mapping can be written as

$$w = f(v) = f\left(\sum_{i=1}^{n} \alpha_i \tilde{v}_i\right) = \sum_{i=1}^{n} \alpha_i f(\tilde{v}_i) = \sum_{i=1}^{n} \alpha_i w_i, \tag{5.4}$$

which is just the decomposition over the elements w_i of Eq. (5.1). Substituting the expressions for these elements from Eq. (5.2), we get the formula

$$w = \sum_{i=1}^{n} \alpha_i \sum_{j=1}^{m} f_{ji}\tilde{w}_j = \sum_{i=1}^{n}\sum_{j=1}^{m} \alpha_i f_{ji}\tilde{w}_j. \tag{5.5}$$

On the other hand, from the "$\tilde{w} - basis$" point of view, Eq. (5.5) means nothing more than a decomposition of the element w over the elements of the "$\tilde{w}$ basis". To make this clear, we change the order of the summations,

$$w = \sum_{j=1}^{m}\sum_{i=1}^{n} \alpha_i f_{ji}\tilde{w}_j = \sum_{j=1}^{m} \tilde{w}_j \sum_{i=1}^{n} f_{ji}\alpha_i = \sum_{j=1}^{m} \gamma_j \tilde{w}_j, \tag{5.6}$$

where the coefficients γ_j just constitute an m-tuple representing the element w in C^m-space.

Thus, in fact, the formula of Eq. (5.6) expresses relations between coefficients α_i of the decomposition over the "$\tilde{v}$ basis" and coefficients γ_i of the decomposition over the "$\tilde{w}$-basis",

$$\gamma_j = \sum_{i=1}^{n} f_{ji}\alpha_i \tag{5.7}$$

or, in other words, this formula defines the mapping $C^n \to C^m$.

It is convenient to write in Eq. (5.7) the two-index entity f_{ji} as a table (called a *matrix*) having m rows and n columns, so that the formula itself takes the form

$$\begin{pmatrix} f_{11} & f_{12} & \cdots & f_{1n} \\ f_{21} & f_{22} & \cdots & f_{2n} \\ \vdots & & & \\ f_{m1} & f_{m2} & \cdots & f_{mn} \end{pmatrix} \begin{pmatrix} \alpha_1 \\ \alpha_2 \\ \vdots \\ \alpha_n \end{pmatrix} = \begin{pmatrix} \gamma_1 \\ \gamma_2 \\ \vdots \\ \gamma_m \end{pmatrix}. \qquad (5.8)$$

Actually, the formula of Eq. (5.8) gives the rule for obtaining the *vector-by-matrix product*, which can be formulated as the following:

The number of components in the resulting column vector is the same as the number of rows in the matrix; an i-th component of the resulting vector is obtained as a linear combination of the components of the original vector with coefficients that are the elements of the i-th row of the matrix.

We will denote matrices as $\hat{F}$, or, sometimes $\hat{F}^{m \times n}$, to emphasize their dimensions. Correspondingly, the elements of matrix $\hat{F}$ will be denoted as f_{ij}, or, sometimes, as $(\hat{F})_{ij}$. As we have shown in this section, *any* linear mapping F can be completely characterized by $n \times m$ elements of the related matrix $\hat{F}$.

5.3 THE MATRIX PRODUCT

Since matrices represent linear mappings, any modification of a linear mapping means some modification of its matrix, and vice versa. In particular, we can multiply a matrix by a number, α, so that $\hat{B} = \alpha \hat{A}$ just means $b_{ij} = \alpha \cdot a_{ij}$ for all indices i and j. This corresponds to modification of a mapping $v \to a(v)$ into mapping $v \to \alpha a(v)$. We can also add matrices, provided they are of the same dimensions $(n \times m)$. The sum of two matrices $\hat{C} = \hat{A} + \hat{B}$ just means the matrix with elements $c_{ij} = a_{ij} + b_{ij}$. This addition of matrices corresponds to the union of two mappings, $v \to a(v)$ and $v \to b(v)$, into mapping $v \to (a(v) + b(v))$. These operations being established, all matrices of a given dimension $(n \times m)$ constitute a linear space, in which the role of the zero element is played by the matrix

$$\hat{O} = \begin{pmatrix} 0 & 0 & \cdots & 0 \\ 0 & 0 & \cdots & 0 \\ \vdots & & & \\ 0 & 0 & \cdots & 0 \end{pmatrix}. \qquad (5.9)$$

A much less trivial fact is that for matrices there exists a very specific operation of *matrix multiplication*. To introduce this operation properly and to realize its meaning, let us consider two consecutive mappings, $V \to W$ and $W \to U$, where U is yet another vector space, of dimension l (see Fig. 5.1). The first mapping transforms an element $v \in V$ into element $w = f(v) \in W$. The second mapping transforms

98 LINEAR MAPPINGS AND MATRICES

the element w into element $u = g(w) \in U$. On the other hand, instead of these two consecutive mappings we can consider one *composite* mapping $V \to U$, which eventually transforms the element v into element $u = h(v) = g(f(v))$. In the realm of column-vector spaces these consecutive mappings correspond to mappings

$$C^n \to C^m \to C^l. \tag{5.10}$$

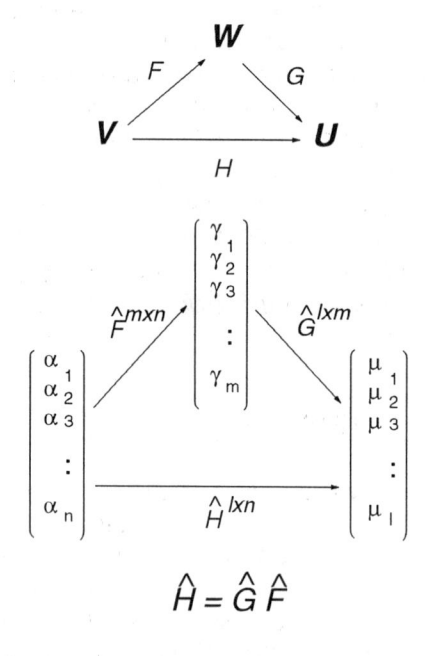

Fig. 5.1 Matrix multiplication.

Let $\{\tilde{u}_1, \tilde{u}_2, \tilde{u}_3, ..., \tilde{u}_n\}$ be a basis in U space. We, then, represent the vectors under consideration, v, w, and u, as decompositions over bases in their respective linear spaces,

$$v = \sum_{i=1}^{n} \alpha_i \tilde{v}_i; \quad w = \sum_{j=1}^{m} \gamma_j \tilde{w}_j; \quad u = \sum_{k=1}^{l} \mu_k \tilde{u}_k. \tag{5.11}$$

If the mapping $v \to f(v)$ corresponds to the matrix $\hat{F}$, and the mapping $w \to g(w)$ corresponds to the matrix $\hat{G}$, we get from Eq.(5.7)

$$\gamma_j = \sum_{i=1}^{n} f_{ji} \alpha_i \tag{5.12}$$

and

$$\mu_k = \sum_{j=1}^{m} g_{kj}\gamma_j = \sum_{j=1}^{m} g_{kj} \sum_{i=1}^{n} f_{ji}\alpha_i. \tag{5.13}$$

On the other hand, for the *composite* mapping h the same formula (5.7) gives

$$\mu_k = \sum_{i=1}^{n} h_{ki}\alpha_i. \tag{5.14}$$

Comparing Eqs. (5.13) and (5.14), and changing the order of the summations in Eq. (5.13), we obtain the following formula for the matrix $\hat{H}$ of the composite mapping $v \to h(v)$:

$$h_{ki} = \sum_{j=1}^{m} g_{kj} f_{ji}. \tag{5.15}$$

In fact, Eq. (5.15) defines a new operation called *matrix multiplication*:

$$\hat{H} = \hat{G}\hat{F}. \tag{5.16}$$

As we have seen, this operation is possible only if the number of columns in $\hat{G}$ is equal to the number of rows in $\hat{F}$. The rule of matrix multiplication is thus formulated as follows:
The product of matrices $\hat{G}$ and $\hat{F}$ is a matrix $\hat{H}$ which has as many rows as matrix $\hat{G}$ and as many columns as matrix $\hat{F}$; any column of the resulting matrix $\hat{H}$ is obtained by multiplying the corresponding column of the matrix $\hat{F}$ by the matrix $\hat{G}$.

5.4 MATRIX ALGEBRA

As we see, matrix multiplication requires *a priori* matching of the numbers of rows and columns of the matrices involved in the multiplication. We can simplify the situation by restricting our attention to *square* matrices of a given dimension, say $n \times n$. (E.g., all possible matrices of this type correspond to all possible linear mappings of an n-dimensional space V into itself). In this case, matrix multiplication is always possible. However, even in this case it is a *non-commutative operation*. This means that, in general,

$$\hat{A}\hat{B} \neq \hat{B}\hat{A}. \tag{5.17}$$

Example of the non-commutativity of matrix multiplication
Let

$$\hat{A} = \begin{pmatrix} 1 & 0 \\ 1 & 1 \end{pmatrix}, \quad \hat{B} = \begin{pmatrix} 1 & 1 \\ 0 & 1 \end{pmatrix}. \tag{5.18}$$

100 LINEAR MAPPINGS AND MATRICES

Then

$$\hat{A}\hat{B} = \begin{pmatrix} 1 & 0 \\ 1 & 1 \end{pmatrix} \begin{pmatrix} 1 & 1 \\ 0 & 1 \end{pmatrix} = \begin{pmatrix} 1 & 1 \\ 1 & 2 \end{pmatrix}; \quad (5.19)$$

$$\hat{B}\hat{A} = \begin{pmatrix} 1 & 1 \\ 0 & 1 \end{pmatrix} \begin{pmatrix} 1 & 0 \\ 1 & 1 \end{pmatrix} = \begin{pmatrix} 2 & 1 \\ 1 & 1 \end{pmatrix}. \quad (5.20)$$

Clearly, in this case $\hat{A}\hat{B} \neq \hat{B}\hat{A}$.

There exists also a specific $n \times n$ matrix which plays the role of the identity element with respect to matrix multiplication. This is the so-called *identity matrix*. It has the form

$$\hat{I} = \begin{pmatrix} 1 & 0 & \cdots & 0 \\ 0 & 1 & \cdots & 0 \\ \vdots & & & \\ 0 & 0 & \cdots & 1 \end{pmatrix}. \quad (5.21)$$

Overall, square $n \times n$ matrices seem promising to constitute a group (non-Abelian!) with respect to matrix multiplication. To complete the construction of this group, we need to introduce the operation of inversion, that is, to define for each matrix $\hat{A}$ its inverse $\hat{A}^{-1}$, such that

$$\hat{A}\hat{A}^{-1} = \hat{A}^{-1}\hat{A} = \hat{I}. \quad (5.22)$$

However, not every square matrix has an inverse. There exists a whole class of matrices, called *singular* matrices, which have no inverse. As you should know from an elementary algebra course, for a matrix to have an inverse (that is, to be *non-singular*) its determinant must be non-zero,

$$\det(\hat{A}) \neq 0. \quad (5.23)$$

We will shortly make use of this criterion and discuss it in more detail. Thus, if we restrict the scope of our consideration to *all non-singular square matrices of given dimension,* we obtain a non-Abelian group with respect to matrix multiplication.

Since we now have two operations, matrix addition and matrix multiplication, it is tempting to combine them in a comprehensive *matrix algebra* structure, similar to that which we constructed for complex numbers in Chapter 4. However, an unexpected obstacle lurks in this path: the operation of addition does not preserve non-singularity of matrices, that is, a sum of two non-singular matrices can be a singular matrix.

Example
Let

$$\hat{A} = \hat{I} = \begin{pmatrix} 1 & 0 \\ 0 & 1 \end{pmatrix} \quad \hat{B} = \begin{pmatrix} -1 & 1 \\ 0 & -1 \end{pmatrix}. \quad (5.24)$$

Then

$$\hat{A} + \hat{B} = \begin{pmatrix} 1 & 0 \\ 0 & 1 \end{pmatrix} + \begin{pmatrix} -1 & 1 \\ 0 & -1 \end{pmatrix} = \begin{pmatrix} 0 & 1 \\ 0 & 0 \end{pmatrix}. \tag{5.25}$$

Thus, $\hat{A} + \hat{B}$ is a singular matrix different from $\hat{O}$. This means that the inclusion of the addition operation irrevocably destroys the fragile construction of the non-Abelian group we just attempted.

Finally, the set A of square matrices of given dimension, $\hat{A}^{n \times n}$, taken along with operations of matrix addition and matrix multiplication, constitute *the matrix algebra* having signature $(A, +, \cdot, -, \hat{A}^{-1}, \hat{O}, \hat{I})$. This structure is an Abelian group with respect to addition, however it is not a group with respect to multiplication. Nevertheless, the property of associativity of multiplication,

$$(\hat{A}_1 \hat{A}_2) \hat{A}_3 = \hat{A}_1 (\hat{A}_2 \hat{A}_3), \tag{5.26}$$

and the property of distributivity,

$$(\hat{A}_1 + \hat{A}_2) \hat{A}_3 = \hat{A}_1 \hat{A}_3 + \hat{A}_2 \hat{A}_3, \tag{5.27}$$

hold in this case (and are left for the reader to prove). As we see, this new mathematical structure is really quite different from both the algebra of complex numbers we considered in Chapter 4 and the Boolean Algebra we dealt with in Chapter 3.

5.5 TRANSPOSE OF A MATRIX

In this section we return for a while (i.e., for Sections 5.5 through 5.8) to the general case of rectangular $n \times m$ matrices. For any rectangular matrix A, there exists a special associated matrix, which will play an essential role in further considerations involving matrices, in Chapters 7, 9, and 11. This is the so called *transpose matrix*, denoted as $\hat{A}^T$. It is obtained by interchanging rows and columns of the initial matrix, so that i-th row of $\hat{A}$ becomes the i-th column of $\hat{A}^T$ (and vice versa). Thus, if $\hat{A}$ is an $n \times m$ matrix, than $\hat{A}^T$ is an $m \times n$ matrix with the elements:

$$(\hat{A}^T)_{ij} = a_{ji}. \tag{5.28}$$

According to this definition, for two matrices of the same dimensions, $\hat{A}$ and $\hat{B}$,

$$(\hat{A} + \hat{B})^T = \hat{A}^T + \hat{B}^T. \tag{5.29}$$

Less obvious is the following relation holding when the multiplication of the matrices $\hat{A}$ and $\hat{B}$ makes sense:

$$(\hat{A}\hat{B})^T = \hat{B}^T \hat{A}^T. \tag{5.30}$$

102 LINEAR MAPPINGS AND MATRICES

To prove this relation, consider matrices $\hat{A}^{n \times m}$ and $\hat{B}^{m \times l}$. According to the rule of matrix multiplication given by Eq. (5.15), the ij-element of the matrix on the left-hand side in Eq. (5.30) is obtained as

$$((\hat{A}\hat{B})^T)_{ij} = (\hat{A}\hat{B})_{ji} = \sum_{k=1}^{m} a_{jk} b_{ki}. \tag{5.31}$$

On the other hand, the ij-element of the right-hand-side matrix in Eq. (5.30) is

$$(\hat{B}^T \hat{A}^T)_{ij} = \sum_{k=1}^{m} (\hat{B}^T)_{ik} (\hat{A}^T)_{kj} = \sum_{k=1}^{m} b_{ki} a_{jk}, \tag{5.32}$$

which is exactly the same expression as that of Eq. (5.31). Thus, formula (5.30) is established.

Based on formula (5.30), we can easily show that for any square nonsingular matrix $\hat{A}$ the following important relation holds:

$$(\hat{A}^{-1})^T = (\hat{A}^T)^{-1}. \tag{5.33}$$

In order to do so, we merely transpose both sides of the relation (5.22),

$$(\hat{A}\hat{A}^{-1})^T = I^T, \tag{5.34}$$

and note that $I^T = I$. Thus,

$$(\hat{A}\hat{A}^{-1})^T = (\hat{A}^{-1})^T \hat{A}^T = \hat{I}, \tag{5.35}$$

which establishes $(\hat{A}^{-1})^T$ as the inverse of $\hat{A}^T$ and, therefore, proves formula (5.33).

5.6 FUNDAMENTAL SUBSPACES OF COLUMN VECTORS ASSOCIATED WITH A GIVEN MATRIX

We have introduced matrices as a tool for mapping a linear space of column vectors C^n into another linear space of column vectors C^m. Following our general approach to mappings formulated in Chapter 2, we should then indicate in the codomain C^m the subspace which is the *range* of the mapping under consideration. This is the subspace of all vectors resulting from the mapping,

$$\vec{\gamma} = \hat{A}\vec{\alpha}, \tag{5.36}$$

where $\vec{\alpha}$ is an arbitrary vector in C^n, and $\vec{\gamma} \in C^m$. This range subspace is denoted $\mathcal{R}(\hat{A})$. Of course, its dimension and basis are determined by the structure of the matrix $\hat{A}$.

In fact, the relation of $\mathcal{R}(\hat{A})$ to the matrix $\hat{A}$ turns out to be very simple: the range $\mathcal{R}(\hat{A})$ is just the span of the set of vectors which are columns of the matrix $\hat{A}$. To show this, it is sufficient to note that vector $\vec{\gamma}$ in the formula (5.36) is produced as a linear combination of the columns of matrix $\hat{A}$ with coefficients that are the components of the vector $\vec{\alpha}$. According to our discussion in Section 4.8, this means that all *linearly independent* columns of $\hat{A}$ constitute a basis of the range space, and the number of these linearly independent columns is the dimension of this space. This dimension is called the *rank* of matrix $\hat{A}$ and denoted as $\rho(\hat{A})$. It is a key characteristic of a matrix, whose usefulness we will shortly demonstrate in this chapter. Intuitively, the rank shows how far a mapping under consideration is from being *a surjection*. Indeed, if $\rho(\hat{A})$ were equal to m, the range would cover all the codomain space, and the mapping defined by matrix $\hat{A}$ would be a surjection.

There is another important subspace associated with matrix $\hat{A}$. This is the so-called *nullspace* of $\hat{A}$, denoted as $\mathcal{N}(\hat{A})$ and determined as the linear space of all those vectors α in $\mathcal{C}^n$, which are mapped by the matrix $\hat{A}$ to the zero vector of $\mathcal{C}^m$,

$$\hat{A}\vec{\alpha} = \theta. \tag{5.37}$$

Note that the nullspace belongs to the domain $\mathcal{C}^n$, while the range space belongs to the codomain $\mathcal{C}^m$!

The dimension of the nullspace is called the *nullity* and denoted as $\nu(\hat{A})$. This characteristic of a matrix is also of great importance; it shows how far the matrix-related mapping is from being an injection. If $\nu(\hat{A}) = 0$, that is, the equality (5.37) holds only for $\vec{\alpha} = \theta$, the matrix $\hat{A}$ performs a one-to-one mapping. To prove this statement, suppose $\vec{\alpha_1} \neq \vec{\alpha_2}$ but $\hat{A}\vec{\alpha_1} = \hat{A}\vec{\alpha_2}$. The latter equality implies

$$\hat{A}\vec{\alpha_1} - \hat{A}\vec{\alpha_2} = \hat{A}(\vec{\alpha_1} - \vec{\alpha_2}) = \theta, \tag{5.38}$$

which implies $\vec{\alpha_1} = \vec{\alpha_2}$, which contradicts our assumption. Thus, the statement is proven by *reductio ad absurdum*.

As we will discover in the following sections, the rank and nullity are not independent characteristics of a matrix. They are connected by a remarkable relation which is sometimes referred to as the *Fundamental Theorem of Linear Algebra*. However, before discussing this statement we need to learn more about the properties of a matrix.

5.7 THE ECHELON MATRIX. LU-DECOMPOSITION

The range space and nullspace of a matrix $\hat{A}$ introduced in the previous section are intimately connected with solving systems of linear algebraic equations. If the components of vector $\vec{\gamma}$ are given numbers and the components of vector $\vec{\alpha}$ are unknown, Eq. (5.36) represents a system of linear algebraic equations in a *matrix form*. Rewriting this system in a more traditional way, we get:

$$\begin{cases} a_{11}\alpha_1 + a_{12}\alpha_2 + \ldots + a_{1n}\alpha_n = \gamma_1 \\ a_{21}\alpha_1 + a_{22}\alpha_2 + \ldots + a_{2n}\alpha_n = \gamma_2 \\ \quad \vdots \\ a_{m1}\alpha_1 + a_{m2}\alpha_2 + \ldots + a_{mn}\alpha_n = \gamma_m \end{cases} \quad (5.39)$$

Considering the range space and the nullspace from the standpoint of solving the system of Eqs. (5.39), one can say that the system (5.39) has a solution only if the vector $\vec{\gamma}$ belongs to the range space of the matrix $\hat{A}$, and that this solution is unique if the nullity of the matrix $\hat{A}$ is equal to zero.

To pursue our goal of establishing a relation between $\rho(\hat{A})$, $\nu(\hat{A})$, and the matrix structure, we use Gaussian elimination to solve (5.39). Let $a_{11} \neq 0$ (if this is not true, we reshuffle the equations of the system so that the first equation is one which has a nonzero coefficient for α_1; there is for sure at least one such coefficient). Then we subtract from the second equation the first one multiplied by a_{21}/a_{11}, subtract from the third equation the first one multiplied by a_{31}/a_{11}, and so on, up to the very last equation. As a result, we obtain a system having zero coefficients at α_1 in all the rows but the first one:

$$\begin{cases} a_{11}\alpha_1 + a_{12}\alpha_2 + \ldots + a_{1n}\alpha_n = \gamma_1 \\ 0 + a'_{22}\alpha_2 + \ldots + a'_{2n}\alpha_n = \gamma'_2 \\ \quad \vdots \\ 0 + a'_{m2}\alpha_2 + \ldots + a'_{mn}\alpha_n = \gamma'_m \end{cases}, \quad (5.40)$$

where

$$a'_{ij} = a_{ij} - \frac{a_{i1}a_{1j}}{a_{11}}; \quad \gamma'_i = \gamma_i - \frac{a_{i1}}{a_{11}}\gamma_1. \quad (5.41)$$

Then we reshuffle all the rows but the first to get a second row having nonzero coefficient at α_2 (if this is impossible, i.e., if all the coefficients at α_2 are zero, we set the second row with nonzero coefficient at α_3, etc.). Repeating the elimination procedure for the second to m-th equations of this new system gives:

$$\begin{cases} a_{11}\alpha_1 + a_{12}\alpha_2 + \ldots + a_{1n}\alpha_n = \gamma_1 \\ 0 + a'_{22}\alpha_2 + \ldots + a'_{2n}\alpha_n = \gamma'_2 \\ 0 + 0 + \ldots + a''_{2n}\alpha_n = \gamma''_2 \\ \quad \vdots \\ 0 + 0 + a'''_{m3}\alpha_2 + \ldots + a'''_{mn}\alpha_n = \gamma'''_m \end{cases}, \quad (5.42)$$

where

$$a''_{ij} = a'_{ij} - \frac{a'_{i2}a'_{2j}}{a'_{22}}; \quad \gamma''_i = \gamma'_i - \frac{a'_{i2}}{a'_{22}}\gamma'_1, \tag{5.43}$$

provided that after the second reshuffling $a'_{22} \neq 0$. Then we repeat the elimination procedure, starting from the third row down, and so on. Eventually, this process of eliminating coefficients leads to a system of equations being equivalent to system (5.39) (i.e., having the same solution) but having the matrix of a much-simplified, so called upper-triangular structure. In this upper-triangular matrix all the elements lying beneath the *main diagonal* (i.e., beneath the elements having the same number of row and column) are zeros. This new matrix is usually denoted $\hat{U}$ (for "upper-triangular") and called the *echelon matrix* corresponding to matrix $\hat{A}$.

The important fact here is that we obtain the $\hat{U}$-matrix by using a finite number of operations on the equations of system (5.39), or, equivalently, on the rows of matrix $\hat{A}$. Each of these operations, in turn, can be represented as an action on matrix $\hat{A}$ by a special (and very simple) auxiliary square matrix. For instance, the subtraction of the first row multiplied by (a_{21}/a_{11}) from the second row is equivalent to multiplying $\hat{A}$ from the left by the $m \times m$ matrix

$$\hat{D}^{(2,1)}\left(-\frac{a_{21}}{a_{11}}\right) = \begin{pmatrix} 1 & 0 & 0 & \cdots & 0 \\ -\frac{a_{21}}{a_{11}} & 1 & 0 & \cdots & 0 \\ 0 & 0 & 1 & \cdots & 0 \\ \vdots & \vdots & \vdots & \ddots & \vdots \\ 0 & 0 & \cdots & \cdots & 1 \end{pmatrix}. \tag{5.44}$$

To generalize, each time we eliminate some "vulnerable" coefficients, we multiply the intermediate-result matrix from the left by a matrix having all main-diagonal elements as 1 and only one non-zero off-diagonal element lying just below the diagonal. The interchange of, say, the second and fourth rows of a matrix is produced by multiplying the matrix by:

$$\hat{E}^{(2,4)} = \begin{pmatrix} 1 & 0 & 0 & 0 & \cdots & 0 \\ 0 & 0 & 0 & 1 & \cdots & 0 \\ 0 & 0 & 1 & 0 & \cdots & 0 \\ 0 & 1 & 0 & 0 & \cdots & 0 \\ \vdots & \vdots & \vdots & \vdots & \ddots & \vdots \\ 0 & 0 & 0 & 0 & \cdots & 1 \end{pmatrix}. \tag{5.45}$$

Thus, we obtain the matrix $\hat{U}$ from the original matrix $\hat{A}$ by consecutively multiplying the latter by some $\hat{D}$ and $\hat{E}$ matrices. Correspondingly, we can represent $\hat{A}$ as the matrix $\hat{U}$ multiplied by $\hat{D}^{-1}$ and $\hat{E}^{-1}$ matrices taken in inverse order. What is especially nice is that these inverse matrices can be easily obtained as

$$(\hat{E}^{(i,j)})^{-1} = \hat{E}^{(i,j)}; \quad (\hat{D}^{(i,j)}(-\beta))^{-1} = \hat{D}^{(i,j)}(\beta). \tag{5.46}$$

We can establish the relations of Eq. (5.46) by direct calculation. On the other hand, we can see that the action of the matrices $\hat{D}^{-1}$ and $\hat{E}^{-1}$ on matrix $\hat{A}$ must simply cancel the actions of the matrices $\hat{D}$ and $\hat{E}$, respectively. This observation directly leads to the relations of Eq. (5.46).

The net result of these lengthy considerations is that we have established an algorithm for representing a given matrix $\hat{A}$ in factored form,

$$\hat{A} = \hat{R}\hat{U}, \qquad (5.47)$$

with straightforward rules for finding the $\hat{R}$ and $\hat{U}$ matrices.

The matrix $\hat{R}$ becomes especially simple and easy-to-find if in the process of obtaining the $\hat{U}$-matrix no row exchanges are required. In this case matrix $\hat{R}$ is just a product of several $\hat{D}^{(i,j)}(\beta)$ matrices:

$$\begin{aligned}\hat{R} &= \hat{D}^{(2,1)}\left(\frac{a_{21}}{a_{11}}\right)\hat{D}^{(3,1)}\left(\frac{a_{31}}{a_{11}}\right)\hat{D}^{(4,1)}\left(\frac{a_{41}}{a_{11}}\right)\cdot\ldots\cdot\hat{D}^{(m,1)}\left(\frac{a_{m1}}{a_{11}}\right)\\ &\times \hat{D}^{(3,2)}\left(\frac{a'_{32}}{a'_{22}}\right)\hat{D}^{(4,2)}\left(\frac{a'_{42}}{a'_{22}}\right)\cdot\ldots\cdot\hat{D}^{(m,2)}\\ &\times \left(\frac{a'_{m2}}{a'_{22}}\right)\cdot\ldots\cdot\hat{D}^{(m,m-1)}(\ldots).\end{aligned} \qquad (5.48)$$

Directly computing such a product (which is left for the reader's own entertainment) shows that in this case the matrix $\hat{R}$ has a 1 in all places on the main diagonal, zeros for all the above diagonal elements, and all the elements β of the factor $\hat{D}$ matrices being put in their respective (i,j) places. Thus, $\hat{R}$ turns out to be a *lower-triangular* matrix. In this, not uncommon, case, we correspondingly call it the $\hat{L}$-matrix and refer to the whole process as to LU-decomposition (or LU-factorization) of matrix $\hat{A}$.

5.8 THE FUNDAMENTAL THEOREM OF LINEAR ALGEBRA

Now we have all necessary means to formulate and prove the Fundamental Theorem of Linear Algebra. It reads, the rank of a matrix is equal to the rank of its transpose, $\rho(\hat{A}) = \rho(\hat{A}^T)$. In other words, the number of independent columns equals the number of independent rows.

We will prove this statement through a somewhat lengthy and academic step-by-step procedure.
1) The number of linearly independent rows of $n \times m$ matrix $\hat{A}$ equals the number of independent rows of the corresponding echelon matrix $\hat{U}$. This is obvious, because the rows of $\hat{U}$ are linear combinations of the rows of $\hat{A}$.
2) The independent rows of $\hat{U}$ are just all the non-zero rows of this matrix, which follows from the construction of matrix $\hat{U}$. Since rows of $\hat{U}$ are columns of $\hat{U}^T$, the linearly independent rows of $\hat{U}$ are linearly independent columns of $\hat{U}^T$; their number is $\rho(\hat{U}^T)$.

THE FUNDAMENTAL THEOREM OF LINEAR ALGEBRA   107

3) If vector $\vec{\alpha}$ is a solution of the equation $\hat{U}\vec{\alpha} = \theta$, $\rho(\hat{U}^T)$ components of this vector can be determined in terms of the other, *free*, components.
4) The number of *free* components, which is actually the dimension of the *nullspace* of matrix $\hat{U}$, $\nu(\hat{U})$, is thus equal to $n - \rho(\hat{U}^T)$.
5) All of the solutions of equation $\hat{U}\vec{\alpha} = \theta$ are solutions of equation $\hat{A}\vec{\alpha} = \theta$ and vice versa (indeed, this was the reason for constructing $\hat{U}$ through the elimination process). Hence, $\nu(\hat{U}) = \nu(A)$, and $\nu(\hat{A}) = n - \rho(\hat{A}^T)$.
6) A corollary of the previous statement is that if a linear combination of some columns of the matrix $\hat{A}$ gives the zero vector, then a linear combination of the corresponding columns of matrix $\hat{U}$ with the same coefficients also gives the zero vector, and vice versa. This means, that the number of *linearly* independent columns of matrix $\hat{A}$ is equal to that of matrix $\hat{U}$, $\rho(\hat{A}) = \rho(\hat{U})$.
7) We can always choose a basis $\{\vec{\alpha}_1, \vec{\alpha}_2, ..., \vec{\alpha}_n\}$ in n-dimensional "α-space", so that its first ν vectors $\{\vec{\alpha}_1, \vec{\alpha}_2, ...\vec{\alpha}_\nu\}$ constitute a basis of the nullspace $\mathcal{N}(\hat{A})$.
8) Then the vectors $\{\vec{\gamma}_{\nu+1} = \hat{A}\vec{\alpha}_{\nu+1}, \vec{\gamma}_{\nu+2} = \hat{A}\vec{\alpha}_{\nu+2}, ...\vec{\gamma}_n = \hat{A}\vec{\alpha}_n\}$ constitute a basis of subspace $\mathcal{R}(\hat{A})$ in m-dimensional "γ-space". To confirm this statement, we just check whether the set $\{\vec{\gamma}_{\nu+1}, ..., \vec{\gamma}_n\}$ has all the properties of a basis. First, by construction, these are definitely non-zero vectors. Second, they are linearly independent, which we easily prove by contradiction. Suppose the vectors are linearly dependent, that is, we can construct a linear combination with all non-zero coefficients $\sum_{i=\nu+1}^{n} \beta_i \vec{\gamma}_i = 0$. This means that

$$\sum_{i=\nu+1}^{n} \beta_i \left(\hat{A}\vec{\alpha}_i\right) = \hat{A}\left(\sum_{i=\nu+1}^{n} \beta_i \vec{\alpha}_i\right) = \theta, \tag{5.49}$$

that is, that vector

$$\vec{\alpha}^* = \sum_{i=\nu+1}^{n} \beta_i \vec{\alpha}_i \tag{5.50}$$

belongs to $\mathcal{N}(\hat{A})$. Thus, we can express this vector as a linear combination of the vectors $\{\vec{\alpha}_1, ..., \vec{\alpha}_\nu\}$,

$$\vec{\alpha}^* = \sum_{j=1}^{\nu} \mu_j \vec{\alpha}_j. \tag{5.51}$$

Comparing Eqs. (5.50) and (5.51), we get:

$$\sum_{j=1}^{\nu} \mu_j \vec{\alpha}_j - \sum_{i=\nu+1}^{n} \beta_i \vec{\alpha}_i = \theta. \tag{5.52}$$

This means that we have constructed some *trivial* linear combination of the vectors of basis $\{\vec{\alpha}_1, ..., \vec{\alpha}_n\}$. Thus, these vectors turn out to be linearly dependent, a

108 LINEAR MAPPINGS AND MATRICES

contradiction to their constituting a basis. So, our initial assumption was wrong, and the vectors $\{\vec{\gamma}_{\nu+1}, ...\vec{\gamma}_n\}$ are really linearly independent.

Finally, the set $\{\vec{\gamma}_{\nu+1}, ..., \vec{\gamma}_n\}$ spans the entire range space $\mathcal{R}(\hat{A})$, because every vector $\vec{\alpha}$ giving non-zero $\vec{\gamma} = \hat{A}\vec{\alpha}$ can be represented as a linear combination of vectors $\{\vec{\alpha}_{\nu+1}, ..., \vec{\alpha}_n\}$.

9) The set $\{\vec{\gamma}_{\nu+1}, ..., \vec{\gamma}_n\}$ being a basis of $\mathcal{R}(\hat{A})$, the dimension of the range, $\rho(\hat{A}) = n - \nu(\hat{A})$.

10) Finally: comparing statement 9 to statement 5, we conclude that $\rho(\hat{A}) = \rho(\hat{A}^T)$.

5.9 SAMPLE PROBLEM I

Find the LU decomposition of the following matrix:

$$\begin{pmatrix} 1 & 3 & 2 & 4 \\ 2 & 7 & 5 & 9 \\ 3 & 8 & 6 & 12 \\ 1 & 2 & 3 & 6 \end{pmatrix}. \tag{5.53}$$

Solution

We represent the steps of the Gaussian elimination process by matrix multiplications:

1)
$$\begin{pmatrix} 1 & 0 & 0 & 0 \\ -2 & 1 & 0 & 0 \\ 0 & 0 & 1 & 0 \\ 0 & 0 & 0 & 1 \end{pmatrix} \begin{pmatrix} 1 & 3 & 2 & 4 \\ 2 & 7 & 5 & 9 \\ 3 & 8 & 6 & 12 \\ 1 & 2 & 3 & 6 \end{pmatrix} = \begin{pmatrix} 1 & 3 & 2 & 4 \\ 0 & 1 & 1 & 1 \\ 3 & 8 & 6 & 12 \\ 1 & 2 & 3 & 6 \end{pmatrix}. \tag{5.54}$$

2)
$$\begin{pmatrix} 1 & 0 & 0 & 0 \\ 0 & 1 & 0 & 0 \\ -3 & 0 & 1 & 0 \\ 0 & 0 & 0 & 1 \end{pmatrix} \begin{pmatrix} 1 & 3 & 2 & 4 \\ 0 & 1 & 1 & 1 \\ 3 & 8 & 6 & 12 \\ 1 & 2 & 3 & 6 \end{pmatrix} = \begin{pmatrix} 1 & 3 & 2 & 4 \\ 0 & 1 & 1 & 1 \\ 0 & -1 & 0 & 0 \\ 1 & 2 & 3 & 6 \end{pmatrix}. \tag{5.55}$$

3)
$$\begin{pmatrix} 1 & 0 & 0 & 0 \\ 0 & 1 & 0 & 0 \\ 0 & 0 & 1 & 0 \\ -1 & 0 & 0 & 1 \end{pmatrix} \begin{pmatrix} 1 & 3 & 2 & 4 \\ 0 & 1 & 1 & 1 \\ 0 & -1 & 0 & 0 \\ 1 & 2 & 3 & 6 \end{pmatrix} = \begin{pmatrix} 1 & 3 & 2 & 4 \\ 0 & 1 & 1 & 1 \\ 0 & -1 & 0 & 0 \\ 0 & -1 & 1 & 2 \end{pmatrix}. \tag{5.56}$$

4)
$$\begin{pmatrix} 1 & 0 & 0 & 0 \\ 0 & 1 & 0 & 0 \\ 0 & 1 & 1 & 0 \\ 0 & 0 & 0 & 1 \end{pmatrix} \begin{pmatrix} 1 & 3 & 2 & 4 \\ 0 & 1 & 1 & 1 \\ 0 & -1 & 0 & 0 \\ 0 & -1 & 1 & 2 \end{pmatrix} = \begin{pmatrix} 1 & 3 & 2 & 4 \\ 0 & 1 & 1 & 1 \\ 0 & 0 & 1 & 1 \\ 0 & -1 & 1 & 2 \end{pmatrix}. \tag{5.57}$$

5)
$$\begin{pmatrix} 1 & 0 & 0 & 0 \\ 0 & 1 & 0 & 0 \\ 0 & 0 & 1 & 0 \\ 0 & 1 & 0 & 1 \end{pmatrix} \begin{pmatrix} 1 & 3 & 2 & 4 \\ 0 & 1 & 1 & 1 \\ 0 & 0 & 1 & 1 \\ 0 & -1 & 1 & 2 \end{pmatrix} = \begin{pmatrix} 1 & 3 & 2 & 4 \\ 0 & 1 & 1 & 1 \\ 0 & 0 & 1 & 1 \\ 0 & 0 & 2 & 3 \end{pmatrix}. \quad (5.58)$$

6)
$$\begin{pmatrix} 1 & 0 & 0 & 0 \\ 0 & 1 & 0 & 0 \\ 0 & 0 & 1 & 0 \\ 0 & 0 & -2 & 1 \end{pmatrix} \begin{pmatrix} 1 & 3 & 2 & 4 \\ 0 & 1 & 1 & 1 \\ 0 & 0 & 1 & 1 \\ 0 & 0 & 2 & 3 \end{pmatrix} = \begin{pmatrix} 1 & 3 & 2 & 4 \\ 0 & 1 & 1 & 1 \\ 0 & 0 & 1 & 1 \\ 0 & 0 & 0 & 1 \end{pmatrix} = \hat{U}. \quad (5.59)$$

Matrix $\hat{L}$ is given as the inverse of the product of the $\hat{D}$-matrices used in steps 1) through 6):

$$\hat{L} = \begin{pmatrix} 1 & 0 & 0 & 0 \\ -2 & 1 & 0 & 0 \\ 0 & 0 & 1 & 0 \\ 0 & 0 & 0 & 1 \end{pmatrix}^{-1} \begin{pmatrix} 1 & 0 & 0 & 0 \\ 0 & 1 & 0 & 0 \\ -3 & 0 & 1 & 0 \\ 0 & 0 & 0 & 1 \end{pmatrix}^{-1} \begin{pmatrix} 1 & 0 & 0 & 0 \\ 0 & 1 & 0 & 0 \\ 0 & 0 & 1 & 0 \\ -1 & 0 & 0 & 1 \end{pmatrix}^{-1}$$

$$\times \begin{pmatrix} 1 & 0 & 0 & 0 \\ 0 & 1 & 0 & 0 \\ 0 & 1 & 1 & 0 \\ 0 & 0 & 0 & 1 \end{pmatrix}^{-1} \begin{pmatrix} 1 & 0 & 0 & 0 \\ 0 & 1 & 0 & 0 \\ 0 & 0 & 1 & 0 \\ 0 & 1 & 0 & 1 \end{pmatrix}^{-1} \begin{pmatrix} 1 & 0 & 0 & 0 \\ 0 & 1 & 0 & 0 \\ 0 & 0 & 1 & 0 \\ 0 & 0 & -2 & 1 \end{pmatrix}^{-1}$$

$$= \begin{pmatrix} 1 & 0 & 0 & 0 \\ 2 & 1 & 0 & 0 \\ 3 & -1 & 1 & 0 \\ 1 & -1 & 2 & 1 \end{pmatrix} = \hat{L}. \quad (5.60)$$

Check the result:

$$\hat{L}\hat{U} = \begin{pmatrix} 1 & 0 & 0 & 0 \\ 2 & 1 & 0 & 0 \\ 3 & -1 & 1 & 0 \\ 1 & -1 & 2 & 1 \end{pmatrix} \begin{pmatrix} 1 & 3 & 2 & 4 \\ 0 & 1 & 1 & 1 \\ 0 & 0 & 1 & 1 \\ 0 & 0 & 0 & 1 \end{pmatrix}$$

$$= \begin{pmatrix} 1 & 3 & 2 & 4 \\ 2 & 7 & 5 & 9 \\ 3 & 8 & 6 & 12 \\ 1 & 2 & 3 & 6 \end{pmatrix} = \hat{A}. \quad (5.61)$$

5.10 EIGENVALUES, EIGENVECTORS, AND THE CHARACTERISTIC EQUATION

Let us now return to consideration of square matrices, $\hat{A}^{n \times n}$. As mentioned previously, such a matrix performs some mapping of the linear space C^n into itself, that is, transforms each of the n-component column vectors into some other n-component vector, $\hat{A}\vec{\alpha} = \vec{\alpha}'$. There exists, however, a vector which stays unaffected under this mapping. This is the vector θ, $\hat{A}\theta = \theta$. The question is whether we can find *other* "mapping-resistant" vectors. To make the task easier, we slightly relax this requirement and look for *non-zero* vectors which are transformed by matrix $\hat{A}$ into *proportional* vectors:

$$\hat{A}\vec{\alpha} = \lambda \cdot \vec{\alpha}. \tag{5.62}$$

The non-zero vectors satisfying the condition of Eq. (5.62) for some λ are called *eigenvectors* of the matrix $\hat{A}$, and the corresponding scalars λ are called *eigenvalues*. Both $\vec{\alpha}$ and $\lambda = \lambda(\vec{\alpha})$ being unknown, Eq. (5.62) is *non-linear*. We can, however, separate the problems of finding λ and $\vec{\alpha}$.

Let us rewrite Eq. (5.62) as

$$(\hat{A} - \lambda \hat{I})\vec{\alpha} = \theta. \tag{5.63}$$

In this form, it is a homogeneous system of linear algebraic equations in the components of the vector $\vec{\alpha}$. We know that the condition for this system to have a non-zero solution is

$$\det(\hat{A} - \lambda \hat{I}) = 0. \tag{5.64}$$

This equation determines the eigenvalues. It is called the *characteristic equation*.

The determinant in Eq. (5.64) is a polynomial in λ of degree n. This equation has in total n solutions: $\lambda_1, \lambda_2, \ldots, \lambda_n$. Although, in general, some of these λ_i might be the same, for now we will restrict our attention to the case when all eigenvalues are distinct. If we substitute each of these λ_i in Eq. (5.63), we obtain each time an equation for $\vec{\alpha}$ with a singular matrix, that is, an equation having a non-zero solution. These n non-zero solutions $\vec{\alpha}^{(1)}, \vec{\alpha}^{(2)}, \ldots, \vec{\alpha}^{(n)}$ are just the eigenvectors of the matrix $\hat{A}$, corresponding to eigenvalues $\lambda_1, \lambda_2, \ldots, \lambda_n$, respectively. (Note, that we changed the problem in a sneaky way. Instead of finding $\vec{\alpha}$ and $\lambda(\vec{\alpha})$, we have found λ_i and $\vec{\alpha}(\lambda_i)$. However, this does not affect the result being pairs $(\lambda_1, \vec{\alpha}^{(1)}), (\lambda_2, \vec{\alpha}^{(2)}), \ldots, (\lambda_n, \vec{\alpha}^{(n)})$.)

We will show that in the case of distinct eigenvalues all the n eigenvectors are also distinct and, moreover, linearly independent. Using our favorite method, *reductio ad absurdum*, we assume that the eigenvectors are linearly dependent. This means that there exists a trivial linear combination,

$$\sum_{i=1}^{n} \beta_i \vec{\alpha}^{(i)} = \theta \tag{5.65}$$

with some non-zero coefficients β_i. Let the matrix $\hat{A}$ act on both sides of Eq. (5.65):

$$\hat{A}\left(\sum_{i=1}^{n} \beta_i \vec{\alpha}^{(i)}\right) = \sum_{i=1}^{n} \beta_i(\hat{A}\vec{\alpha}^{(i)}) = \sum_{i=1}^{n} \beta_i \lambda_i \vec{\alpha}^{(i)} = \theta. \tag{5.66}$$

Then we multiply Eq. (5.65) by λ_1 and subtract it from Eq. (5.66), obtaining

$$\sum_{i=1}^{n} \beta_i \lambda_i \vec{\alpha}^{(i)} - \lambda_1 \sum_{i=1}^{n} \beta_i \vec{\alpha}^{(i)}$$

$$= \sum_{i=1}^{n} \beta_i (\lambda_i - \lambda_1) \vec{\alpha}^{(i)}$$

$$= \sum_{i=2}^{n} \beta_i (\lambda_i - \lambda_1) \vec{\alpha}^{(i)} = \theta. \tag{5.67}$$

In Eq. (5.67) we have another trivial linear combination, this time of the $(n-1)$ vectors $\vec{\alpha}^{(2)}, \vec{\alpha}^{(3)}, ..., \vec{\alpha}^{(n)}$:

$$\sum_{i=2}^{n} \beta'_i \vec{\alpha}^{(i)} = \theta \tag{5.68}$$

with coefficients $\beta'_i = \beta_i \cdot (\lambda_i - \lambda_1)$. Note that if $\beta_i \neq 0$, then also $\beta'_i \neq 0$, because all λ's are distinct.

Then, we repeat the procedure. We act on Eq. (5.68) by matrix $\hat{A}$, then subtract this equation multiplied by λ_2 from the result obtained. Thus, we get another linear combination, of the form:

$$\sum_{i=3}^{n} \beta'_i (\lambda_i - \lambda_2) \vec{\alpha}^{(i)} = \theta. \tag{5.69}$$

So, now it is a combination of the $(n-2)$ remaining vectors, $\vec{\alpha}^{(3)}, \vec{\alpha}^{(4)}, ..., \vec{\alpha}^{(n)}$, with the coefficients still being non-zero.

Certainly, we can continue reducing the number of eigenvectors participating in the trivial combination further and further, up to the moment when we come to the last remaining vector, $\vec{\alpha}^{(n)}$ (or to the vector corresponding to last non-zero original coefficient β_i). The coefficient of this latter vector being non-zero, the vector itself must be the zero vector, $\vec{\alpha}^{(n)} = \theta$. But this is a contradiction, the eigenvectors are non-zero *by definition*. Thus, our initial assumption of the eigenvectors being linearly dependent was wrong, and they are linearly independent.

5.11 MATRIX DIAGONALIZATION

The fact that the eigenvectors are linearly independent provides an interesting opportunity for decomposition (or factorization) of the original matrix $\hat{A}$. Let us construct an auxiliary $n \times n$ matrix, whose columns are the eigenvectors $\vec{\alpha}^{(1)}, \vec{\alpha}^{(2)}, ..., \vec{\alpha}^{(n)}$:

$$\hat{S} = (\vec{\alpha}^{(1)} \vec{\alpha}^{(2)} \ldots \vec{\alpha}^{(n)}). \tag{5.70}$$

These columns are linearly independent, so $\rho(\hat{S}) = n$. Now, let us find the product matrix $\hat{A}\hat{S}$. The action of $\hat{A}$ on each of the columns of $\hat{S}$ is described by Eq. (5.62). Thus, the result of the multiplication is just a matrix whose columns are the columns of $\hat{S}$ multiplied by the corresponding λ_i,

$$\hat{A}\hat{S} = (\lambda_1 \vec{\alpha}^{(1)} \lambda_2 \vec{\alpha}^{(2)} \lambda_3 \vec{\alpha}^{(3)} \ldots \lambda_n \vec{\alpha}^{(n)}). \tag{5.71}$$

We can rewrite (5.71) as a matrix product of the following form:

$$(\lambda_1 \vec{\alpha}^{(1)} \lambda_2 \vec{\alpha}^{(2)} \ldots \lambda_n \vec{\alpha}^{(n)})$$

$$= (\vec{\alpha}^{(1)} \vec{\alpha}^{(2)} \ldots \vec{\alpha}^{(n)}) \begin{pmatrix} \lambda_1 & 0 & 0 \ldots & 0 \\ 0 & \lambda_2 & 0 \ldots & 0 \\ 0 & 0 & \lambda_3 \ldots & 0 \\ \vdots & & & \\ 0 & 0 & 0 \ldots & \lambda_n \end{pmatrix} = \hat{S}\hat{\Lambda}, \tag{5.72}$$

where $\hat{\Lambda}$ is the diagonal matrix of eigenvalues. Thus,

$$\hat{A}\hat{S} = \hat{S}\hat{\Lambda}. \tag{5.73}$$

The fact that $\rho(\hat{S}) = n$ means that $\hat{S}$ is a non-singular matrix. So, we can find $\hat{S}^{-1}$ and multiply both sides of Eq. (5.73) by this $\hat{S}^{-1}$ from the right:

$$(\hat{A}\hat{S})\hat{S}^{-1} = \hat{S}\hat{\Lambda}\hat{S}^{-1}. \tag{5.74}$$

Since $\hat{S}\hat{S}^{-1} = \hat{I}$, we obtain the decomposition of the matrix $\hat{A}$ in the following form:

$$\hat{A} = \hat{S}\hat{\Lambda}\hat{S}^{-1}. \tag{5.75}$$

Thus, we have succeeded in expressing $\hat{A}$ in terms of the diagonal matrix $\hat{\Lambda}$. Correspondingly, this procedure is referred as *diagonalization*, and the matrix $\hat{S}$ as the *diagonalizing matrix*.

5.12 FUNCTIONS OF MATRICES

The diagonal form of a matrix and the related transformation (5.75) are indispensable in numerous applications. As an example of the usefulness of decomposition (5.75), consider the following problem.

Suppose we are given a matrix $\hat{A}$ and asked to find some power of this matrix, say, $\hat{A}^4$. Of course, we can obtain the result by performing consecutive matrix multiplications $\hat{A}\hat{A}\hat{A}\hat{A}$. This would be, however, a rather tedious procedure, which grows

FUNCTIONS OF MATRICES 113

unbearable as the power increases. No person in their right mind would volunteer to calculate in this way, say, $\hat{A}^{25}$.

Using the factorization (5.75), we can suggest a sensible alternative to this indiscriminate multiplication. We just write $\hat{A}^4$ and make use of (5.75):

$$\begin{aligned}\hat{A}^4 &= (\hat{S}\hat{\Lambda}\hat{S}^{-1})(\hat{S}\hat{\Lambda}\hat{S}^{-1})(\hat{S}\hat{\Lambda}\hat{S}^{-1})(\hat{S}\hat{\Lambda}\hat{S}^{-1}) \\ &= \hat{S}\hat{\Lambda}(\hat{S}^{-1}\hat{S})\hat{\Lambda}(\hat{S}^{-1}\hat{S})\hat{\Lambda}(\hat{S}^{-1}\hat{S})\hat{\Lambda}\hat{S}^{-1} \\ &= \hat{S}\hat{\Lambda}\hat{I}\hat{\Lambda}\hat{I}\hat{\Lambda}\hat{I}\hat{\Lambda}\hat{S}^{-1} = \hat{S}\hat{\Lambda}^4\hat{S}^{-1}.\end{aligned} \quad (5.76)$$

We have thus reduced finding $\hat{A}^4$ to finding $\hat{\Lambda}^4$. The latter task is easy, since

$$\hat{\Lambda}^4 = \begin{pmatrix} \lambda_1 & 0 & 0\dots & 0 \\ 0 & \lambda_2 & 0\dots & 0 \\ 0 & 0 & \lambda_3 \dots & 0 \\ \vdots & & & \\ 0 & 0 & 0\dots & \lambda_n \end{pmatrix}^4 = \begin{pmatrix} \lambda_1^4 & 0 & 0\dots & 0 \\ 0 & \lambda_2^4 & 0\dots & 0 \\ 0 & 0 & \lambda_3^4 \dots & 0 \\ \vdots & & & \\ 0 & 0 & 0\dots & \lambda_n^4 \end{pmatrix}. \quad (5.77)$$

As can be seen, there is now no major difference between calculating $\hat{A}^4$ or $\hat{A}^{25}$.

We can proceed even further and consider arbitrary functions of a matrix, $f(\hat{A})$. Of course, this is not just idle curiosity; such functions are often required in applications. In particular, the exponential function of a matrix, $e^{\hat{A}}$, appears in solving systems of linear differential equations.

First, we need to suggest a proper definition of a function of a matrix. The available operations on matrices are addition, multiplication by a number, and matrix multiplication. Thus, a natural way to define a function of a matrix is to use Taylor's expansion of a function of scalar argument x:

$$f(x) = \sum_{m=0}^{\infty} a_m x^m \quad (5.78)$$

(we assume $f(x)$ to be a regular function allowing the expansion). Substituting the matrix $\hat{A}$ in place of x, we obtain both the definition and a way of computing the function $f(\hat{A})$:

$$f(\hat{A}) = \sum_{m=0}^{\infty} a_m \hat{A}^m. \quad (5.79)$$

Now, we make use of our results for powers of matrices. We substitute in formula (5.79) the matrix $\hat{A}$ in factored form, $\hat{A} = \hat{S}\hat{\Lambda}\hat{S}^{-1}$, and take the following transformations:

$$\begin{aligned} f(\hat{A}) &= \sum_{m=0}^{\infty} a_m (\hat{S}\hat{\Lambda}\hat{S}^{-1})^m = \sum_{m=0}^{\infty} a_m \hat{S}\hat{\Lambda}^m \hat{S}^{-1} \\ &= \hat{S}\left(\sum_{m=0}^{\infty} a_m \hat{\Lambda}^m\right)\hat{S}^{-1} = \hat{S}f(\hat{\Lambda})\hat{S}^{-1}. \end{aligned} \quad (5.80)$$

114 LINEAR MAPPINGS AND MATRICES

Moreover,

$$f(\hat{\Lambda}) = \sum_{m=0}^{\infty} a_m \hat{\Lambda}^m = \sum_{m=0}^{\infty} a_m \begin{pmatrix} \lambda_1^m & 0 & 0 & \cdots & 0 \\ 0 & \lambda_2^m & 0 & \cdots & 0 \\ \vdots & & & & \\ 0 & 0 & 0 & \cdots & \lambda_n^m \end{pmatrix}$$

$$= \begin{pmatrix} \sum_{m=0}^{\infty} a_m \lambda_1^m & 0 & 0 & \cdots & 0 \\ 0 & \sum_{m=0}^{\infty} a_m \lambda_2^m & 0 & \cdots & 0 \\ \vdots & & & & \\ 0 & 0 & 0 & \cdots & \sum_{m=0}^{\infty} a_m \lambda_2^m \end{pmatrix}$$

$$= \begin{pmatrix} f(\lambda_1) & 0 & 0 & \cdots & 0 \\ 0 & f(\lambda_2) & 0 & \cdots & 0 \\ \vdots & & & & \\ 0 & 0 & 0 & \cdots & f(\lambda_n) \end{pmatrix}. \qquad (5.81)$$

Thus, the algorithm for computing $f(\hat{A})$ is as follows:
1) Diagonalize matrix $\hat{A}$, that is, find matrices $\hat{\Lambda}$ and $\hat{S}$;
2) Find $f(\hat{\Lambda})$, according to Eq. (5.81);
3) Finally, find $f(\hat{A}) = \hat{S} f(\hat{\Lambda}) \hat{S}^{-1}$.

Example of calculating a function of a matrix
Suppose, we are given the following system of equations:

$$\left[\exp \begin{pmatrix} \frac{\pi}{2} & -\frac{\pi}{2} \\ \frac{\pi}{2} & \frac{\pi}{2} \end{pmatrix} \right] \begin{pmatrix} x \\ y \end{pmatrix} = \begin{pmatrix} 1 \\ 1 \end{pmatrix}. \qquad (5.82)$$

To solve this system,

$$\left[\exp \left(\frac{\pi}{2} \cdot \hat{A} \right) \right] \begin{pmatrix} x \\ y \end{pmatrix} = \begin{pmatrix} 1 \\ 1 \end{pmatrix}, \qquad (5.83)$$

with

$$\hat{A} = \begin{pmatrix} 1 & -1 \\ 1 & 1 \end{pmatrix}, \qquad (5.84)$$

we first need to find explicitly

$$\exp \left(\frac{\pi}{2} \cdot \hat{A} \right) = \exp \left(\frac{\pi}{2} \cdot \hat{S} \hat{\Lambda} \hat{S}^{-1} \right) = \hat{S} \left(\exp \left(\frac{\pi}{2} \cdot \hat{\Lambda} \right) \right) \hat{S}^{-1}, \qquad (5.85)$$

where $\hat{\Lambda}$ is the diagonal matrix of eigenvalues of matrix $\hat{A}$, and $\hat{S}$ is the matrix of eigenvectors of $\hat{A}$.

Calculating the eigenvalues and eigenvectors of $\hat{A}$

a) *eigenvalues*

$$\det(\hat{A} - \lambda \hat{I}) = \begin{vmatrix} 1-\lambda & -1 \\ 1 & 1-\lambda \end{vmatrix} = (\lambda - 1)^2 + 1 = 0, \tag{5.86}$$

$$\lambda_{1,2} = 1 \pm i. \tag{5.87}$$

b) *eigenvectors*

The eigenvector $\vec{\alpha}_1$ corresponding to λ_1 is found from the system,

$$\begin{pmatrix} 1-\lambda_1 & -1 \\ 1 & 1-\lambda_1 \end{pmatrix} \begin{pmatrix} \alpha_1^{(1)} \\ \alpha_2^{(1)} \end{pmatrix} = \begin{pmatrix} -i & -1 \\ 1 & -i \end{pmatrix} \begin{pmatrix} \alpha_1^{(1)} \\ \alpha_2^{(1)} \end{pmatrix} = \begin{pmatrix} 0 \\ 0 \end{pmatrix}. \tag{5.88}$$

Taking the first equation,

$$-i\alpha_1^{(1)} - \alpha_2^{(1)} = 0, \tag{5.89}$$

we obtain

$$\alpha_2^{(1)} = -i\alpha_1^{(1)}. \tag{5.90}$$

Thus, we can take the first eigenvector as

$$\vec{\alpha}^{(1)} = \begin{pmatrix} 1 \\ -i \end{pmatrix}. \tag{5.91}$$

Similarly, the eigenvector $\vec{\alpha}_2$ corresponding to λ_2 is found from the system

$$\begin{pmatrix} 1-\lambda_2 & -1 \\ 1 & 1-\lambda_2 \end{pmatrix} \begin{pmatrix} \alpha_1^{(2)} \\ \alpha_2^{(2)} \end{pmatrix} = \begin{pmatrix} i & -1 \\ 1 & i \end{pmatrix} \begin{pmatrix} \alpha_1^{(2)} \\ \alpha_2^{(2)} \end{pmatrix} = \begin{pmatrix} 0 \\ 0 \end{pmatrix}. \tag{5.92}$$

We again take the first equation,

$$i\alpha_1^{(2)} - \alpha_2^{(2)} = 0, \tag{5.93}$$

and get

$$\alpha_2^{(2)} = i\alpha_1^{(2)}. \tag{5.94}$$

Thus,

$$\vec{\alpha}^{(2)} = \begin{pmatrix} 1 \\ i \end{pmatrix}. \tag{5.95}$$

116 LINEAR MAPPINGS AND MATRICES

Therefore, we have obtained the matrix $\hat{S}$ as

$$\begin{pmatrix} 1 & 1 \\ -i & i \end{pmatrix}. \tag{5.96}$$

Its inverse is given as

$$\hat{S}^{-1} = \frac{1}{\det \hat{S}} \cdot \begin{pmatrix} s_{22} & -s_{12} \\ -s_{21} & s_{11} \end{pmatrix} = \frac{1}{2i} \cdot \begin{pmatrix} i & -1 \\ i & 1 \end{pmatrix} = \frac{1}{2} \begin{pmatrix} 1 & i \\ 1 & -i \end{pmatrix}. \tag{5.97}$$

Now, for the exponential function of the matrix $\hat{A}$ we have

$$\exp\left(\frac{\pi}{2} \cdot \hat{A}\right) = \begin{pmatrix} 1 & 1 \\ -i & i \end{pmatrix} \exp\left(\frac{\pi}{2} \cdot \begin{pmatrix} 1+i & 0 \\ 0 & 1-i \end{pmatrix}\right) \frac{1}{2} \begin{pmatrix} 1 & i \\ 1 & -i \end{pmatrix}$$

$$= \frac{1}{2} \cdot \begin{pmatrix} 1 & 1 \\ -i & i \end{pmatrix} \begin{pmatrix} e^{\frac{\pi}{2}+i\frac{\pi}{2}} & 0 \\ 0 & e^{\frac{\pi}{2}-i\frac{\pi}{2}} \end{pmatrix} \begin{pmatrix} 1 & i \\ 1 & -i \end{pmatrix}$$

$$= \frac{1}{2} e^{\frac{\pi}{2}} \cdot \begin{pmatrix} 1 & 1 \\ -i & i \end{pmatrix} \begin{pmatrix} e^{i\frac{\pi}{2}} & 0 \\ 0 & e^{-i\frac{\pi}{2}} \end{pmatrix} \begin{pmatrix} 1 & i \\ 1 & -i \end{pmatrix}$$

$$= \frac{1}{2} e^{\frac{\pi}{2}} \cdot \begin{pmatrix} 1 & 1 \\ -i & i \end{pmatrix} \begin{pmatrix} i & 0 \\ 0 & -i \end{pmatrix} \begin{pmatrix} 1 & i \\ 1 & -i \end{pmatrix}$$

$$= \frac{1}{2} e^{\frac{\pi}{2}} \cdot \begin{pmatrix} 1 & 1 \\ -i & i \end{pmatrix} \begin{pmatrix} i & -1 \\ -i & -1 \end{pmatrix}$$

$$= \frac{1}{2} e^{\frac{\pi}{2}} \cdot \begin{pmatrix} 0 & -2 \\ 2 & 0 \end{pmatrix} = e^{\frac{\pi}{2}} \cdot \begin{pmatrix} 0 & -1 \\ 1 & 0 \end{pmatrix}. \tag{5.98}$$

Thus, the original equation now reads

$$e^{\frac{\pi}{2}} \cdot \begin{pmatrix} 0 & -1 \\ 1 & 0 \end{pmatrix} \begin{pmatrix} x \\ y \end{pmatrix} = \begin{pmatrix} 1 \\ 1 \end{pmatrix}. \tag{5.99}$$

It has the obvious solution:

$$x = e^{-\frac{\pi}{2}}; \quad y = -e^{-\frac{\pi}{2}}. \tag{5.100}$$

5.13 SAMPLE PROBLEM II

Find

$$\cosh\begin{pmatrix} 2+i\pi & i\pi \\ i\pi & 2+i\pi \end{pmatrix}. \tag{5.101}$$

Solution
According to the general procedure, we first have to diagonalize the matrix

$$\hat{A} = \begin{pmatrix} 2+i\pi & i\pi \\ i\pi & 2+i\pi \end{pmatrix}. \tag{5.102}$$

Let

$$(\hat{A} - \lambda\hat{I}) = (2+i\pi-\lambda)^2 - (i\pi)^2 = 0. \tag{5.103}$$

This means:

$$(2+i\pi-\lambda)^2 = (i\pi)^2, \tag{5.104}$$

so that

$$\lambda_{1,2} = 2+i\pi \pm i\pi; \tag{5.105}$$

$$\lambda_1 = 2+2\pi i; \quad \lambda_2 = 2. \tag{5.106}$$

The next step should be that of finding the eigenvectors of matrix $\hat{A}$. In this particular case, however, we proceed directly to computing $\cosh(\hat{\Lambda})$; you will see why. We write

$$\cosh(\hat{\Lambda}) = \begin{pmatrix} \cosh(2+2\pi i) & 0 \\ 0 & \cosh(2) \end{pmatrix}. \tag{5.107}$$

But $\cosh(z)$ is a periodic function of the complex variable z along the imaginary axis, with period $2\pi i$. So, $\cosh(2+2\pi i) = \cosh(2)$, and

$$\cosh(\hat{\Lambda}) = \cosh(2)\begin{pmatrix} 1 & 0 \\ 0 & 1 \end{pmatrix} = \cosh(2) \cdot \hat{I}. \tag{5.108}$$

Now, we see that we really have no need for matrix $\hat{S}$, because:

$$\cosh(\hat{A}) = \hat{S}\cosh(\hat{\Lambda})\hat{S}^{-1} = \hat{S}(\cosh(2)\hat{I})\hat{S}^{-1}$$
$$= \cosh(2) \cdot \hat{S}\hat{I}\hat{S}^{-1} = \cosh(2) \cdot \hat{S}\hat{S}^{-1} = \cosh(2) \cdot \hat{I}. \tag{5.109}$$

This is the final result, which was hardly obvious at the beginning of this exercise:

$$\cosh(\hat{A}) = \cosh(2) \cdot \hat{I}. \tag{5.110}$$

Of course, this situation which allowed us to avoid computing the matrix $\hat{S}$ is rather exceptional. Usually calculating functions of matrices takes more time and effort.

5.14 MULTIPLE EIGENVALUES. THE JORDAN FORM

Now we can return and examine our requirement for the eigenvalues to be distinct. This requirement provided the existence of exactly n linearly independent eigenvectors and, thus, the existence and non-singularity of the diagonalizing matrix $\hat{S}$. Though most applications (even the vast majority) satisfy this requirement, the question is still open, what should we do if the characteristic equation has multiple roots. Here we briefly discuss this situation and how to deal with it.

First of all, the fact that some eigenvalues are equal does not necessary mean that we cannot find several independent solutions of Eq. (5.63) for this eigenvalue. Indeed, if for the eigenvalue in question, $\tilde{\lambda}$, the rank of the matrix $(\hat{A} - \tilde{\lambda}\hat{I})$ is less, than $n-1$, we can construct several independent vectors satisfying Eq. (5.63) for this $\tilde{\lambda}$: we just pick the additional vectors from the nullspace of the matrix $(\hat{A} - \tilde{\lambda}\hat{I})$. We illustrate this case by the following simple example.

Example of diagonalizable matrix with multiple eigenvalues
Consider the matrix

$$\hat{A} = \begin{pmatrix} 1 & 0 & 2 \\ 0 & 1 & 0 \\ 0 & 0 & 3 \end{pmatrix}. \tag{5.111}$$

The characteristic equation in this case is

$$(1-\lambda)^2(3-\lambda) = 0. \tag{5.112}$$

It has roots

$$\lambda_1 = \lambda_2 = 1; \qquad \lambda_3 = 3. \tag{5.113}$$

The equation for the eigenvectors corresponding to the eigenvalues λ_1 and λ_2 then takes the form

$$\begin{pmatrix} 0 & 0 & 2 \\ 0 & 0 & 0 \\ 0 & 0 & 2 \end{pmatrix} \begin{pmatrix} s_1 \\ s_2 \\ s_3 \end{pmatrix} = \theta. \tag{5.114}$$

The rank of matrix in this equation is 1, and we easily find two independent solutions as, for example:

$$\vec{s}^{(1)} = \begin{pmatrix} 1 \\ 0 \\ 0 \end{pmatrix}; \qquad \vec{s}^{(2)} = \begin{pmatrix} 0 \\ 1 \\ 0 \end{pmatrix}. \tag{5.115}$$

The equation for the third eigenvector reads

$$\begin{pmatrix} -2 & 0 & 2 \\ 0 & -2 & 0 \\ 0 & 0 & 0 \end{pmatrix} \begin{pmatrix} s_1^{(3)} \\ s_2^{(3)} \\ s_3^{(3)} \end{pmatrix} = \theta. \tag{5.116}$$

Here, the rank of the matrix is two, and the eigenvector is found as

$$\vec{s}^{(3)} = \begin{pmatrix} 1 \\ 0 \\ 1 \end{pmatrix}. \tag{5.117}$$

Thus, the matrix of eigenvectors is

$$\hat{S} = \begin{pmatrix} 1 & 0 & 1 \\ 0 & 1 & 0 \\ 0 & 0 & 1 \end{pmatrix}. \tag{5.118}$$

By inspection and in comparison with the matrices we used for $\hat{L}\hat{U}$ decomposition, we guess (and check) that

$$\hat{S}^{-1} = \begin{pmatrix} 1 & 0 & -1 \\ 0 & 1 & 0 \\ 0 & 0 & 1 \end{pmatrix}. \tag{5.119}$$

Thus,

$$\hat{A} = \begin{pmatrix} 1 & 0 & 1 \\ 0 & 1 & 0 \\ 0 & 0 & 1 \end{pmatrix} \begin{pmatrix} 1 & 0 & 0 \\ 0 & 1 & 0 \\ 0 & 0 & 3 \end{pmatrix} \begin{pmatrix} 1 & 0 & -1 \\ 0 & 1 & 0 \\ 0 & 0 & 1 \end{pmatrix}. \tag{5.120}$$

In this particular case, the multiplicity of eigenvalues did not cause too much trouble.

In a more general and less favorable case, when it is impossible to find m independent eigenvectors for an eigenvalue repeated m times, the diagonalization fails. All what we can do in this case is to put the matrix to a form as *close* to the diagonal as possible. This "close to the diagonal form" of the matrix and the decomposition procedure are established by the following fundamental theorem which we discuss here very briefly, without proof and even without much rationale.

The Jordan Theorem
If an $n \times n$ matrix $\hat{A}$ has $s \leq n$ linearly independent eigenvectors, it can always be decomposed as

$$\hat{A} = \hat{M}\hat{J}\hat{M}^{-1}, \tag{5.121}$$

where the matrix $\hat{J}$, called the *Jordan form*, is a block-diagonal matrix,

$$\hat{J} = \begin{pmatrix} \hat{J}_1 & 0 & 0 & \cdots & 0 \\ 0 & \hat{J}_2 & 0 & \cdots & 0 \\ 0 & 0 & \hat{J}_3 & \cdots & 0 \\ \vdots & \vdots & \vdots & \ddots & \vdots \\ 0 & 0 & 0 & \cdots & \hat{J}_s \end{pmatrix}, \tag{5.122}$$

with its blocks being simple triangular matrices of the form

$$\hat{J}_i = \begin{pmatrix} \lambda_i & 1 & 0 & \cdots & 0 & 0 \\ 0 & \lambda_i & 1 & \cdots & 0 & 0 \\ 0 & 0 & \lambda_i & \cdots & 0 & 0 \\ \vdots & \vdots & \vdots & \ddots & \vdots & \vdots \\ 0 & 0 & 0 & \cdots & \lambda_i & 1 \\ 0 & 0 & 0 & \cdots & 0 & \lambda_i \end{pmatrix}. \tag{5.123}$$

Each such *Jordan block* corresponds to an independent eigenvector. Since a multiple eigenvalue *can* correspond to several independent eigenvectors (as we saw in the preceding example), there is no universal rule determining the dimensions of the Jordan block. For the same reason, the same eigenvalue λ_i may appear in several blocks. To determine the matrix $\hat{M}$, we just rewrite Eq. (5.121) as

$$\hat{A}\hat{M} = \hat{M}\hat{J} \tag{5.124}$$

and from this obtain the set of equations on the columns of the matrix $\hat{M}$. We explain this procedure on the following example.

Example on the Jordan Form
Consider the matrix

$$\hat{B} = \begin{pmatrix} 1 & 4 & 2 \\ 0 & 1 & 0 \\ 0 & 0 & 3 \end{pmatrix}, \tag{5.125}$$

which is only slightly different from the matrix $\hat{A}$ of the preceding example. The characteristic equation in this case is the same as we had for matrix $\hat{A}$,

$$(1 - \lambda)^2(3 - \lambda) = 0, \tag{5.126}$$

and it produces the same roots,

$$\lambda_1 = \lambda_2 = 1; \qquad \lambda_3 = 3. \tag{5.127}$$

However, the equation for the eigenvectors corresponding to the eigenvalues λ_1 and λ_2 is now different,

$$\begin{pmatrix} 0 & 4 & 2 \\ 0 & 0 & 0 \\ 0 & 0 & 2 \end{pmatrix} \begin{pmatrix} s_1 \\ s_2 \\ s_3 \end{pmatrix} = \theta. \tag{5.128}$$

The rank of matrix in this equation is now 2. This means, that the equation gives us only one eigenvector,

$$\vec{s}^{(1)} = \begin{pmatrix} 1 \\ 0 \\ 0 \end{pmatrix}. \tag{5.129}$$

Taken with the eigenvector corresponding to the eigenvalue λ_3, we have only two eigenvectors. This means, that we cannot diagonalize the matrix, and we must resort to the Jordan form. In this case, the block-diagonal matrix of the Jordan form contains one 2×2 Jordan block corresponding to the eigenvalues $\lambda_1 = \lambda_2$, and one 1×1 block (i.e., just the diagonal element), for the eigenvalue λ_3. Eq. (5.124) for the matrix

$$\hat{M} = (\vec{m}^{(1)}, \vec{m}^{(2)}, \vec{m}^{(3)}) = \begin{pmatrix} m_1^{(1)} & m_1^{(2)} & m_1^{(3)} \\ m_2^{(1)} & m_2^{(2)} & m_2^{(3)} \\ m_3^{(1)} & m_3^{(2)} & m_3^{(3)} \end{pmatrix} \quad (5.130)$$

yields the equation for its first column,

$$\hat{B}\vec{m}^{(1)} = \lambda_1 \vec{m}^{(1)}. \quad (5.131)$$

Thus, this first column is just the eigenvector we have already found,

$$\vec{m}^{(1)} = \vec{s}^{(1)} = \begin{pmatrix} 1 \\ 0 \\ 0 \end{pmatrix}. \quad (5.132)$$

For the second column, Eq. (5.124) yields

$$\hat{B}\vec{m}^{(2)} = \lambda_1 \vec{m}^{(2)} + \vec{m}^{(1)}, \quad (5.133)$$

that is,

$$\begin{pmatrix} 0 & 4 & 2 \\ 0 & 0 & 0 \\ 0 & 0 & 2 \end{pmatrix} \begin{pmatrix} m_1^{(2)} \\ m_2^{(2)} \\ m_3^{(2)} \end{pmatrix} = \begin{pmatrix} 1 \\ 0 \\ 0 \end{pmatrix}. \quad (5.134)$$

From this equation, we find

$$\vec{m}^{(2)} = \begin{pmatrix} m_1^{(2)} \\ m_2^{(2)} \\ m_3^{(2)} \end{pmatrix} = \begin{pmatrix} 0 \\ 1/4 \\ 0 \end{pmatrix}. \quad (5.135)$$

Note that we made the simplest choice, $m_1^{(2)} = 0$ for the *arbitrary* first component. As we see, the equation for the third column, $\vec{m}^{(3)}$, happens to be the same as Eq. (5.116) for the third eigenvector of the previous example. Thus, we can write at once,

$$\vec{m}^{(3)} = \begin{pmatrix} 1 \\ 0 \\ 1 \end{pmatrix}. \quad (5.136)$$

Finally, the matrix $\hat{M}$ is obtained as

$$\hat{M} = \begin{pmatrix} 1 & 0 & 1 \\ 0 & 1/4 & 0 \\ 0 & 0 & 1 \end{pmatrix}. \quad (5.137)$$

Its careful inspection and comparison with the matrices (5.44) of the $\hat{L}\hat{U}$ decomposition result in

$$\hat{M}^{-1} = \begin{pmatrix} 1 & 0 & -1 \\ 0 & 4 & 0 \\ 0 & 0 & 1 \end{pmatrix}. \tag{5.138}$$

Thus,

$$\hat{B} = \begin{pmatrix} 1 & 0 & 1 \\ 0 & 1/4 & 0 \\ 0 & 0 & 1 \end{pmatrix} \begin{pmatrix} 1 & 1 & 0 \\ 0 & 1 & 0 \\ 0 & 0 & 3 \end{pmatrix} \begin{pmatrix} 1 & 0 & -1 \\ 0 & 4 & 0 \\ 0 & 0 & 1 \end{pmatrix}. \tag{5.139}$$

The reader is invited to check this by direct calculation.

The Jordan form of a matrix having being established, the question is whether we can use it to express a function of the matrix, in the same manner as we did it with the diagonal matrix $\hat{\Lambda}$ in the previous section. Certainly, we can, and the only difference is that now the powers of the Jordan form are more complicated that the powers of the diagonal matrix. Namely, the l-th power of the matrix $\hat{J}$ is

$$\hat{J}^l = \begin{pmatrix} \hat{J}_1^l & 0 & 0 & \cdots & 0 \\ 0 & \hat{J}_2^l & 0 & \cdots & 0 \\ 0 & 0 & \hat{J}_3^l & \cdots & 0 \\ \vdots & \vdots & \vdots & \ddots & \vdots \\ 0 & 0 & 0 & \cdots & \hat{J}_s^l \end{pmatrix}, \tag{5.140}$$

when the powers of the Jordan blocks are given as

$$\hat{J}_i^l = \begin{pmatrix} \lambda_i^l & l\lambda_i^{l-1} & l(l-1)\lambda_i^{l-2} & \cdots & 0 \\ 0 & \lambda_i^l & l\lambda_i^{l-1} & \cdots & 0 \\ 0 & 0 & \lambda_i^l & \cdots & 0 \\ \vdots & \vdots & \vdots & \ddots & \vdots \\ 0 & 0 & 0 & \cdots & 1 \\ 0 & 0 & 0 & \cdots & \lambda_i^l \end{pmatrix}. \tag{5.141}$$

With the formulas (5.140) and (5.141) one can compute a regular function of *any* matrix, though the computations are more complex compared to the case of a diagonalizable matrix.

For more details on the Jordan form and its applications we refer the reader to the literature mentioned in the end of this chapter.

5.15 SUMMARY OF CHAPTER 5

1) A linear mapping of an n-dimensional linear space into an m-dimensional linear space corresponds to a linear transformation of an n-component column vector into

an m-component column vector; this transformation is performed using an $n \times m$ table called a *matrix*. The elements of the matrix correspond to transformations of the vectors of the basis in the first space, expressed in terms of the basis in the second space.

2) Two consecutive mappings correspond to a *matrix product*. For the multiplication to be possible, the factor matrices must satisfy certain matching conditions.

3) The set of square matrices of given dimension, taken along with operations of matrix addition and matrix multiplication, constitute a *matrix algebra*.

4) The *transpose matrix* is defined as the matrix obtained from a given matrix by transforming its rows into columns (and vice versa). The transpose of a product of two matrices is equal to the product of their transposes taken in inverse order.

5) All the linearly independent columns of a given matrix constitute a basis in its *range space*. The dimension of the range space is called the *rank* of the matrix, it is also equal to the number of linearly independent *rows*.

6) Any square matrix can be easily reduced to an upper-triangular matrix $\hat{U}$ called the *echelon matrix*. The rank of the echelon matrix is equal to the rank of the original matrix.

7) The eigenvalues of a square matrix are roots of its *characteristic equation*. The eigenvectors corresponding to distinct eigenvalues are linearly independent. If all the n eigenvalues of an $n \times n$ matrix are distinct, the corresponding linearly independent eigenvectors constitute a convenient basis in the space of n-component column vectors.

8) In the case of distinct eigenvalues, the *diagonalization* of a matrix is achieved by the diagonalizing matrix whose columns are the n eigenvectors.

9) By means of diagonalization, any regular function of a matrix can be computed straightforwardly.

10) For repeated eigenvalues, the Jordan form plays the same role as the diagonal matrix.

5.16 PROBLEMS

I. Write down the matrix which transforms the cartesian coordinates, x, y, of a given point, when the coordinate axes rotate at an angle α counterclockwise.

II. In a complex linear space of four-component column vectors, find the dimension of the linear subspace generated by the vectors,

$$\begin{pmatrix} 1 \\ i \\ i \\ 1 \end{pmatrix}, \begin{pmatrix} -2i \\ 1 \\ 1 \\ -2i \end{pmatrix}, \begin{pmatrix} 1 \\ 2i \\ 2i \\ -1 \end{pmatrix}, \text{ and } \begin{pmatrix} i \\ 1 \\ 1 \\ -i \end{pmatrix}.$$

III. Answer the same question for the linear subspace generated by the vectors,

$$\begin{pmatrix} 1 \\ 2i \\ 2i \\ 1 \end{pmatrix}, \begin{pmatrix} -3i \\ 1 \\ 1 \\ -3i \end{pmatrix}, \begin{pmatrix} 1 \\ 3i \\ 3i \\ -1 \end{pmatrix}, \text{ and } \begin{pmatrix} 2i \\ 1 \\ 1 \\ -2i \end{pmatrix}.$$

IV. Solve the following system by means of LU decomposition:

$$\begin{cases} w + 3x + 2y + z = 7 \\ 2w + 7x + 7y + 4z = 20 \\ 3w + 10x + 10y + 8z = 31 \\ 5w + 16x + 12y + 5z = 38 \end{cases}.$$

V. Find

$$\cos \begin{pmatrix} \frac{\pi}{2} & \frac{\pi}{2} \\ 0 & \pi \end{pmatrix} + \sin \begin{pmatrix} \frac{\pi}{2} & 0 \\ \frac{\pi}{2} & \pi \end{pmatrix}.$$

VI. In the linear space of two-component column vectors, find explicitly the element,

$$\begin{pmatrix} \alpha_1 \\ \alpha_2 \end{pmatrix} = \cos \begin{pmatrix} \pi & \pi \\ 0 & 2\pi \end{pmatrix} \begin{pmatrix} 1 \\ -1 \end{pmatrix} + \cos \begin{pmatrix} \pi & 0 \\ \pi & 2\pi \end{pmatrix} \begin{pmatrix} -1 \\ 1 \end{pmatrix}.$$

VII. Solve the following system of differential equations by using the Jordan form of its matrix:

$$\frac{d}{dt} \begin{pmatrix} u_1 \\ u_2 \end{pmatrix} = \begin{pmatrix} 2 & 1 \\ -1 & 4 \end{pmatrix} \begin{pmatrix} u_1 \\ u_2 \end{pmatrix}; \quad \begin{pmatrix} u_1(0) \\ u_2(0) \end{pmatrix} = \begin{pmatrix} 1 \\ 2 \end{pmatrix}.$$

5.17 FURTHER READING

The first book to be recommended to everybody interested in matrices is a classic two-volume compendium,

F. R. Gantmacher, *The Theory of Matrices*, v. 1,2, Chelsea, New York, 1960.
More recent theoretical results can be found in the following books:
P. Lancaster, M. Tismenetsky, *The Theory of Matrices*, 2nd ed., Academic Press, Orlando, FL, 1985.

The authors endeavor to bridge the gap between the treatments of matrix theory and linear algebra to be found in most current textbooks and the mastery of these topics required in important areas of application.

J. M. Ortega, *Matrix Theory – A Second Course*, Plenum, New York, 1987.

The book is concentrated on the aspects of matrix theory which are essential for various graduate courses. It intends to give the student adequate working knowledge of the subject.

In the following book, matrix theory is considered as an essential part of linear algebra, similar to our approach in this chapter.

I. N. Herstein, D. J. Winter, *Matrix Theory and Linear Algebra*, Macmillan, New York, 1988.

The book is aimed at those students who not only must acquire a solid understanding of matrix calculations but also need deeper results, knowledge, and familiarity with the theoretical aspects of the subject. This will allow them to adapt arguments or extend results to the particular problems they are considering.

Computational aspects of matrix theory and matrix computations in modern computer codes are thoroughly discussed in the following two books:

G. H. Golub, C. F. Van Loan, *Matrix Computations*, Johns Hopkins University Press, Baltimore, 1983.

D. S. Watkins, *Fundamentals of Matrix Computations*, Wiley, New York, 1991.

6
Metrics and Topological Properties

6.1 NECESSITY OF DISTANCE CONCEPT

So far, we have treated vector spaces as algebraic structures, that is, in relation to group operations and to their linear properties. We have obtained numerous very important results, but our development still needs one essential aspect to make it useful for real-world problems. The reason is that in the framework of linear spaces we cannot discuss *proximity* of the elements. The only thing we can say about two vectors v_1 and v_2 is whether they are equal of not. We cannot say whether v_2 is closer to v_1 than v_3, and we cannot say how we should modify v_2 for it to be closer to v_1. Thus, we are unable to treat such matters as distance, passing to the limit, continuity, and so on, that are called *topological properties* and are very important in the real world. Therefore, we urgently need to introduce a concept enabling us to describe *distance* between elements of a space.

6.2 METRIC MAPPING AND DISTANCE: DEFINITION AND AXIOMS

Although the three-dimensional space of the real world to which we unavoidably refer in our mathematical constructions is a linear space and has a distance, these two properties are completely different. In fact, the concept of Metric Distance we are going to study has nothing in common with the properties of linearity. This concept can be applied to sets which are not linear spaces. Let us consider an arbitrary set V, and let us introduce a specific mapping from the cross-product set V^2 into the set of real numbers, $V \times V \to \mathcal{R}$, that is, a mapping which transforms a pair of elements of the set V into some real number,

$$(v_1, v_2) \longrightarrow \rho(v_1, v_2). \tag{6.1}$$

This mapping is called a *metric mapping* and, correspondingly, the real number $\rho(v_1, v_2)$ is called a *metric distance* (or just distance) between the elements v_1 and v_2 if the following three axioms hold:

1) The metric distance between two elements is zero if and only if they are the same element,

$$\forall v_1, v_2 \in V: \ \rho(v_1, v_2) = 0 \leftrightarrow v_1 = v_2. \tag{6.2}$$

2) *The axiom of symmetry*: the metric distance between elements v_1 and v_2 is the same as that between v_2 and v_1:

$$\forall v_1, v_2 \in V: \rho(v_1, v_2) = \rho(v_2, v_1). \tag{6.3}$$

3) *The triangle axiom* or *the triangle inequality*:

$$\forall v_1, v_2, v_3 \in V: \rho(v_1, v_3) \leq \rho(v_1, v_2) + \rho(v_2, v_3). \tag{6.4}$$

The choice of a metric mapping is generally not unique, since the restrictions imposed by these three axioms are not too difficult to meet. In particular cases this choice is mainly determined by convenience (or just convention). The three axioms nevertheless stipulate some particular features of metric distance. The most important of these features is that the *metric distance is always a non-negative number*,

$$\rho(v_1, v_2) \geq 0. \tag{6.5}$$

To prove this, we substitute in the triangle inequality (6.4) $v_3 = v_1$, and obtain

$$\rho(v_1, v_1) \leq \rho(v_1, v_2) + \rho(v_2, v_1). \tag{6.6}$$

Then, from the axiom of symmetry the right-hand side of inequality (6.6) is just $2\rho(v_1, v_2)$. As for the left-hand side of this inequality, the first axiom, Eq. (6.2), requires $\rho(v_1, v_1) = 0$. The result,

$$0 = \rho(v_1, v_1) \leq 2 \cdot \rho(v_1, v_2), \tag{6.7}$$

implies $\rho(v_1, v_2) \geq 0$, which completes our proof.

6.3 METRIC SPACES

The set V taken along with a properly defined (i.e., satisfying the axioms) metric distance, that is (V, ρ), is called a *metric space* (often denoted just as V). This name is somewhat misleading, because, as already noted, the set V *is not* supposed to be a *linear space*. Here, the word "space" is just used for "set", for historical reasons.

Examples of metric spaces
a) A trivial choice of metric distance. For any set V we can choose a metric mapping in the form

$$\rho(v_1, v_2) = \begin{cases} 0 & v_1 = v_2 \\ 1 & v_1 \neq v_2 \end{cases}. \quad (6.8)$$

This choice satisfies all the three axioms:
1) $\rho(v_1, v_1) = 0$, by definition;
2) $\rho(v_1, v_2) = \rho(v_2, v_1) = 1$ when $v_1 \neq v_2$;
3) $\rho(v_1, v_3) = \rho(v_1, v_2) = \rho(v_2, v_3) = 1$. Therefore, $\rho(v_1, v_2) + \rho(v_2, v_3) = 2$, and $\rho(v_1, v_3) \leq \rho(v_1, v_2) + \rho(v_2, v_3)$.

Such a metric distance, however, is not too useful.
b) Let set V be the set of natural numbers $\mathcal{N}$ (note that this is not a linear space, as we mentioned in Chapter 4). In this case, the "natural" choice of metric distance is

$$\rho(n_1, n_2) = |n_1 - n_2|, \quad (6.9)$$

which clearly satisfies the three axioms:
1) $\rho(n_1, n_1) = |n_1 - n_1| = 0$;
2) $\rho(n_2, n_1) = |n_2 - n_1| = |n_1 - n_2| = \rho(n_1, n_2)$;
3) It is known from arithmetic that for any three numbers, n_1, n_2, and n_3, the inequality $|n_1 - n_3| \leq |n_1 - n_2| + |n_2 - n_3|$ holds. According to the definition of Eq. (6.9), this means $\rho(v_1, v_3) \leq \rho(v_1, v_2) + \rho(v_2, v_3)$.
c) The same natural choice of metric distance that we used for natural numbers also works for a wider set of real numbers:

$$\rho(r_1, r_2) = |r_1 - r_2|. \quad (6.10)$$

Here, it is also easy to show that the axioms are satisfied.
d) Let $V = \mathcal{C}^n$, the set of ordered n-tuples of complex numbers or n-component column vectors with complex-number components. This set, as we know from Chapter 4 (Section 4.9), *is* a linear space. Moreover, this is a generic linear space representing all the linear spaces of a given dimension n. So, this example is especially important.

For any two elements of V,

$$v_1 = \begin{pmatrix} \alpha_1 \\ \alpha_2 \\ \alpha_3 \\ \cdot \\ \cdot \\ \cdot \\ \alpha_n \end{pmatrix} \text{ and } v_2 = \begin{pmatrix} \beta_1 \\ \beta_2 \\ \beta_3 \\ \cdot \\ \cdot \\ \cdot \\ \beta_n \end{pmatrix}, \quad (6.11)$$

we define metric distance as

$$\rho(v_1, v_2) = \left(\sum_{j=1}^{n} |\alpha_j - \beta_j|^2 \right)^{1/2}. \tag{6.12}$$

This definition clearly satisfies the first two axioms of the metric distance:

1)

$$\rho(v_1, v_1) = \left(\sum_{j=1}^{n} |\alpha_j - \alpha_j|^2 \right)^{1/2} = \left(\sum_{j=1}^{n} |0|^2 \right)^{1/2} = 0; \tag{6.13}$$

2)

$$\rho(v_2, v_1) = \left(\sum_{j=1}^{n} |\beta_j - \alpha_j|^2 \right)^{1/2} = \left(\sum_{j=1}^{n} |\alpha_j - \beta_j|^2 \right)^{1/2} = \rho(v_1, v_2). \tag{6.14}$$

3) It would need, however, much effort and a very tricky consideration to show that the third axiom is satisfied. We will not do this here. Later, in Chapter 7, we will obtain the proof in the particular case of a Hilbert space, as a corollary of another fundamental inequality.

In the particular case of the three-dimensional *real* linear space of the real world, this definition of the metric distance gives for the real-component vectors

$$v_1 = \begin{pmatrix} x_1 \\ y_1 \\ z_1 \end{pmatrix} \quad \text{and} \quad v_2 = \begin{pmatrix} x_2 \\ y_2 \\ z_2 \end{pmatrix} \tag{6.15}$$

the metric distance,

$$\rho(v_1, v_2) = \sqrt{(x_1 - x_2)^2 + (y_1 - y_2)^2 + (z_1 - z_2)^2}. \tag{6.16}$$

Of course, this space is just the three-dimensional *Euclidian space*, the space encountered in high-school geometry. The components of column vectors v_1 and v_2 are then the cartesian coordinates of two three-dimensional vectors.

e) The case of set V being an *infinite-dimensional* linear space of functions on the interval $[0, 1]$, which we have also considered in Chapter 4 (Section 4.4). In this case, we have to construct a metric mapping which transforms *functions* into *real numbers*. We can come to a proper definition in the following heuristic way. Let us symbolically imagine a function $f(x)$, determined by its values at *all* the points in the interval $[0, 1]$, as some limit case of an n-tuple whose components are the values of $f(x_i)$ at a *large but finite number* of equidistant points $x_i = i/n$,

$i = 1, \ldots, n$ in the interval. Then, for two functions, $f_1(x)$ and $f_2(x)$, the metric distance between the n-tuples $f_1 = (f_1(x_1), f_1(x_2), f_1(x_3), \ldots, f_1(x_n))$ and $f_2 = (f_2(x_1), f_2(x_2), f_2(x_3), \ldots, f_2(x_n))$ can be written as

$$\rho(f_1, f_2) = \left(\frac{1}{n} \sum_{i=1}^{i=n} |f_1(x_i) - f_2(x_i)|^2 \right)^{1/2}. \tag{6.17}$$

(We add here the factor $1/n$, compared to Eq. (6.12), so that $\rho(f_1, f_2)$ does not become too large for large n). Then, we generalize this definition of metric distance to the case of $n \to \infty$, the sum of Eq. (6.17) becomes an integral,

$$\rho(f_1(x), f_2(x)) = \lim_{n \to \infty} \rho(f_1, f_2) = \left(\int_0^1 dx |f_1(x) - f_2(x)|^2 \right)^{1/2}. \tag{6.18}$$

Thus, in this infinite-dimensional linear space the metric distance is defined via a specific *definite integral*.

The fact that the definition of Eq. (6.18) is obtained as a limit of (6.12) means that it also satisfies the axioms of metric distance. However, the limit transition to the integral implies that the functions f_1 and f_2 satisfy some conditions, elaborated in calculus. We do not intend to cast our heuristics in mathematically rigorous form and will not go into these subtleties. In what follows, we will always assume that we deal with "reasonable" functions, so-called *square-integrable* functions, which allow us to use this definition of the metric distance. This assumption makes sense, because this covers virtually all functions encountered in engineering practice.

There is also another useful definition of metric distance in the space of functions. This definition refers to (6.10) and gives the formula

$$\rho_2(f_1(x), f_2(x)) = \int_0^1 dx |f_1(x) - f_2(x)|. \tag{6.19}$$

To show that this definition satisfies the axioms, we would merely need to rewrite our arguments for the formula (6.9), adding the sign of the integral. As we will shortly see, this second definition of metric distance is more convenient in theoretical considerations than that of Eq. (6.18). However, it is usually easier to operate with the *square* than with the *modulus*, so the formula (6.18) is preferable. Also, the metric distance of Eq. (6.19) can be of help in cases of functions which are not square-integrable, for instance, when functions $f_1(x)$ and $f_2(x)$ behave as $1/\sqrt{x}$ at $x \to 0$.

There is yet another known definition of a metric distance in the case of functions,

$$\rho_3(f_1(x), f_2(x)) = \max_{x \in [0,1]} |f_1(x) - f_2(x)|. \tag{6.20}$$

We will not use it in this book, though it is useful in some specific control problems.

6.4 SAMPLE PROBLEM

Which of the polynomials $p_1(t) = 1 - t$ and $p_2(t) = 1 - t^2$ gives a better approximation of $\cos t$ on the interval $[0, \pi/2]$ in the sense of formula (6.18)?
Hint
We need to compare the metric distances $\rho(\cos t, 1 - t)$ and $\rho(\cos t, 1 - t^2)$. That metric distance which is smaller corresponds to the polynomial which is a better approximation.

Solution
In practical calculations, it is easier to compare *squares* of the metric distances, because these squares do not contain square roots. Thus, we calculate the difference

$$(\rho(\cos t, 1 - t))^2 - (\rho(\cos t, 1 - t^2))^2 =$$
$$\int_0^{\pi/2} (\cos t - (1 - t))^2 dt - \int_0^{\pi/2} (\cos t - (1 - t^2))^2 dt$$
$$= \int_0^{\pi/2} \left((\cos t - 1 + t)^2 - (\cos t - 1 + t^2)^2\right) dt$$
$$= \int_0^{\pi/2} \left[(\cos t - 1)^2 + 2t(\cos t - 1) + t^2 \right.$$
$$\left. - (\cos t - 1)^2 - 2t^2(\cos t - 1) - t^4\right] dt =$$
$$= \int_0^{\pi/2} \left(2t \cos t - 2t + t^2 - 2t^2 \cos t + 2t^2 - t^4\right) dt$$
$$= \int_0^{\pi/2} \left(2t \cos t - 2t^2 \cos t - 2t + 3t^2 - t^4\right) dt, \tag{6.21}$$

and to check whether the result is negative or positive.
The resulting expression in Eq. (6.21) contains the following integrals:

$$\int t \cos t\, dt = t \sin t - \int \sin t\, dt = t \sin t + \cos t$$

$$\int t^2 \cos t\, dt = t^2 \sin t - \int 2t \sin t\, dt$$
$$= t^2 \sin t + 2t \cos t - \int 2 \cos t\, dt$$
$$= t^2 \sin t + 2t \cos t - 2 \sin t. \tag{6.22}$$

Making use of the results (6.22) in the expression (6.21), we obtain:

$$(\rho(\cos t, 1 - t))^2 - (\rho(\cos t, 1 - t^2))^2$$
$$= \Big(2\left((t \sin t + \cos t) - (t^2 \sin t + 2t \cos t - 2 \sin t)\right)$$

$$\left. - t^2 + t^3 - \frac{1}{5}t^5 \right)\Big|_0^{\pi/2}$$
$$= \left(2\sin t(t - t^2 + 2) + 2\cos t(1 - 2t) - t^2 + t^3 - \frac{1}{5}t^5 \right)\Big|_0^{\pi/2}$$
$$= 2 + \pi - 3\left(\frac{\pi}{2}\right)^2 + \left(\frac{\pi}{2}\right)^3 - \frac{1}{5}\left(\frac{\pi}{2}\right)^5 \approx -0.3. \tag{6.23}$$

Thus,
$$\rho(\cos t, 1 - t)) < (\rho(\cos t, 1 - t^2), \tag{6.24}$$

which means that for approximating $\cos t$ on the interval $[0, \pi/2]$ the polynomial $p_1(t) = 1 - t$ is better than the polynomial $p_2(t) = 1 - t^2$.

6.5 PROXIMITY. LIMITS IN METRIC SPACES

Having in hand both the concept and a working definition of metric distance, we are able to arrange the elements of a metric space by their distance to a given element. We are also able to investigate such cases when the distances between a given element and the other elements of the metric space approaching it become infinitely small. This *passing to the limit* in the metric space and all the related matters can be considered in strict analogy with those considerations for real numbers which have been made in calculus.

Let the set V be a metric space with metric distance $\rho(v_1, v_2)$. Let us pick elements from V according to some rule, and arrange these elements as the components of an infinite tuple, called a *sequence*,
$$(v_1, v_2, v_3, \ldots, v_n, \ldots). \tag{6.25}$$

The sequence (6.25) is said to *converge* to some element of V, $\tilde{v}$, if the metric distances $\rho(\tilde{v}, v_n)$ go to zero with increasing n. Or, in more formal language, the sequence (6.25) converges to $\tilde{v}$ if for any small real number ε there exists a number $N(\varepsilon)$ such that all the elements of the sequence with indexes greater than N have their distances to $\tilde{v}$ less than ε,
$$\forall \varepsilon > 0, \ \exists N(\varepsilon) \Rightarrow \forall n \geq N, \ \rho(\tilde{v}, v_n) < \varepsilon. \tag{6.26}$$

If this is the case, the element $\tilde{v}$ is called the *limit* of the sequence.

An important fact is that if a given sequence has a limit, this *limit is unique*. We prove it by *reductio ad absurdum*. Suppose that a sequence has two different limits, say, $\tilde{v}$ and $\tilde{\tilde{v}}$. According to (6.26), the $\tilde{v}$ being a limit means that starting with some large number all the elements of the sequence must occur in close vicinity (we can call it ε-vicinity, that is, being at a distance closer than ε) of $\tilde{v}$. But at the same time, these very elements must be in the ε-vicinity of the other limit, $\tilde{\tilde{v}}$. This means that the metric distance $\rho(\tilde{v}, \tilde{\tilde{v}}) \leq 2\varepsilon$. The ε being an arbitrary small parameter, this distance must be zero, which implies, according to the first axiom of metric distance, $\tilde{v} = \tilde{\tilde{v}}$. Thus, our assumption of two different limits was wrong, and the limit is really unique.

6.6 CAUCHY SEQUENCES

The preceding consideration of passing to the limit had one uncomfortable feature: it referred to the limiting element $\tilde{v}$ itself. Indeed, according to (6.26), to determine whether a sequence converges we had to use its limiting element, the very result of that alleged convergence. Definitely, it would be much more useful (and much more consistent) if we could answer the question of convergence, making use of only the elements of the sequence. Fortunately, a criterion of convergence which is based only on the elements of a sequence does exist. This criterion was formulated by the French mathematician Cauchy, and it is so useful and important that now convergent sequences are often called *Cauchy sequences*.

The Cauchy criterion looks very much like the limit definition of Eq. (6.26), with one essential distinction. It reads: a sequence is convergent if for any small real number ε there exists a number $N(\varepsilon)$ such that the metric distance between any two of the elements having numbers greater than N is less than ε,

$$\forall \epsilon > 0, \ \exists N(\varepsilon) \Rightarrow \forall n, m \geq N, \ \rho(v_m, v_n) < \epsilon. \tag{6.27}$$

This statement operates only on elements of the sequence, and does not refer to the limiting element.

We will not rigorously prove the equivalence of the Cauchy criterion and the limit definition. We note only that this equivalence is quite apparent if we use the concept of an ε-ball.

6.7 OPEN BALL: A GEOMETRIC INTERPRETATION

While considering uniqueness of the limiting element, we used a vague image of "ε-vicinity" of an element of the metric space V. Let us now define an *open ε-ball* (or just an *ε-ball*) centered at $\bar{v}$, $B_\varepsilon(\bar{v})$ as a subset of the space V, whose distance to $\bar{v}$ is less than ε:

$$v \in B_\varepsilon(\bar{v}) \leftrightarrow \rho(v, \bar{v}) < \varepsilon. \tag{6.28}$$

In the customary three-dimensional Euclidean space this is really a ball with radius ε centered at $\bar{v}$.

From the standpoint of this new concept, the definition of a convergent sequence can be reformulated as follows: the sequence $(v_1, \ldots, v_n, \ldots)$ converges to an element $\tilde{v}$ if for any small real number ε there exists a number $N(\varepsilon)$ such that all the

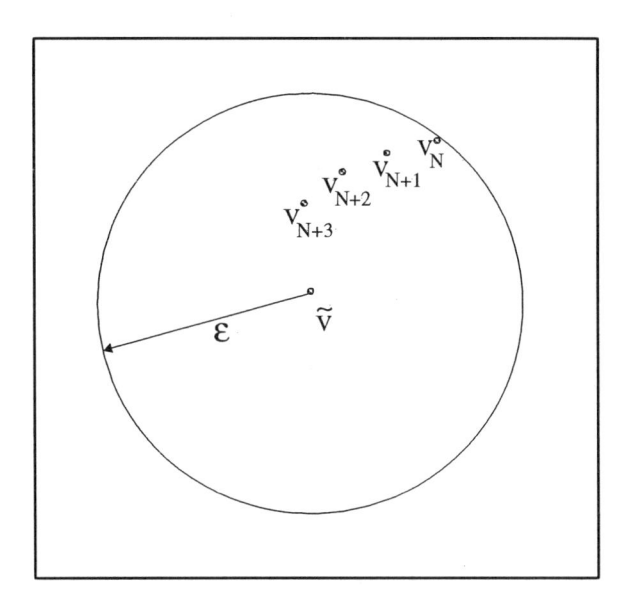

Fig. 6.1 Necessity of the Cauchy criterion.

elements of the sequence with indexes greater than $N(\varepsilon)$ fall into the ε-ball centered at $\tilde{v}$,

$$\forall \varepsilon > 0, \ \exists N(\varepsilon) \Rightarrow n \geq N(\varepsilon), \ v_n \in B_\varepsilon(\tilde{v}). \tag{6.29}$$

In fact, this is merely a restatement of Eq. (6.26). However, this new formulation invokes an image which is very useful for our next step, the demonstration of the equivalence of the Cauchy criterion (6.27) and the definition of Eq. (6.29).

To show the *necessity* of the Cauchy criterion, we make the observation that starting with some $N(\varepsilon)$ all the elements of the convergent sequence are packed in the ε-ball. This means that distances between them are no greater than 2ε, see Fig. 6.1. The ε being an arbitrary small number, this implies that the Cauchy condition is satisfied.

To show the *sufficiency* of the Cauchy criterion, we note that if starting with v_N the distances between *any* two of the elements of the sequence are less than ε, then all of these elements are packed in the ε-ball centered at v_N. Thus, they are not further than $2 \cdot \varepsilon$ from some limiting element lying somewhere inside this ball, see Fig. 6.2.

6.8 EXAMPLES OF CAUCHY SEQUENCES

a) Let the set V be the set of *positive numbers*:

$$V = \{v \in \mathcal{R} \land v > 0\}. \tag{6.30}$$

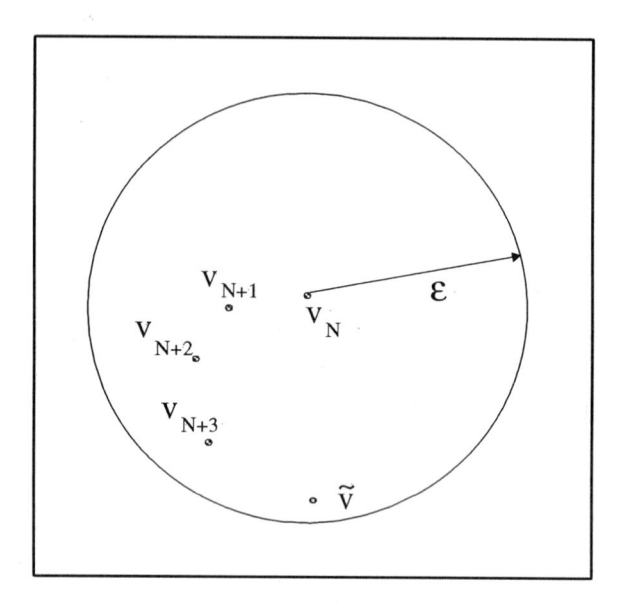

Fig. 6.2 Sufficiency of the Cauchy criterion.

Consider a sequence of elements of this set,

$$(1, 1/2, 1/4, 1/8, 1/16, \ldots, 1/2^n, \ldots). \tag{6.31}$$

According to definition (6.9), the metric distance between the n-th and m-th elements of this sequence is

$$\rho(v_n, v_m) = |1/2^n - 1/2^m|. \tag{6.32}$$

To be precise, suppose $m > n$. Then we can rewrite the metric distance (6.32) as

$$\rho(v_n, v_m) = (1/2^n)(1 - 2^{n-m}) \leq 1/2^n. \tag{6.33}$$

Therefore, this distance will be less than a given small number ε, if

$$n > N(\varepsilon) = [\log_2 (1/\varepsilon)], \tag{6.34}$$

where the brackets, [...], mean *integer part* of the expression. Thus, we have found the number $N(\varepsilon)$ for the criterion of Eq. (6.27), and have proven the sequence (6.31) to be a Cauchy sequence.

b) Let the set V be the set of *continuous functions* on the interval $[0, 1]$. Consider a sequence of elements $v_n = \phi_n(x)$ of this set, the functions

$$(\phi_1(x), \phi_2(x), \phi_3(x), \phi_4(x), \ldots \phi_n(x), \ldots), \tag{6.35}$$

where

$$\phi_n(x) = \begin{cases} (2x)^n & \text{if } x < 1/2 \\ 1 & \text{if } x > 1/2 \end{cases} \qquad (6.36)$$

(the choice seems a bit strange, but we will also need this sequence in the following). According to the basic definition of Eq. (6.18), the metric distance between the n-th and m-th elements of this sequence is

$$\rho(\phi_n(x), \phi_m(x)) = \sqrt{\int_0^1 dx (\phi_n(x) - \phi_m(x))^2}$$

$$= \sqrt{\int_0^{1/2} dx ((2x)^n - (2x)^m)^2}. \qquad (6.37)$$

We again suppose $m > n$, and get the estimate as

$$\rho(\phi_n(x), \phi_m(x)) = \sqrt{\int_0^{1/2} dx (2x)^{2n} (1 - (2x)^{m-n})^2}. \qquad (6.38)$$

Since the argument x is always less than 1/2, the value of the integral in (6.38) becomes larger when we drop the term $(2x)^{m-n}$; thus

$$\rho(\phi_n(x), \phi_m(x)) = \sqrt{\int_0^{1/2} dx (2x)^{2n} (1 - (2x)^{m-n})^2}$$

$$\leq \sqrt{\int_0^{1/2} dx (2x)^{2n}} = \frac{1}{\sqrt{2(2n+1)}}. \qquad (6.39)$$

This metric distance will be less than a prescribed ε for any

$$n \geq N(\varepsilon) = \left[\frac{1}{2} \left(\frac{1}{2\varepsilon^2} - 1 \right) \right]. \qquad (6.40)$$

So, we again have found the number $N(\varepsilon)$ for the Cauchy criterion (6.27), and thus have shown the sequence (6.35) to be a Cauchy sequence.

6.9 OPEN AND CLOSED SETS. COMPLETENESS VS. CLOSEDNESS

Complete sets
Typically the set from which the elements of a Cauchy sequence are picked is a subset

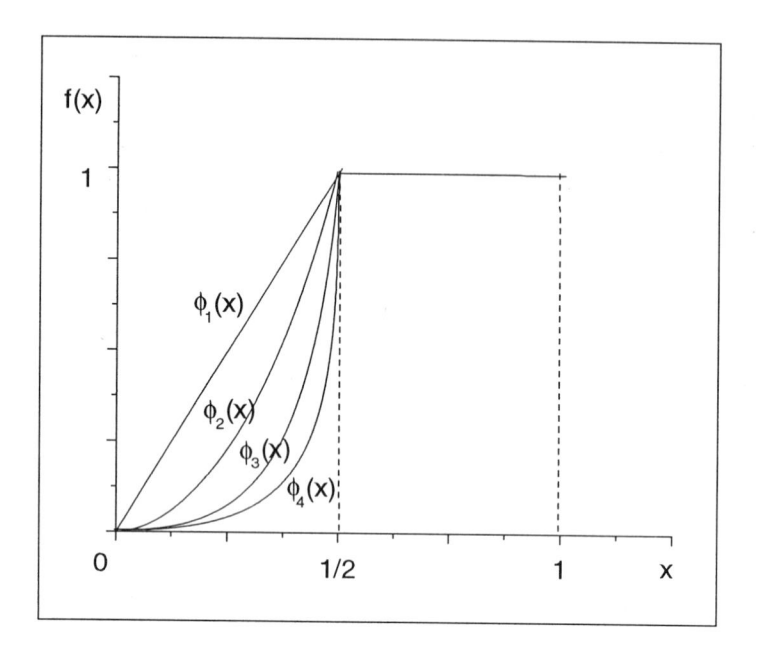

Fig. 6.3 The sequence of functions $\phi_n(x)$.

of some larger set. In particular, this was the case in both of the examples of Cauchy sequences that we considered in the previous section. In example a), the set of positive numbers is a subset of the set of real numbers, $\mathcal{R}$. In example b), the set of continuous functions is a subset of the set of all functions on a given interval.

In such situations, the fact that a Cauchy sequence necessarily converges to some limiting element does not mean that this limiting element must belong to the same subset as the elements of the sequence itself. In fact, in both examples considered above, this *was not* the case. In example a) the limiting element is 0, which clearly does not belong to the subset of positive numbers. In example b) the limit is a very special function of argument x on the interval $[0, 1]$ which is equal to 0 when $x < 1/2$ and equal to 1 when $x > 1/2$ (see Fig. 6.3). This is certainly not a continuous function.

In practical instances, it is much better to deal with such situations when the limiting element *belongs* to the subset on which the Cauchy sequence is defined. In this case, the limiting element has the same attributes as the elements of the sequence and can be treated with the same specific techniques. Usually, it is possible to construct not just one, but many Cauchy sequences from the elements of a given set. Thus, instead of checking whether the limit element of any individual Cauchy sequence belongs to the sequence, it is worth considering the problem with respect to the whole set. Namely, a subset V_1 of a metric space V is called *complete* if *every* Cauchy sequence of elements of this subset converges to an element $\tilde{v} \in V_1$. (One might notice that in some sense the situation with a complete subset is similar to that of a linear subspace.

The latter must contain all the linear combinations of its elements; the former must contain the limiting elements of all its Cauchy sequences.)

For instance, in example a) we can make the set complete simply by adding to it one element, the number 0, that is, by the transition from the set of *positive numbers* to the set of *non-negative numbers*. Well-known examples of complete sets are the set of real numbers, $\mathcal{R}$, and the set of complex numbers, $\mathcal{C}$.

Closed sets
When thinking of applications, the concept of complete set might seem basically useless. Indeed, in most cases it is virtually impossible to check all the imaginable Cauchy sequences to determine whether a given subset is complete or not. For this reason, we consider another characteristic of a set, which is related to completeness, but operates with different concepts, those of *interior* and *closure points*

A point v_i of a subset $V_1 \subset V$ is called an *interior point* of this subset if there exists a small positive number ε such that all the elements of V constituting the ε-ball centered at v_i belong to the subset V_1,

$$\exists \epsilon : B_\varepsilon(v_i) \subseteq V_1. \tag{6.41}$$

Then, if *all* the points of some subset are interior points, such a subset is called an *open subset*.

A point v_c of a set V is called a *closure point* of a subset $V_1 \subset V$, if for *any given* number ε there is a point $v_j \in V_1$ such that the metric distance $\rho(v_j, v_c) < \epsilon$,

$$\forall \epsilon \ (B_\varepsilon(v_c) \cap V_1 \neq \emptyset). \tag{6.42}$$

Note the significant difference between *interior point* and *closure point*. The definition of interior point requires that starting with some small ε *all* the elements of the ε-ball belong to subset V_1, while the definition of closure point merely demands that *some* elements of V_1 belong to ε-ball $B_\varepsilon(v_c)$. In this sense, all the interior points of a subset are its closure points, but a closure point of the subset is not necessarily its interior point. Moreover, a closure point might not belong to the subset at all. Subset V_1 is called a *closed subset* if it contains all of its closure points. If this is not so, it is always possible to construct a *closure* of the subset V_1,

$$\bar{V}_1 = V_1 \cup (\text{ all its closure points}). \tag{6.43}$$

Examples of open and closed sets
a) According to their definitions (Chapter 1, Section 1.6), the interval $[0, 1]$ is a closed subset of $\mathcal{R}$, while the interval $(0, 1)$ is an open subset of $\mathcal{R}$.
b) The set of all positive numbers is an open set. On the other hand, the set of all non-negative numbers is a closed set, because it contains the closure point 0.
c) A set containing only one element is always a closed set, because the ε-ball centered at this element contains just only the element itself.

140 METRICS AND TOPOLOGICAL PROPERTIES

d) An interesting example is the *empty set*, which is open as well as closed. It is open because it does not contain any elements which are not interior points. It is closed, because the set of its closure points is also closed and thereby belongs to it.

e) Other sets which demonstrate the feature of being both open and closed at the same time, are the set of real numbers, $\mathcal{R}$, and the set of complex numbers, $\mathcal{C}$.

Comparison of example b) with the examples of complete sets given earlier indicates that these two properties, completeness and closedness, are connected. And this is really so, the general statement holds: *if a subset is closed, it is complete, and vice versa.*

To prove this statement, we use the second definition of limit, that of Eq. (6.29). Juxtaposing this definition with the definition of closure point, Eq. (6.42), we see, that the limit point of a Cauchy sequence whose elements belong to some subset is necessarily a closure point of this subset. Thus, the statements "all the closure points belong to the subset" and "the limit points of all Cauchy sequences belong to the subset" mean the same thing.

6.10 CONTINUOUS MAPPING

The last of the important concepts related to metric distance we need to know is *continuous mapping*. This concept relates metric distances in different metric spaces.

Consider two complete metric spaces, V and W. Generally speaking, these metric spaces can be of quite different nature and have their metric distances defined quite differently; these details do not matter for the following. Consider a mapping F from the space V into the space W, which transforms an element $v \in V$ into an element $w = f(v) \in W$.

Let the sequence of elements of the set V,

$$(v_1, v_2, v_3, ... v_n, ...), \qquad (6.44)$$

be a Cauchy sequence with the limit $\tilde{v}$. Under the mapping F the elements of this sequence are being transformed into some elements of set W, constituting a sequence

$$(w_1 = f(v_1), w_2 = f(v_2), w_3 = f(v_3), ..., w_n = f(v_n), ...). \qquad (6.45)$$

In general, this new sequence is not necessarily a Cauchy sequence. Consider, for instance, the sequence of Eq. (6.31). The mapping from the set of positive numbers into set of all real numbers, given as $w = log_2(v)$, transforms this sequence into the sequence

$$(0, -1, -2, -3, ..., -n, ...), \qquad (6.46)$$

which does not converge.

In some instances, however, the new sequence also converges to some limiting element,

$$\tilde{w} = \lim_{n \to \infty} w_n. \qquad (6.47)$$

If this is the case for all Cauchy sequences in V, and if for each of these sequences the limiting element $\tilde{w}$ is the value of $\tilde{v}$ under the mapping F, that is,

$$\lim_{n\to\infty} f(v_n) = f(\lim_{n\to\infty} v_n), \tag{6.48}$$

then the mapping is called *continuous*.

6.11 LIPSCHITZ (BOUNDED) MAPPING

It is generally very hard to check whether a mapping is continuous or not, since we cannot try all the possible Cauchy sequences in V. There is, however, a criterion, which can be used to easily verify whether a mapping is continuous. At the same time, this criterion elucidates the meaning of a continuous mapping. This is the so-called *Lipschitz criterion*.

A mapping F is said to be a Lipschitz (bounded) mapping, if

$$\forall v_1, v_2 \in V : \rho(f(v_1), f(v_2)) \leq \mu \cdot \rho(v_1, v_2), \tag{6.49}$$

where the *Lipschitz constant* μ is some positive number, independent of the choice of elements v_1 and v_2. This means that close points of set V are being transformed into close points of set W.

Since the Lipschitz condition (6.49) bounds the metric distance between the values of two elements by the distance between the elements themselves, it is evident that if the distance between two elements, v_1 and v_2 of the sequence (6.44) is less than ε, then the distance between their values, w_1 and w_2 is less than $\mu\varepsilon$. If this ε is the arbitrary small number of the definition of a Cauchy sequence, Eq. (6.29), then we conclude that if the sequence of Eq. (6.44) is a Cauchy sequence with $\tilde{v}$ being its limit, this implies the sequence of Eq. (6.45) is also a Cauchy sequence with $f(\tilde{v})$ as its limit. Thus, *if the mapping is bounded, then it is continuous*.

If for some mapping F the Lipschitz constant $\mu < 1$, such a mapping is called a *contraction*. The notion of a contraction mapping will be very useful when treating operator equations in Chapter 9.

6.12 EXAMPLE OF A CONTRACTION MAPPING

Consider the mapping of the interval of real numbers $[0, 1]$ into itself,

$$x \to \cos x, \quad x \in [0, 1]. \tag{6.50}$$

We want to show that this mapping is a contraction. To do so, we need to express the metric distance between the values of two elements in terms of the distance between the elements themselves. According to the definition of Eq. (6.9), the metric distance between two real-number values of Eq. (6.50) is

$$\rho(f(x_1), f(x_2)) = |\cos(x_1) - \cos(x_2)|. \tag{6.51}$$

To express this through the metric distance between the two real-number elements, x_1 and x_2,

$$\rho(x_1, x_2) = |x_1 - x_2|, \tag{6.52}$$

we transform the trigonometric expression in the right-hand side of Eq. (6.51) as

$$\cos(x_1) - \cos(x_2) = 2 \sin\left(\frac{x_1 - x_2}{2}\right) \cdot \sin\left(\frac{x_1 + x_2}{2}\right). \tag{6.53}$$

Then, recalling that

$$|\sin \alpha| \leq |\alpha|, \tag{6.54}$$

we have

$$\left|\sin\left(\frac{x_1 - x_2}{2}\right)\right| \leq \frac{1}{2} |x_1 - x_2|, \tag{6.55}$$

On the other hand, since both x_1 and x_2 belong to the interval $[0, 1]$,

$$\sin\left(\frac{x_1 + x_2}{2}\right) \leq \sin\left(\frac{1 + 1}{2}\right) = \sin 1. \tag{6.56}$$

Thus, we get the estimate,

$$\begin{aligned}\rho(f(x_1), f(x_2)) &= |\cos(x_1) - \cos(x_2)| \\ &\leq |\sin 1| \cdot |x_1 - x_2| = |\sin 1| \rho(x_1, x_2).\end{aligned} \tag{6.57}$$

Therefore, the mapping of Eq. (6.50) is a Lipschitz mapping with Lipschitz constant $\mu = |\sin 1|$. The fact that $\mu < 1$ means that the mapping in question is a contraction.

6.13 SUMMARY OF CHAPTER 6

1) The concept of *metric mapping* is introduced for an arbitrary set V regardless of its algebraic properties; it is a mapping $V \times V \to \mathcal{R}$ which transforms a pair of elements v_1, v_2 of the set V into a real number $\rho(v_1, v_2)$ called the *metric distance* between the elements v_1 and v_2.
2) The metric distance satisfies three axioms:
a) $\rho(x, y) = 0 \leftrightarrow x = y$;
b) $\rho(x, y) = \rho(y, x)$ — the axiom of symmetry;
c) $\rho(x, z) \leq \rho(x, y) + \rho(y, z)$ — the triangle axiom.

3) The set V taken along with the metric mapping is called a *Metric Space*. In particular,
a) the set of n-tuples (or n-component column vectors) with complex-number components, constitutes a metric space with metric distance,

$$\rho(x,y) = \left(\sum_{i=1}^{n} |y_i - x_i|^2\right)^{1/2};$$

b) the set of square-integrable functions of real argument on a given interval $[a, b]$ is a metric space with metric distance,

$$\rho = \left(\int_a^b |f_1(x) - f_2(x)|^2 dx\right)^{1/2}.$$

4) *Convergence.* A sequence of elements of a metric space, $(v_1, v_2, v_3, \ldots, v_n, \ldots)$, converges to some *limiting element*, $\tilde{v}$, if with growing n the metric distance between v_n and $\tilde{v}$ goes to zero. the limit of a convergent sequence is unique. Convergent sequences are also called *Cauchy sequences*, they satisfy the Cauchy criterion of Eq. (6.27).

5) *Interior and closure points.* An element of a subset $V_1 \subset V$ is called an interior point of this subset, if there is a close vicinity of the element which contains only elements of V_1. A subset for which all elements are interior points is called an *open* set. An element $v \in V$ is called a *closure point* of a subset V_1, if in however close vicinity of this element there exist some elements of V_1. Each interior point of a subset is its closure point; the inverse statement is not true.

6) *Completeness and closedness.* A subset V_1 is called *complete* if every Cauchy sequence of its elements converges to some element of the same subset, V_1. A subset is called *closed* if it contains all its *closure points*. A closed set is necessarily complete and vice versa.

7) *Continuous and bounded mappings.* A mapping $V \to W$ is called *continuous*, if a Cauchy sequence in the metric space V, $(v_1, v_2, v_3, \ldots v_n, \ldots)$ is being transformed into a Cauchy sequence in the metric space W, $(w_1, w_2, w_3, \ldots, w_n, \ldots)$, and $\lim_{n\to\infty} v_n$ is transformed into $\lim_{n\to\infty} w_n$. A mapping $V \to W$ is called a *bounded* or *Lipschitz* mapping if the Lipschitz condition is satisfied: for any two elements, $v_1, v_2 \in V$, and their values, $w_1, w_2 \in W$, $\rho(w_1, w_2) \leq \mu\rho(v_1, v_2)$, where μ is a constant (called the Lipschitz constant). If a mapping is a Lipschitz mapping, it is continuous. A Lipschitz mapping with the constant $\mu < 1$ is called a *contraction*.

6.14 PROBLEMS

I. Check whether the metric distance on the set $\mathcal{N}$, defined as

$$\rho(n_1, n_2) = (n_1 - n_2)^2,$$

is really a metric distance (i.e., satisfying the axioms).

II. Which of the polynomials $p_1(t) = t$ and $p_2(t) = t - t^2/2$ gives a better approximation of the function $\sin t$ on the interval $[0, \pi/2]$ and why?

III. At what value of the parameter α does the function $p(t) = \alpha t$ give the best approximation of $\sin(\pi t/3)$ on the interval $[0, 1]$.

IV. Consider a mapping $\mathbf{R} \to \mathbf{C}$,
$$x \mapsto e^{ix}, x \in [0, 2\pi].$$
Show the mapping to be continuous. Is it a contraction?

V. Consider the mapping of the space of continuous functions $f(t)$ on the interval $[0,1]$ into itself: $f(t) \to f(\ln(1+t))$. Is this a Lipschitz mapping? Is it a contraction?

VI. A mapping $V \to V$, where V is the linear space of functions $f(x)$ of real variable $x \in [0, 1]$, is defined as
$$f(x) \longrightarrow \frac{1}{2} f\left(\tan\left(\frac{\pi}{4}x\right)\right).$$
Is this a Lipschitz mapping? Is it a contraction?

6.15 FURTHER READING

In this chapter, we have only considered those basics of set topology which are necessary for understanding of the topics to be covered in the following chapters. The books devoted to topology proper are numerous, their level ranging from advanced undergraduate to highly specialized. Probably the best of simple texts is

M. Mansfield, *Introduction to Topology*, Van Nostrand, Princeton, NJ, 1965.
The scope of this book is almost the same as of this chapter, but the presentation is much more detailed.

At the next stage, the following two books can be recommended:
D. W. Hall, G. L. Spencer, *Elementary Topology*, Wiley, New York, 1955;
C. W. Baker, *Introduction to Topology*, Wm. C. Brown, Dubuque, IA, 1991.

Much higher level, broader scope, and more details the reader can find in
W. J. Pervin, *Foundations of General Topology*, Academic Press, New York, 1964, and
J. Kelley, *General Topology*, Van Nostrand, New York, 1955.

The concepts related to metric spaces are primarily developed in and applied to real analysis and function theory. In these areas, the general topological treatment tends to be relatively non-technical, while the abstraction to the general topological settings often clarifies how proofs are constructed. Moreover, the interplay between topology and analysis is a harbinger of the interplay between abstraction and particularization that is so important in mathematics and its applications. In this regard, the books closest to our approach in this chapter are:

D. L. Stancl, M. L. Stancl, *Real Analysis with Point-Set Topology*, Marcel Dekker, New York, 1987;

A. N. Kolmogorov, S. V. Fomin, *Introductory Real Analysis*, Dover, New York, 1970.

A more in-depth study can be assisted by:

J. Foran, *Fundamentals of Real Analysis*, Marcel Dekker, New York, 1991;

E. Hewitt, K. Stromberg, *Real and Abstract Analysis*, Springer-Verlag, New York, 1965;

J. P. Aubin, *Applied Abstract Analysis*, Wiley, New York, 1977.

Approximation of functions by polynomials and other functions of simple constructions is exhaustively treated in the classical texts,

I. P. Natanson, *Theory of Functions of a Real Variable*, Volume I, Ungar, New York, 1955;

A. F. Timan, *Theory of Approximation of Functions of a Real Variable*, Pergamon, New York, 1963.

7
Banach and Hilbert Spaces

7.1 INTRODUCTION: THE GREAT ALLIANCE OF LINEARITY AND METRICS

In Chapters 4 and 6 the concepts of linear space and of metric mapping were introduced as completely independent and fully self-sufficient. And so they are. However, most practical applications deal with systems which are linear spaces and metric spaces at the same time. This is not surprising, because, as we mentioned, both the mathematical concepts of linearity and metrics represent some features of the real world. Now, if a system is both a metric and linear space, the questions can be asked, which additional features does it possess, and how can one profit from these additional features in the analysis of particular situations.

In this chapter, we are going to follow the implications of the union of these two basic ideas, and to use the fruits of this union for solving the general problem which was left hanging from Chapter 4, namely, that of construction of a basis in an infinite-dimensional linear space. This problem will be elegantly and handily solved using the new notions, which are to be developed here.

7.2 NORM MAPPING. BANACH SPACE

If the set V is a linear space, there is an easier way to describe its topological properties, compared to the general approach of metric distance. Any two elements of the space V, v_1 and v_2, can be combined in a linear combination providing a unique result, say

$v = v_1 - v_2$. This means that, instead of considering a metric mapping $V \times V \to \mathcal{R}$, for all "metric" purposes it is sufficient to define the mapping $V \to \mathcal{R}$ which establishes a correspondence between each element v of the space V and some *real* number, denoted as $\|v\|$. This mapping is called the *norm mapping* or just the *norm* on the space V (and, correspondingly, the number $\|v\|$ is called the *norm* of an element v), if it satisfies the following four axioms:

1) The norm is a non-negative number: for any element v of the space V,
$$\|v\| \geq 0. \tag{7.1}$$

2) The norm of an element v being zero is equivalent to this element being the zero element in the space V,
$$\|v\| = 0 \quad \leftrightarrow \quad v = \theta. \tag{7.2}$$

3) Norm stretching: for any element v of the space V and for any complex number α,
$$\|\alpha \cdot v\| = |\alpha| \cdot \|v\|. \tag{7.3}$$

4) The triangle axiom for the norm: for any two elements, v and v', of the space V,
$$\|v + v'\| \leq \|v\| + \|v'\|. \tag{7.4}$$

A linear space with norm mapping satisfying Eqs. (7.1) – (7.4) is called a *normed space*.

By its origin and construction, the concept of norm is purposely similar to the concept of metric distance, compare the axioms of norm with those of metric distance, Eqs. (6.2 – 6.4). This means that norm mapping yields a metric mapping (called the *induced metric*) under the definition
$$\rho(v, v') = \|v - v'\|. \tag{7.5}$$

Thus, a normed space is always a metric space. A complete (or closed) normed space is called a *Banach space*.

Examples of Banach spaces
Linear spaces of one-, two-, three-, or more dimensional column vectors, $\mathcal{R}, \mathcal{R}^2, \mathcal{R}^3$, and so on, are all Banach spaces with the norm of an n-dimensional column vector,
$$\vec{v} = \begin{pmatrix} \alpha_1 \\ \alpha_2 \\ \alpha_3 \\ \vdots \\ \alpha_n \end{pmatrix}, \tag{7.6}$$

naturally defined as
$$\|\vec{v}\| = \sqrt{\alpha_1^2 + \alpha_2^2 + \alpha_3^2 + \ldots + \alpha_n^2}. \tag{7.7}$$

A one-dimensional subspace in the three-dimensional Euclidean space, and the corresponding three-dimensional real column vectors with only one, say, the first, non-zero component constitute a Banach subspace, $\mathcal{R} \subset \mathcal{R}^3$. (Visually, this subspace is just the x axis of the cartesian coordinate system).

The linear space of continuous functions on $[0, 1]$ *is not* a Banach space, because it is not complete (as it was shown in Chapter 6 (Section 6.9), this space contains a Cauchy sequence of functions $\phi(x)$, whose limit is not a continuous function, that is, it lies outside the space).

7.3 SAMPLE PROBLEM I

Consider the vector

$$v = \begin{pmatrix} 1 \\ i \\ 1 \end{pmatrix} \tag{7.8}$$

in the three-dimensional Banach space of column vectors and the vector

$$f = e^x \tag{7.9}$$

in the infinite-dimensional Banach space of functions on the interval $[0, 1]$. Which of these vectors has the least norm?

Solution
Note that the question is resonable, because in all cases the norm is just a real number, so we can choose to compare elements of different Banach spaces by their norms.

The norm of the vector v is

$$\|v\| = \sqrt{v_1 v_1^* + v_2 v_2^* + v_3 v_3^*} = \sqrt{3}. \tag{7.10}$$

The norm of the vector (function) $f(x)$ is

$$\|f\| = \sqrt{\int_0^1 dx\, e^{2x}} = \sqrt{\frac{e^2 - 1}{2}}. \tag{7.11}$$

Since

$$\sqrt{\frac{e^2 - 1}{2}} > \sqrt{3}, \tag{7.12}$$

the vector v has the least norm.

7.4 FINITE-DIMENSIONAL SUBSPACES IN BANACH SPACE

The following considerations show the benefits of working with a Banach space where the properties of linearity can be used for describing its topological structure.

Consider some Cauchy sequence in a finite-dimensional Banach space V:

$$(v_1, v_2, \ldots, v_m, \ldots). \tag{7.13}$$

Let this sequence have the limit $\bar{v} \in V$,

$$\bar{v} = \lim_{m \to \infty} v_m. \tag{7.14}$$

This means, that for an arbitrarily small positive number ϵ there exists a natural number $\mathcal{N}(\epsilon)$ such that for all the elements v_m with indexes $m > \mathcal{N}(\epsilon)$,

$$\rho(v_m, \bar{v}) \equiv \|v_m - \bar{v}\| < \epsilon \tag{7.15}$$

This implies that

$$\rho(v_m - \bar{v}, \theta) \equiv \|v_m - \bar{v}\| < \epsilon, \tag{7.16}$$

that is, that the zero element of space V is the limiting element of the sequence

$$(v_1 - \bar{v}, v_2 - \bar{v}, v_3 - \bar{v}, \ldots, v_m - \bar{v}, \ldots), \tag{7.17}$$

which, therefore, is also a Cauchy sequence.

Given a basis in the linear space V,

$$\{\tilde{v}_1, \tilde{v}_2, \ldots, \tilde{v}_n\}, \tag{7.18}$$

each of the elements of the original sequence (7.13) can be represented as the decomposition over the elements of this basis with some complex coefficients,

$$v_m = \sum_{i=1}^{n} \alpha_i^{(m)} \cdot \tilde{v}_i. \tag{7.19}$$

In the same way, the limiting element (7.14) is represented as

$$\bar{v} = \sum_{i=1}^{n} \bar{\alpha}_i \cdot \tilde{v}_i, \tag{7.20}$$

and the elements of the sequence (7.17) are represented as

$$v_m - \bar{v} = \sum_{i=1}^{n} (\alpha_i^{(m)} - \bar{\alpha}_i) \cdot \tilde{v}_i. \tag{7.21}$$

Now, for the limiting element of this latter sequence, that is, for the zero element of the space V, we can write

$$\lim_{m \to \infty} (v_m - \bar{v}) = \left(\lim_{m \to \infty} \sum_{i=1}^{n} (\alpha_i^{(m)} - \bar{\alpha}_i) \cdot \tilde{v}_i \right)$$

$$= \sum_{i=1}^{n} \left(\lim_{m \to \infty} (\alpha_i^{(m)} - \bar{\alpha}_i) \right) \cdot \tilde{v}_i = \theta. \tag{7.22}$$

The elements of the basis (7.18) being linearly independent implies that all the coefficients of the linear combination (7.22) are zeroes,

$$\lim_{m \to \infty} (\alpha_i^{(m)} - \bar{\alpha}_i) = 0. \tag{7.23}$$

This, in turn, means that for any i,

$$\lim_{m \to \infty} (\alpha_i^{(m)}) = \bar{\alpha}_i. \tag{7.24}$$

The result (7.24) is a fundamental property of any finite-dimensional Banach space: in this space, if a sequence of elements converges to some limiting element, then all the sequences of the coefficients of decompositions of these elements over a given basis converge to corresponding coefficients of the decomposition of that limiting element, and vice versa.

As a corollary, any finite-dimensional subspace of a Banach space is complete, that is, it is a Banach space. Indeed, for any Cauchy sequence in the subspace, all the sequences of coefficients $\alpha_i^{(m)}$ are also Cauchy sequences, with limits $\bar{\alpha}_i$, respectively. The element $\bar{v}$, reconstructed from these limits through formula (7.20), is the limit of the initial Cauchy sequence. On the other hand, this element belongs to the subspace, because it is a linear combination of the elements of a basis in this subspace. Thus, the subspace contains the limiting elements of all its Cauchy sequences, that is, it is complete.

The application of the concept of limit to linear spaces also makes it possible to extend the use of decomposition over a basis to infinite-dimensional cases. Namely, if the linear space V under consideration is a Banach space, it is possible to define a basis in this space as an infinite *countable set* or an infinite *sequence* of elements,

$$(\tilde{v}_1, \tilde{v}_2, \tilde{v}_3, \ldots, \tilde{v}_n, \ldots), \tag{7.25}$$

such that any element v of the space V can be represented as

$$v = \sum_{i=1}^{\infty} \alpha_i \cdot \tilde{v}_i = \lim_{n \to \infty} \left(\sum_{i=1}^{n} \alpha_i \cdot \tilde{v}_i \right). \tag{7.26}$$

This definition implies that when n grows, the partial sums approach the limit of the infinite sum. In other words, in this infinite sum the contributions of the elements $\tilde{v}_i$ with large i are negligibly small.

Note, however, that we still have no reliable method to decompose an element of a linear (even Banach) space over its basis, that is, to explicitly find the coefficients of the decomposition. We have no regular procedure for doing this decomposition even in finite-dimensional spaces, let alone infinite-dimensional cases. To resolve this problem, we need to introduce yet another important concept.

7.5 INNER-PRODUCT MAPPING. HILBERT SPACE

The ultimate purpose of the concept we are going to develop here is to establish a quantitative characteristic related to linear independence. Once established, this characteristic can be (and will be) of indispensable help in constructing a basis. However, its scope is much broader; we will meet it again and again in many of the following chapters.

We construct a mapping $V \times V \to \mathcal{C}$ which transforms any pair (v_1, v_2) of elements of a linear space V into some complex number called the *inner product* of the elements v_1 and v_2 and denoted $<v_1, v_2>$:

$$(v_1, v_2) \to <v_1, v_2> \in \mathcal{C}. \tag{7.27}$$

The mapping itself is called the *inner-product mapping*, provided it satisfies the following five axioms:

1) The inner product of any element $v \in V$ and itself is a non-negative real number,

$$<v, v> \geq 0. \tag{7.28}$$

2) The inner product of an element $v \in V$ with itself being zero is equivalent to this element being the zero element in the space V,

$$<v, v> = 0 \leftrightarrow v = \theta. \tag{7.29}$$

3) *Axiom of asymmetry*: the inversion of order of the elements leads to complex conjugation of the inner product,

$$<v_2, v_1> = (<v_1, v_2>)^*. \tag{7.30}$$

The following two axioms express the properties of *linearity* of the inner-product mapping.

4) For any three elements v_1, v_2, and v_3 of the space V,

$$<v_1 + v_2, v_3> = <v_1, v_3> + <v_2, v_3>. \tag{7.31}$$

5) For any two elements v_1 and v_2 of the space V and for any complex number α,

$$<\alpha \cdot v_1, v_2> = \alpha \cdot <v_1, v_2>. \tag{7.32}$$

It is instructive to compare these axioms with those of metric distance and of norm mapping. Basically, the three sets of axioms relate to similar properties. Note, however, that the axiom 3) establishes a definite asymmetry between the first and the second elements in the inner product, in contrast to the axiom of symmetry which holds for metric distance. The comparison of the sets of axioms shows also that, in fact, the inner product yields a norm. Indeed, the definition of norm through the inner product,

$$\|v\| = (<v, v>)^{\frac{1}{2}}, \tag{7.33}$$

clearly satisfies all four axioms of the norm.

A complete linear space with inner product satisfying (7.28) – (7.32) is called a *Hilbert space*. As we will see, among all mathematical constructions considered so far, this is closest to the real world, and, therefore, to most applications.

Examples of Hilbert spaces
The most familiar example of a Hilbert space is the three-dimensional Euclidean space encountered in high-school geometry. Since this is a *real* linear space, the inner product of two its elements is a real number. It is just the dot product of two vectors:

$$< \vec{v}_1, \vec{v}_2 > = |\vec{v}_1| \cdot |\vec{v}_2| \cdot \cos\phi, \qquad (7.34)$$

where $|\vec{v}_1|, |\vec{v}_2|$ are the magnitudes of the vectors, and ϕ is the angle between them. In the cartesian coordinate system with axes x, y, and z, the inner product (7.34) is written as:

$$< \vec{v}_1, \vec{v}_2 > = v_{1x} \cdot v_{2x} + v_{1y} \cdot v_{2y} + v_{1z} \cdot v_{2z}. \qquad (7.35)$$

Another important example of a Hilbert space is the n-dimensional linear space of column vectors with complex-number components, taken along with the inner product defined in analogy to formula (7.35): for two such column vectors,

$$\vec{\alpha} = \begin{pmatrix} \alpha_1 \\ \alpha_2 \\ \vdots \\ \alpha_n \end{pmatrix} ; \quad \vec{\beta} = \begin{pmatrix} \beta_1 \\ \beta_2 \\ \vdots \\ \beta_n \end{pmatrix}, \qquad (7.36)$$

their inner product is given as

$$< \vec{\alpha}, \vec{\beta} > = \alpha_1 \cdot \beta_1^* + \alpha_2 \cdot \beta_2^* + \ldots + \alpha_n \cdot \beta_n^*. \qquad (7.37)$$

Note, that the complex conjugation must be used here to satisfy the Axiom of asymmetry of the inner product.

Note also, that the linear space of n-component column vectors was established in Chapter 4 as representing *any* n-dimensional linear space. Correspondingly, the *Hilbert space of n-component column vectors* described above represents *any* n-dimensional Hilbert space. Thus, a thorough understanding of the properties of this Hilbert space is quite sufficient for understanding any finite-dimensional Hilbert space.

As for *infinite-dimensional* Hilbert spaces, a basic example of such a space is obtained from the linear space of complex-valued functions $f(x)$ of real variable x on the given interval $[0, 1]$ which was also considered in Chapter 4. Referring to the metric distance (6.12), it is natural in this case to define the inner product as

$$< f_1(x), f_2(x) > = \int_0^1 dx\, f_1(x) \cdot f_2^*(x). \qquad (7.38)$$

154 BANACH AND HILBERT SPACES

This definition assumes that we will consider only *square-integrable* functions, that is, those for which

$$\int_0^1 dx |f(x)|^2 < \infty. \tag{7.39}$$

7.6 SAMPLE PROBLEMS II

I. In the four-dimensional Hilbert space of column vectors with complex components, find the inner product of the vectors

$$v_1 = \begin{pmatrix} 1 \\ i \\ 2 \\ 3i \end{pmatrix} \quad \text{and} \quad v_2 = \begin{pmatrix} 3i \\ 2 \\ i \\ 1 \end{pmatrix}. \tag{7.40}$$

Solution
According to definition of Eq. (7.37),

$$\begin{aligned} &< v_1, v_2 > \\ &= (v_1)_1 \cdot (v_2)_1^* + (v_1)_2 \cdot (v_2)_2^* + (v_1)_3 \cdot (v_2)_3^* + (v_1)_4 \cdot (v_2)_4^* \\ &= 1 \cdot (-3i) + i \cdot 2 + 2 \cdot (-i) + (3i) \cdot 1 = 0, \end{aligned} \tag{7.41}$$

where the notation, $(v_1)_j$ and $(v_2)_j$, $j = 1, 2, 3, 4$, was used for components of the vectors v_1 and v_2, respectively.

II. In the Hilbert space of square-integrable complex-valued functions of real argument x on the interval $[0, 1]$, find the inner product of the vectors

$$v_1 = \sin(\pi x) \quad \text{and} \quad v_2 = e^{2i\pi x}. \tag{7.42}$$

Solution
According to definition of Eq. (7.38),

$$< v_1, v_2 > = \int_0^1 dx v_1(x) \cdot v_2^*(x)$$

$$= \int_0^1 dx \sin(\pi x) \cdot e^{-2i\pi x} = \frac{1}{2i} \int_0^1 dx \left(e^{i\pi x} - e^{-i\pi x} \right) \cdot e^{-2i\pi x}$$

$$= \frac{1}{2i}\int_0^1 dx\, (e^{-i\pi x} - e^{-3i\pi x}) = \frac{1}{2\pi}(e^{-\pi i} - 1) - \frac{1}{6\pi}(e^{-3\pi i} - 1)$$

$$= \frac{1}{2\pi}(-1-1) - \frac{1}{6\pi}(-1-1)$$

$$= -\left(\frac{1}{\pi} - \frac{1}{3\pi}\right) = -\frac{2}{3\pi}. \tag{7.43}$$

7.7 THE CAUCHY-SCHWARZ INEQUALITY

Comparing the axioms for metric distance, norm, and inner product, we see that there is no analog of the *triangle axiom* for the inner product. In fact, such an analog exists and counts as a fundamental property of the inner product mapping. The reason for its absence among the axioms merely means that in this case the analogous inequality *can be proved*.

This inequality, known as the *Cauchy-Schwarz inequality*, is formulated as follows. For any two elements v_1 and v_2 on a Hilbert space V the following relation holds:

$$|<v_1, v_2>| \leq (<v_1, v_1>)^{\frac{1}{2}} \cdot (<v_2, v_2>)^{\frac{1}{2}}. \tag{7.44}$$

This inequality can be also rewritten in terms of the norm as defined in Eq. (7.33):

$$|<v_1, v_2>| \leq \|v_1\| \cdot \|v_2\|. \tag{7.45}$$

In Eqs. (7.44) and (7.45) the equality only holds if either at least one of the vectors v_1 and v_2 is the zero vector, θ, in the space V, or if one of these vectors is proportional to the other, say,

$$v_1 = \alpha \cdot v_2, \tag{7.46}$$

with an arbitrary complex number $\alpha \neq 0$ being the coefficient of proportionality.

Proof
First, we easily check that in both of the special cases mentioned above the strict equality really holds. Then, for the rest of this proof, we assume that the vectors v_1 and v_2 do not satisfy the condition (7.46) and neither of them is the zero vector.

To prove the strict inequality in (7.44), we start with the only inequality present in the axioms of inner-product mapping, that is, with axiom 1). We write this axiom for the special linear combination of vectors v_1 and v_2,

$$v_{spec} = v_1 - \beta \cdot v_2 \tag{7.47}$$

(a non-zero vector, within our assumptions), to obtain an expression containing various inner products of the vectors v_1 and v_2. The trick here is to properly choose the arbitrary complex coefficient β.

Thus, we start with

$$< v_{spec}, v_{spec} > = < (v_1 - \beta \cdot v_2), (v_1 - \beta \cdot v_2) > \; > 0. \qquad (7.48)$$

The vector v_{spec} being non-zero, the strict inequality holds. Then, we use axiom 4) of the inner product to rewrite the left-hand side of (7.48) as:

$$< v_{spec}, v_{spec} >$$
$$= < v_1, v_1 > - < \beta \cdot v_2, v_1 >$$
$$- < v_1, \beta \cdot v_2 > + < \beta \cdot v_2, \beta \cdot v_2 >. \qquad (7.49)$$

Now, we use axioms 3) and 5) of the inner product to transform the expression (7.49) into

$$< v_{spec}, v_{spec} >$$
$$= < v_1, v_1 > - \beta < v_2, v_1 >$$
$$- \beta^* < v_1, v_2 > + |\beta|^2 < v_2, v_2 >$$
$$= < v_1, v_1 > - \beta (< v_1, v_2 >)^*$$
$$- \beta^* < v_1, v_2 > + |\beta|^2 < v_2, v_2 >. \qquad (7.50)$$

Comparing Eqs. (7.50) and (7.44), the proper choice of coefficient β is the following:

$$\beta = \frac{< v_1, v_2 >}{< v_2, v_2 >}, \qquad (7.51)$$

which is always possible because $v_2 \neq \theta$. Substituting this β into the expression (7.50), we obtain

$$< v_{spec}, v_{spec} > = < v_1, v_1 > - \frac{< v_1, v_2 > (< v_1, v_2 >)^*}{< v_2, v_2 >}$$
$$- \frac{(< v_1, v_2 >)^* < v_1, v_2 >}{< v_2, v_2 >} + \frac{|< v_1, v_2 >|^2 \cdot < v_2, v_2 >}{|< v_2, v_2 >|^2}. \qquad (7.52)$$

After the obvious cancellation, this results in

$$< v_{spec}, v_{spec} > = < v_1, v_1 > - \frac{|< v_1, v_2 >|^2}{< v_2, v_2 >}, \qquad (7.53)$$

which gives in inequality (7.48),

$$\frac{< v_1, v_1 > < v_2, v_2 > - |< v_1, v_2 >|^2}{< v_2, v_2 >} > 0, \qquad (7.54)$$

that is,

$$(< v_1, v_1 > \cdot < v_2, v_2 >) > |< v_1, v_2 >|^2, \qquad (7.55)$$

both sides of which are definitely positive. Taking the square root of both sides of inequality (7.54), we arrive at inequality (7.44), which is thus proven.

7.8 INNER PRODUCT AND MATRICES. HERMITIAN MATRICES

In this section we are going to develop some important relations between the inner product and mappings. Consider two finite-dimensional Hilbert spaces, an n-dimensional space V and an m-dimensional space W, and a linear mapping transforming element $v \in V$ into element $w \in W$. This mapping corresponds to a mapping between the spaces of n- and m-component column vectors with complex components,

$$\vec{v} \to \hat{A}\vec{v}, \tag{7.56}$$

with some $n \times M$ matrix $\hat{A}$. Then, for *any* element $v \in V$ and an element $w \in W$, the following important relation holds:

$$<\vec{w}, \hat{A}\vec{v}> = <(\hat{A}^T)^*\vec{w}, \vec{v}>. \tag{7.57}$$

Note, that left-hand side of this relation is the inner product in m-dimensional W-space, while the right-hand side is the inner-product in n-dimensional V-space; the dimensions m and n of these two spaces need not be the same.

To *prove* the relation (7.57), we first write the inner products in the form appropriate for column vectors, Eq. (7.37). The left-hand-side inner product then takes the form,

$$<\vec{w}, \hat{A}\vec{v}> = \sum_{i=1}^{m} w_i \cdot (\hat{A}\vec{v})_i^* = \sum_{i=1}^{m} w_i \cdot \left(\sum_{j=1}^{n} a_{ij} v_j\right)^*, \tag{7.58}$$

where a_{ij} are the elements of the matrix $\hat{A}$. We formally transform the latter expression in the following way:

$$\sum_{i=1}^{m} w_i \left(\sum_{j=1}^{n} a_{ij} v_j\right)^* = \sum_{i=1}^{m} w_i \sum_{j=1}^{n} a_{ij}^* v_j^* = \sum_{i=1}^{m} \sum_{j=1}^{n} w_i a_{ij}^* v_j^*$$

$$= \sum_{j=1}^{n} \sum_{i=1}^{m} v_j^* a_{ij}^* w_i = \sum_{j=1}^{n} v_j^* \sum_{i=1}^{m} a_{ij}^* w_i. \tag{7.59}$$

Then, we recall that, by definition,

$$a_{ij} = (\hat{A})_{ij} = (\hat{A}^T)_{ji}. \tag{7.60}$$

So,

$$\sum_{j=1}^{n} v_j^* \sum_{i=1}^{m} a_{ij}^* w_i = \sum_{j=1}^{n} v_j^* \sum_{i=1}^{m} (\hat{A}^T)_{ji}^* w_i$$

$$= \sum_{j=1}^{n} v_j^* ((\hat{A}^T)^* \vec{w})_j = <(\hat{A}^T)^* \vec{w}, \vec{v}>, \tag{7.61}$$

which is just the right-hand side of relation (7.57). Thus, we have proven that relation (7.57) holds.

The matrix $(\hat{A}^T)^*$ which has emerged in Eq. (7.57) is called the *Hermitian conjugate* of matrix $\hat{A}$. In particular (but most interesting) cases when the mapping under consideration is a mapping of a finite-dimensional Hilbert space into itself, that is, $V = W$, the corresponding matrix $\hat{A}$ is a square matrix. Then, it can happen, that the Hermitian conjugate matrix coincides with the original matrix,

$$(\hat{A}^T)^* = \hat{A}. \tag{7.62}$$

Such special matrices are called *Hermitian matrices*. For them, instead of the general relation (7.57), the following formula holds:

$$< \vec{v_1}, \hat{A}\vec{v_2} > = < \hat{A}\vec{v_1}, \vec{v_2} > . \tag{7.63}$$

Based on Eq. (7.63), an obvious feature of a Hermitian matrix is that all its eigenvalues are real numbers. To show this, we take the characteristic equation for the k-th eigenvalue λ_k and the corresponding eigenvector $\vec{v_k}$,

$$\hat{A}\vec{v_k} = \lambda_k \vec{v_k}. \tag{7.64}$$

Then, the inner product is

$$< \vec{v_k}, \hat{A}\vec{v_k} > = < \vec{v_k}, \lambda_k \vec{v_k} > = \lambda_k^* < \vec{v_k}, \vec{v_k} > = \lambda_k^* \|\vec{v_k}\|^2. \tag{7.65}$$

On the other hand, by virtue of formula (7.63), the same inner product can be transformed as

$$< \vec{v_k}, \hat{A}\vec{v_k} > = < \hat{A}\vec{v_k}, \vec{v_k} > = < \lambda_k \vec{v_k}, \vec{v_k} > = \lambda_k < \vec{v_k}, \vec{v_k} > = \lambda_k \|\vec{v_k}\|^2. \tag{7.66}$$

Thus,

$$\lambda_k = \lambda_k^*, \tag{7.67}$$

which just means that λ_k is a real number.

Hermitian matrices play an important role in many applications. We can also generalize this notion for an infinite-dimensional Hilbert space, in a following manner. A mapping of a Hilbert space V into itself, $v' = f(v)$, is called a *Hermitian mapping* (and, correspondingly, f is called *a Hermitian operator*) if for any two elements, $v_1 \in V$ and $v_2 \in V$,

$$< v_2, f(v_1) > = < f(v_2), v_1 > . \tag{7.68}$$

Later in this book we will again meet both Hermitian matrices and general Hermitian operators and make use of their properties.

7.9 CONTINUITY OF THE INNER-PRODUCT MAPPING

Another fundamental feature of an inner-product mapping is that this is a continuous mapping. This property is rigorously formulated in a manner that is a bit awkward. In a Hilbert space V, take two Cauchy sequences,

$$(v_1, v_2, v_3, \ldots, v_n, \ldots) \tag{7.69}$$

and

$$(w_1, w_2, w_3, \ldots, w_n, \ldots), \tag{7.70}$$

which converge to limits $\bar{v}$ and $\bar{w}$, respectively. The sequence of the inner products, the complex numbers

$$(<v_1, w_1>, <v_2, w_2>, \ldots, <v_n, w_n>, \ldots), \tag{7.71}$$

is also a Cauchy sequence, which converges to the number $<\bar{v}, \bar{w}>$. In other words,

$$\lim_{n \to \infty} (<v_n, w_n>) = <\lim_{n \to \infty} v_n, \lim_{n \to \infty} w_n>. \tag{7.72}$$

Proof

To prove the statements of Eq. (7.72), we need only to show that, for an arbitrarily small positive number ε, the metric distance between two complex numbers, $<v_n, w_n>$ and $<\bar{v}, \bar{w}>$, is less than ε:

$$\rho(<v_n, w_n>, <\bar{v}, \bar{w}>) = |<v_n, w_n> - <\bar{v}, \bar{w}>| < \varepsilon, \tag{7.73}$$

for n and m greater than some $\mathcal{N}(\varepsilon)$ (cf. Eq. (6.27)).

To express this metric distance through $\rho(v_n, \bar{v})$ and $\rho(w_n, \bar{w})$, we transform the expression (7.73) as

$$\begin{aligned} &|<v_n, w_n> - <\bar{v}, \bar{w}>| \\ &= |<v_n, w_n> - <v_n, \bar{w}> + <v_n, \bar{w}> - <\bar{v}, \bar{w}>| \\ &= |<v_n, w_n - \bar{w}> + <v_n - \bar{v}, \bar{w}>|. \end{aligned} \tag{7.74}$$

According to the triangle inequality for metric distance,

$$\begin{aligned} &|<v_n, w_n - \bar{w}> + <v_n - \bar{v}, \bar{w}>| \\ &\leq |<v_n, w_n - \bar{w}>| + |<v_n - \bar{v}, \bar{w}>|. \end{aligned} \tag{7.75}$$

Now, we make use of the Cauchy-Schwarz inequality in its form of Eq. (7.45), and get from Eq. (7.74),

$$|<v_n, w_n> - <\bar{v}, \bar{w}>| \leq \|v_n\| \cdot \|w_n - \bar{w}\| + \|\bar{w}\| \cdot \|v_n - \bar{v}\|. \tag{7.76}$$

160   BANACH AND HILBERT SPACES

In the right-hand side of Eq. (7.76), $\|\bar{w}\|$ is a constant with respect to n, the numbers $\|v_n\|$ are bounded, and, according to the assumptions, $\|w_n - \bar{w}\|$ and $\|v_n - \bar{v}\|$ go to zero with increasing n. Thus, for sufficiently large n, this positive right-hand side becomes infinitely small, less than any *a priori* given positive number ε. Thus, the condition of Eq. (7.73) becomes inevitably satisfied, which proves the statement of Eq. (7.72).

7.10   ORTHOGONALITY

Structural properties of Hilbert spaces generalize many of the geometric properties of two- and three-dimensional Euclidean spaces. The most important (and most useful) of such concepts is the concept of *orthogonality*.

In a Hilbert space V, two vectors, v_1 and v_2, are called orthogonal, if

$$< v_1, v_2 > = 0; \tag{7.77}$$

this relation is usually denoted as $v_1 \perp v_2$.

Then, given a vector $v \in V$, all vectors in V which are orthogonal to v constitute a subset $V_1 \subset V$. In this case, the vector v is called orthogonal to the subset V_1,

$$v \perp V_1 \leftrightarrow \forall v_i \in V_1 (< v, v_i > = 0). \tag{7.78}$$

The next natural step in this path is considering *orthogonal subsets*. Two subsets, V_1 and V_2, of the space V, are called orthogonal subsets if all the elements in V_1 are orthogonal to all the elements in V_2,

$$V_1 \perp V_2 \leftrightarrow \forall v_{1i} \in V_1, \forall v_{2j} \in V_2 \quad < v_{1i}, v_{2j} > = 0. \tag{7.79}$$

Thus, for any subset $V_1 \in V$ there exists an *orthogonal complement*, $V_1^\perp$, the set of all vectors of V, each of which is orthogonal to all the vectors of V_1,

$$v \in V_1^\perp \rightarrow v \perp v_i \quad \forall v_i \in V_1. \tag{7.80}$$

It will be shown later, that if the subset V_1 is closed, then the whole space V can be represented as so-called *direct sum* of V_1 and $V_1^\perp$, which means that any vector $v \in V$ is *uniquely* represented as

$$v = v_1 + v_2, \tag{7.81}$$

where $v_1 \in V_1$, $v_2 \in V_1^\perp$.

7.11   SAMPLE PROBLEM III

In the Hilbert space of square-integrable functions on the interval $[0, 1]$, which of the vectors

$$v_1 = x, \quad v_2 = \cos(\pi x), \quad v_3 = \sinh(\pi x), \text{ and } v_4 = \cosh(\pi x) \tag{7.82}$$

SAMPLE PROBLEM III

are orthogonal to the vector $v_0 = \sin(\pi x)$.

Solution

We calculate inner products of v_0 with each of the vectors in (7.82):

$$< v_1, v_0 >$$
$$= \int_0^1 dx\, x \sin(\pi x) = \frac{1}{\pi}\left(-x\cos(\pi x)\Big|_0^1 + \int_0^1 dx \cos(\pi x)\right)$$
$$= \frac{1}{\pi}\left(-x\cos(\pi x)\Big|_0^1 + \frac{1}{\pi}\sin(\pi x)\Big|_0^1\right) = \frac{1}{\pi}(1+0) = \frac{1}{\pi} \neq 0; \quad (7.83)$$

$$< v_2, v_0 >$$
$$= \int_0^1 dx \cos(\pi x) \sin(\pi x) = \frac{1}{2}\int_0^1 dx \sin(2\pi x)$$
$$= \frac{1}{4\pi}\cos(2\pi x)\Big|_0^1 = \frac{1}{4\pi}(1-1) = 0; \quad (7.84)$$

$$< v_3, v_0 >$$
$$= \int_0^1 dx \sinh(\pi x)\sin(\pi x) = -i\int_0^1 dx \sin(\pi i x)\sin(\pi x)$$
$$= -\frac{i}{2}\int_0^1 dx (\cos(\pi x(1-i)) - \cos(\pi x(1+i)))$$
$$= -\frac{i}{2\pi(1-i)}\sin(\pi x(1-i))\Big|_0^1 + \frac{i}{2\pi(1+i)}\sin(\pi x(1+i))\Big|_0^1$$
$$= -\frac{i}{2\pi(1-i)}\sin(\pi(1-i)) + \frac{i}{2\pi(1+i)}\sin(\pi(1+i))$$
$$= -\frac{i}{2\pi(1-i)}(\sin(\pi)\cos(\pi i) - \cos(\pi)\sin(\pi i))$$
$$+ \frac{i}{2\pi(1+i)}(\sin(\pi)\cos(\pi i) + \cos(\pi)\sin(\pi i))$$
$$= -\frac{i}{2\pi(1-i)}\sin(\pi i) - \frac{i}{2\pi(1+i)}\sin(\pi i)$$
$$= \frac{\sinh(\pi)}{2\pi}\left(\frac{1}{1-i} + \frac{1}{1+i}\right) = \frac{\sinh(\pi)}{2\pi} \neq 0; \quad (7.85)$$

162   BANACH AND HILBERT SPACES

$$\begin{aligned}
&< v_4, v_0 > \\
&= \int_0^1 dx \cosh(\pi x) \sin(\pi x) = \int_0^1 dx \cos(\pi i x) \sin(\pi x) \\
&= \frac{1}{2} \int_0^1 dx (\sin(\pi x (1-i)) + \sin(\pi x (1+i))) \\
&= -\frac{1}{2\pi(1-i)} \cos(\pi x(1-i)) \Big|_0^1 - \frac{1}{2\pi(1+i)} \cos(\pi x(1+i)) \Big|_0^1 \\
&= -\frac{1}{2\pi(1-i)} (\cos(\pi(1-i)) - 1) - \frac{1}{2\pi(1+i)} (\cos(\pi(1+i)) - 1) \\
&= -\frac{1}{2\pi(1-i)} (\cos(\pi) \cos(\pi i) + \sin(\pi) \sin(\pi i) - 1) \\
&\quad - \frac{1}{2\pi(1+i)} (\cos(\pi) \cos(\pi i) - \sin(\pi) \sin(\pi i) - 1) \\
&= \frac{1}{2\pi(1-i)} (\cos(\pi i) + 1) + \frac{1}{2\pi(1+i)} (\cos(\pi i) + 1) \\
&= \frac{\cosh(\pi) + 1}{2\pi} \left(\frac{1}{1-i} + \frac{1}{1+i} \right) = \frac{\cosh(\pi) + 1}{2\pi} \neq 0.
\end{aligned} \qquad (7.86)$$

Thus, among all the vectors listed in Eq. (7.82), only vector v_2 is orthogonal to v_0.

7.12 EIGENVECTORS OF HERMITIAN MATRICES

In this section, we use the notion of orthogonality to establish a key property of Hermitian matrices. We have already shown that all eigenvalues of a Hermitian matrix are real numbers. The following theorem states that eigenvectors of a Hermitian matrix are also somewhat special: *The eigenvectors of a Hermitian matrix, corresponding to different eigenvalues, are orthogonal.*

Proof
Consider two different eigenvalues of a Hermitian matrix $\hat{A}$, say, λ_1 and λ_2, and their corresponding eigenvectors, $\vec{v}_1$ and $\vec{v}_2$,

$$\begin{aligned} \hat{A}\vec{v}_1 &= \lambda_1 \vec{v}_1; \\ \hat{A}\vec{v}_2 &= \lambda_2 \vec{v}_2. \end{aligned} \qquad (7.87)$$

Then, according to Eq. (7.67),

$$< \vec{v}_1, \hat{A}\vec{v}_2 > = < \vec{v}_1, \lambda_2 \vec{v}_2 > = \lambda_2^* < \vec{v}_1, \vec{v}_2 > = \lambda_2 < \vec{v}_1, \vec{v}_2 >. \qquad (7.88)$$

On the other hand,

$$< v_1, \hat{A}\vec{v}_2 > = < \hat{A}\vec{v}_1, \vec{v}_2 > = < \lambda_1 \vec{v}_1, \vec{v}_2 > = \lambda_1 < \vec{v}_1, \vec{v}_2 > . \qquad (7.89)$$

Subtracting Eq. (7.89) from Eq. (7.88), we get

$$(\lambda_2 - \lambda_1) < \vec{v}_1, \vec{v}_2 > = 0. \qquad (7.90)$$

From the statement of the theorem,

$$\lambda_2 - \lambda_1 \neq 0. \qquad (7.91)$$

Thus,

$$< \vec{v}_1, \vec{v}_2 > = 0, \qquad (7.92)$$

that is, vectors $\vec{v}_1$ and $\vec{v}_2$ are orthogonal.

7.13 THE PROJECTION THEOREM

The theorem we are going to discuss and prove in this section is the central point of the Hilbert-space-related topics. In its formulation and proof, this theorem is based on the notion of orthogonality. Being particularly important in a fundamental sense, and having far-reaching ramifications in applications, this theorem deserves special attention. For this reason, we will go through its proof in great detail.

The idea of projection is a natural generalization of the well-known fact in high-school geometry, that in a three-dimensional Euclidean space the shortest segment between a given point and a given plane is the perpendicular to the plane (Fig. 7.1.). This fact is generic for any Hilbert space, and results in the following theorem.

The Projection Theorem
Let M be a closed subspace of a Hilbert space V. For any element v of V there exists a unique element m_0 of the subspace M which has the smallest metric distance to v among all the elements of M,

$$\forall v \in V \ \exists m_0 \in M : \ \forall m \in M \ (\|v - m_0\| \leq \|v - m\|); \qquad (7.93)$$

this element m_0 satisfies the condition

$$(v - m_0) \perp M. \qquad (7.94)$$

Proof
The formulation of the theorem includes three statements:
1) The element m_0 exists;

164 BANACH AND HILBERT SPACES

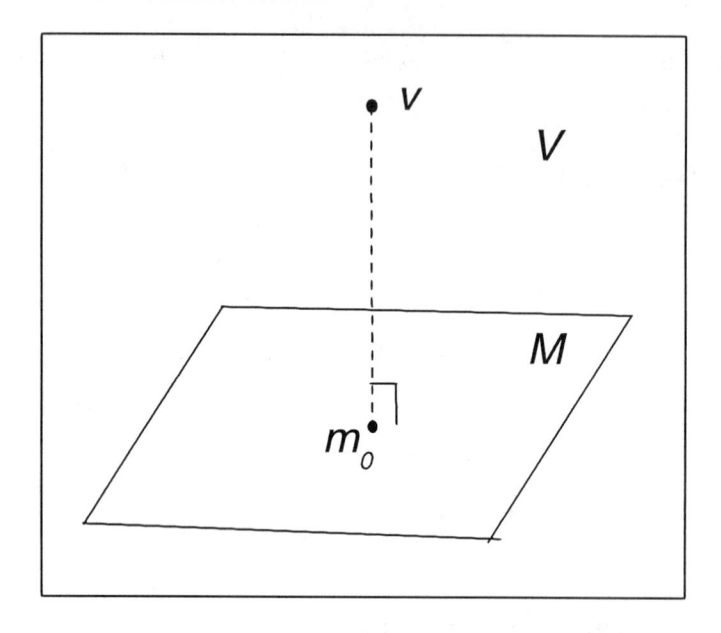

Fig. 7.1 The shortest distance between a point and a plane in three-dimensional Euclidean space.

2) This element is unique;
3) It satisfies the condition of Eq. (7.94).
We will prove these statements from the bottom up.

Statement 3 (condition (7.94) is satisfied):
We prove this statement by the method of *reductio ad absurdum* and by a pervasive trick which is typical in pure mathematics. Suppose that statement 3 is not true, that is, there exists an element $m \in M$ such that the vector $(v - m_0)$ is not orthogonal to m. This means, that the inner product of these two vectors is some nonzero complex number,

$$< (v - m_0), m >= \delta \neq 0. \tag{7.95}$$

Let us construct some other element of the subspace M as the linear combination

$$m_1 = m_0 + \delta \cdot m, \tag{7.96}$$

and find its metric distance to v. The square of this distance is

$$\begin{aligned}\|v - m_1\|^2 &=< (v - m_1), (v - m_1) > \\ &=< (v - m_0 - \delta \cdot m), (v - m_0 - \delta \cdot m) > \\ &=< (v - m_0), (v - m_0) > - < (v - m_0), (\delta \cdot m) > \\ &\quad - < (\delta \cdot m), (v - m_0) > + < \delta \cdot m, \delta \cdot m > .\end{aligned} \tag{7.97}$$

Then, we use the properties of the inner product to transform the latter expression into the form,

$$\begin{aligned}\|v - m_1\|^2 \\ =<(v-m_0),(v-m_0)> -\delta^* <(v-m_0),m> \\ -\delta <(v-m_0),m>^* +\delta\delta^* <m,m>.\end{aligned} \quad (7.98)$$

Substituting the expression for δ of Eq. (7.95), we get

$$\begin{aligned}\|v - m_1\|^2 \\ =<(v-m_0),(v-m_0)> -\delta^*\delta - \delta\delta^* + \delta\delta^* <m,m> \\ =<(v-m_0),(v-m_0)> - \mid\delta\mid^2 (2- <m,m>).\end{aligned} \quad (7.99)$$

Note, that since V is a linear space, the non-orthogonality of Eq. (7.95) holds for all vectors in V proportional to m (of course, the particular value of the number δ depends on the magnitude of m). Thus, (7.99) holds, in particular, for some vector m of unit magnitude,

$$<m,m>=1. \quad (7.100)$$

For this vector, (7.99) reads

$$\|v-m_1\|^2 = \|v-m_0\|^2 - \mid\delta\mid^2, \quad (7.101)$$

which means,

$$\|v-m_1\| < \|v-m_0\|. \quad (7.102)$$

But this is a contradiction, because $\|v - m_0\|$ must be the least distance among those for all the vectors of M. This contradiction means that our initial assumption of Eq. (7.95) was wrong, and the opposite statement of Eq. (7.94) is right. Thus, statement 3 is proven.

Statement 2 (the element m_0 is unique):
We use the same method *reductio ad absurdum*. Suppose there are two different elements, m_0 and m'_0, of the subspace M, which realize the minimum distance to v. Then, we can write

$$\begin{aligned}\|v - m'_0\|^2 = \|v - m_0 + m_0 - m'_0\|^2 \\ =<(v-m_0+m_0-m'_0),(v-m_0+m_0-m'_0)> \\ =<(v-m_0),(v-m_0)> + <(v-m_0),(m_0-m'_0)> \\ + <(m_0-m'_0),(v-m_0)> \\ + <(m_0-m'_0),(m_0-m'_0)>.\end{aligned} \quad (7.103)$$

Since the vector m_0 satisfies the condition of Eq. (7.94),

$$<(v-m_0),(m_0-m'_0)>=0. \quad (7.104)$$

Thus, Eq. (7.103) gives

$$\|v - m_0'\|^2 = \|v - m_0\|^2 + \|m_0 - m_0'\|^2, \qquad (7.105)$$

that is,

$$\|v - m_0'\| > \|v - m_0\|. \qquad (7.106)$$

This is again a contradiction, because these metric distances must be the same minimal distance. Thus, there cannot be two different vectors m_0 and m_0' satisfying the condition of Eq. (7.94), that is, vector m_0 is unique.

Statement 1 (the element m_0 exists):
This part of the theorem is particularly gloomy, especially because Statement 1 seems almost trivial. However, if the subspace M were not closed, we would not be sure to find the desired element m_0, as will be clear from the following.

Consider the metric distances between the vector v and various elements of the subspace m. The values of these distances are bounded by the number 0. So, if we arrange the elements of M in order of decreasing distance to form a sequence, $\{m_n\}$, the sequence of the corresponding distances, $\{\|v - m_n\|\}$, will have some limit,

$$\gamma = \lim_{n \to \infty} \|v - m_n\|. \qquad (7.107)$$

Now, we will produce the element $m_0 \in M$ corresponding to this minimal distance,

$$\|v - m_0\| = \gamma. \qquad (7.108)$$

Consider the sequence $\{m_n\}$ defined above, with $\|v - m_n\| \to \gamma$. We will prove $\{m_n\}$ to be a Cauchy sequence. To do so, we will estimate the metric distances between the elements of this sequence in terms of their distances to the vector v.

For the metric distance between the i-th and j-th elements of the sequence we can write:

$$\begin{aligned}
\|m_i - m_j\|^2 &= \|(m_i - v) + (v - m_j)\|^2 \\
&= <((m_i - v) + (v - m_j)), ((m_i - v) + (v - m_j))> \\
&= <(m_i - v), (m_i - v)> + <(m_i - v), (v - m_j)> \\
&\quad + <(v - m_j), (m_i - v)> + <(v - m_j), (v - m_j)> \\
&= \|m_i - v\|^2 + \|m_j - v\|^2 \\
&\quad - <(m_i - v), (m_j - v)> - <(m_j - v), (m_i - v)>.
\end{aligned} \qquad (7.109)$$

On the other hand, for the auxiliary combination,

$$\begin{aligned}
&\|(m_j - v) - (v - m_i)\|^2 \\
&= <((m_j - v) - (v - m_i)), ((m_j - v) - (v - m_i))>
\end{aligned}$$

$$=< (m_j - v), (m_j - v) > - < (m_j - v), (v - m_i) >$$
$$- < (v - m_i), (m_j - v) > + < (v - m_i), (v - m_i) >$$
$$= \|m_j - v\|^2 + \|m_i - v\|^2$$
$$+ < (m_j - v), (m_i - v) > + < (m_i - v), (m_j - v) > . \quad (7.110)$$

Adding Eqs. (7.109) and (7.110), we get

$$\|m_i - m_j\|^2 + \|(m_j - v) - (v - m_i)\|^2$$
$$= 2\|m_i - v\|^2 + 2\|m_j - v\|^2. \quad (7.111)$$

From Eq. (7.111) we obtain

$$\|m_i - m_j\|^2$$
$$= 2\|m_i - v\|^2 + 2\|m_j - v\|^2 - \|(m_j - v) - (v - m_i)\|^2$$
$$= 2\|m_i - v\|^2 + 2\|m_j - v\|^2 - 4\|v - \frac{m_j + m_i}{2}\|^2. \quad (7.112)$$

The linear combination

$$\frac{m_j + m_i}{2} \in M, \quad (7.113)$$

since M is s linear subspace. According to Eq. (7.107), this means that

$$\|v - \frac{m_j + m_i}{2}\| \geq \gamma. \quad (7.114)$$

Therefore, Eq. (7.112) gives the estimate

$$\|m_i - m_j\|^2 \leq 2\|m_i - v\|^2 + 2\|m_j - v\|^2 - 4\gamma^2. \quad (7.115)$$

When $i \to \infty$ and $j \to \infty$, both of the distances in the right-hand side of Eq. (7.115) go to γ, thereby forcing

$$\|m_i - m_j\|^2 \to 0. \quad (7.116)$$

This affirms that the sequence $\{m_n\}$ is a Cauchy sequence. Since the subspace M is complete, this sequence has as its limit $m_0 \in M$, which is the minimal distance given in Eq. (7.108).

We have thus completed the proof of this most important theorem. The element m_0 playing the central role in this theorem is called the *projection* of vector v onto the subspace M.

7.14 SUMMARY OF CHAPTER 7

1) The application of the concept of metric distance to linear spaces leads to the concept of *norm*. Norm is a mapping from a linear space into the set of real numbers

which satisfies four axioms. The norm of an element v of a linear space V is denoted as $\|v\|$.

2) The norm yields a metric distance as $\rho(v_1, v_2) = \|v_1 - v_2\|$.

3) A complete linear space with a norm is called a Banach space.

4) Any finite-dimensional linear space with a norm is a Banach space.

5) For a linear space V, the inner product is a mapping from $V \times V$ into the set of *complex* numbers which transforms a pair of elements, v_1 and v_2, into a complex number, $< v_1, v_2 >$, and satisfies five axioms.

6) The inner product yields the norm $\|v\| = \sqrt{< v, v >}$.

7) A complete linear space with an inner product is called a Hilbert space. A Hilbert space is automatically a Banach space.

8) For two spaces of column vectors, n-dimensional V and m-dimensional W, and an $n \times m$ matrix $\hat{A}$ the relation $< \hat{A}v, w > = < v, (\hat{A}^T) * w >$ holds. The matrix $(\hat{A}^T)*$ is called *Hermitian conjugate* to the matrix $\hat{A}$.

9) A square matrix which is equal to its own Hermitian conjugate matrix is called a *Hermitian matrix*. All eigenvalues of a Hermitian matrix are *real* numbers.

10) Two nonzero vectors, v_1 and v_2, are called orthogonal, $v_1 \perp v_2$, if their inner product is zero. A vector v is called orthogonal to a set V, $v \perp V$, if it is orthogonal to all vectors of the set.

11) The Projection Theorem states that for any element v of a Hilbert space V and a Hilbert subspace M the element of M closest to v is the *projection* m_0 determined by the condition $(v - m_0) \perp M$.

7.15 PROBLEMS

I. In the real three-dimensional Banach space of column vectors, find the coefficients α and β minimizing the norm of the combination

$$\tilde{v} = v_1 + \alpha v_2 + \beta v_3,$$

where

$$v_1 = \begin{pmatrix} 1 \\ 2 \\ 1 \end{pmatrix}, \quad v_2 = \begin{pmatrix} 2 \\ 1 \\ 1 \end{pmatrix}, \quad \text{and} \quad v_3 = \begin{pmatrix} 1 \\ 1 \\ 2 \end{pmatrix}.$$

II. In a real linear space of square $n \times n$ matrices (see Chapter 5), an inner product is proposed in the form:

$$< \hat{A}, \hat{B} > = \sum_{j=1}^{n} c_{jj},$$

$$\hat{C} = \hat{A}^T \hat{B}.$$

Show that this definition satisfies the axioms of inner product.

III. In the Hilbert space of square-integrable functions on the interval $[0, 1]$, which of the vectors,

$$v_1 = x - \frac{1}{2}, \quad v_2 = \sin(2\pi x), \quad v_3 = e^x - 1, \quad \text{and} \quad v_4 = \sinh(\pi x),$$

are orthogonal to the vector $v_0 = \cos(\pi x)$?

7.16 FURTHER READING

Being pivotal in many application-oriented fields of mathematics, the concepts of Banach and Hilbert spaces get a comprehensive coverage in many books which vary in volume and complexity. As simple introductory textbooks expanding the material of this chapter, the following can be recommended:

S. K. Berberian, *Introduction to Hilbert Space*, Oxford, New York, 1961;

J. L. Soule, *Linear Operators in Hilbert Space*, Gordon and Breach, New York, 1968.

(We prefer these relatively old books, because most newer texts are more specialized and detail-oriented).

For deeper and broader understanding, a classical and ageless book is indispensable,

P. R. Halmos, *Introduction to Hilbert Space*, Chelsea, New York, 1957.

Then, in order of increasing complexity, go the books:

H. L. Hamburger, M. E. Grimshaw, *Linear Transformations in n-dimensional Vector Space*, Cambridge, UK, 1951;

W. Schmeidler, *Linear Operators in Hilbert Space*, Academic Press, New York, 1965;

V. I. Istranescu, *Introduction to Linear Operator Theory*, Marcel Dekker, New York, 1981.

N. I. Akhiezer, I. M. Glazman, *Theory of Linear Operator in Hilbert Space*, v. 1, Pitman, Boston, 1981.

Applications of the Hilbert space concepts and methods to various engineering problems are discussed in the following books:

L. Mate, *Hilbert Space Methods in Science and Engineering*, Adam Hilger, Bristol, UK, 1989;

P. Fuhrmann, *Linear Systems and Operators in Hilbert Space*, McGraw-Hill, New York, 1981;

A. V. Balakrishnan, *Applied Functional Analysis*, Springer-Verlag, New York, 1981.

8

Orthonormal Bases and Fourier Series

8.1 ORTHONORMAL BASIS

To make practical use of the great and mighty Projection Theorem that we have proven in the preceding chapter, it would be nice to have an algorithm for obtaining the projection element m_0 of the subspace M. In the following two sections we construct a special kind of basis, which will be indispensable in developing such an algorithm.

Orthogonal set
A set $\{v_i\}$ of vectors in a Hilbert space V, each of them being orthogonal to all the others,

$$\forall i \neq j \quad <v_i, v_j> = 0, \tag{8.1}$$

is (quite naturally) called an *orthogonal set*. If we add the condition that the vectors of the set all have unit norms,

$$\forall i : \|v_i\| = 1, \tag{8.2}$$

the set is then called an *orthonormal set*. A shorthand way of denoting that a set is orthonormal is

$$<v_i, v_j> = \delta_{i,j}, \tag{8.3}$$

where the special symbol δ_{ij} has the meaning,

$$\delta_{i,j} = \begin{cases} 0 & \text{if } i \neq j \\ 1 & \text{if } i = j \end{cases}. \tag{8.4}$$

171

The most important property of orthogonal sets is formulated as the following theorem.

Theorem of Orthogonal Sets
An orthogonal set of nonzero vectors is a linearly independent set.
Proof
We again (and again) use *reductio ad absurdum*. Suppose, the statement of the theorem is false, that is, the vectors of an orthogonal set,

$$\{e_1, e_2, e_3, \ldots, e_n\}, \tag{8.5}$$

are linearly dependent. According to Section 4.7, this means that it is possible to construct a trivial linear combination,

$$\sum_{i=1}^{n} \alpha_i \cdot e_i = \theta, \tag{8.6}$$

with some nonzero coefficients α_i.

Now, let us take the inner-products of this linear combination with each of the vectors e_j of the set (8.5):

$$<\sum_{i=1}^{n} \alpha_i \cdot e_i, e_j> = \sum_{i=1}^{n} \alpha_i \cdot <e_i, e_j> = 0, \tag{8.7}$$

The set of Eq. (8.5) being an orthogonal set, all the inner-products in (8.7) are zero, except for the case when $i = j$. Thus,

$$\sum_{i=1}^{n} \alpha_i \cdot <e_i, e_j> = \alpha_j \cdot <e_j, e_j> = 0. \tag{8.8}$$

The vectors e_j being non-zero vectors, this means that for all j

$$\alpha_j = 0. \tag{8.9}$$

So, the trivial linear combination (8.6) must have all its coefficients being zero. This contradicts our assumption. Thus, the set (8.5) is really a linearly independent set.

Having established that an orthogonal set is linearly independent makes it a good candidate for a basis.

8.2 THE GRAM - SCHMIDT ORTHONORMALIZATION

As shown in the previous section, if it is possible to find an orthogonal set of n nonzero vectors in n-dimensional Hilbert space, then this set can serve as a basis of the space. However, an even stronger statement is true: starting with an arbitrary basis in a Hilbert space, it is always possible to use this basis to construct an *orthonormal basis*. More precisely, this idea is expressed in the following theorem.

The Gram-Schmidt Theorem
We formulate the statement of this theorem for the more general case of an infinite-dimensional Hilbert space. To formulate it for the case of a finite-dimensional space, one only needs to replace *sequences* with finite n-*tuples*.

Let $\{v_i\}$ be a (*countable*) set of linearly independent vectors in an infinite-dimensional Hilbert space V, and organize them in a sequence,

$$(v_1, v_2, v_3, v_4, \ldots, v_n, \ldots). \tag{8.10}$$

Then, in the space V there exists the sequence of orthonormal vectors,

$$(e_1, e_2, e_3, e_4, \ldots, e_n, \ldots), \tag{8.11}$$

such that for each number n the n-dimensional subspaces of V generated by the first n vectors of the sequences (8.10) and (8.11) are the same:

$$[\{e_1, \ldots e_n\}] = [\{v_1, \ldots v_n\}]. \tag{8.12}$$

Proof
We prove this theorem *constructively*, that is, we produce the elements of the sequence (8.11) by building them from the vectors of the sequence (8.10).

As the first element of the *orthonormal sequence* (8.11) we take the *normalized* first element of the sequence (8.10),

$$e_1 = \frac{v_1}{\|v_1\|}. \tag{8.13}$$

(Note that all the linearly independent vectors of the set $\{v_i\}$ are nonzero vectors.)

Now, to use vectors e_1 and v_2 to produce a vector orthogonal to e_1, we take the specific linear combination,

$$\tilde{v}_2 = v_2 - <v_2, e_1> \cdot e_1. \tag{8.14}$$

Since this is a linear combination of linearly independent vectors with at least one non-zero coefficient, $\tilde{v}_2$ is a non-zero vector. On the other hand, $\tilde{v}_2 \perp e_1$. Indeed,

$$\begin{aligned}
<\tilde{v}_2, e_1> &= <(v_2 - <v_2, e_1> \cdot e_1), e_1> \\
&= <v_2, e_1> - <<v_2, e_1> \cdot e_1, e_1> \\
&= <v_2, e_1> - <v_2, e_1> \cdot <e_1, e_1> \\
&= <v_2, e_1> - <v_2, e_1> \cdot 1 = 0.
\end{aligned} \tag{8.15}$$

Thus, for the second vector of the sequence (8.11) we can take the normalized vector $\tilde{v}_2$,

$$e_2 = \frac{\tilde{v}_2}{\|\tilde{v}_2\|}, \tag{8.16}$$

174 ORTHONORMAL BASES AND FOURIER SERIES

where

$$\begin{aligned}
\|\tilde{v}_2\| &= (<\tilde{v}_2, \tilde{v}_2>)^{1/2} \\
&= (<(v_2-<v_2,e_1>\cdot e_1),(v_2-<v_2,e_1>\cdot e_1)>)^{1/2} \\
&= (<v_2,v_2> - <<v_2,e_1>\cdot e_1, v_2> - <v_2,<v_2,e_1>\cdot e_1> \\
&\quad + <<v_2,e_1>\cdot e_1, <v_2,e_1>\cdot e_1>)^{1/2} \\
&= (<v_2,v_2> - <v_2,e_1>\cdot <e_1,v_2> \\
&\quad - <v_2,e_1>^* \cdot <v_2,e_1> \\
&\quad + <v_2,e_1><v_2,e_1>^* \cdot <e_1,e_1>)^{1/2} \\
&= (<v_2,v_2> - |<v_2,e_1>|^2)^{1/2}.
\end{aligned} \qquad (8.17)$$

Since the linearly independent vectors e_1 and e_2 are linear combinations of the vectors v_1 and v_2,

$$[\{e_1, e_2\}] = [\{v_1, v_2\}]. \qquad (8.18)$$

Now, in the same manner, we construct the vector e_3. First,

$$\tilde{v}_3 = v_3 - <v_3,e_2>\cdot e_2 - <v_3,e_1>\cdot e_1. \qquad (8.19)$$

The vectors e_1 and e_2 being orthogonal,

$$\begin{aligned}
<\tilde{v}_3, e_2> &= <(v_3 - <v_3,e_2>\cdot e_2 - <v_3,e_1>\cdot e_1), e_2> \\
&= <(v_3 - <v_3,e_2>\cdot e_2), e_2> = 0
\end{aligned} \qquad (8.20)$$

and

$$\begin{aligned}
<\tilde{v}_3, e_1> &= <(v_3 - <v_3,e_2>\cdot e_2 - <v_3,e_1>\cdot e_1), e_1> \\
&= <(v_3 - <v_3,e_1>\cdot e_1), e_1> = 0.
\end{aligned} \qquad (8.21)$$

Then, for both Eqs. (8.20) and (8.21) we can continue the calculation as was done in Eq. (8.15), and come to the conclusion that both of these inner products are zero. Thus, the vector $\tilde{v}_3$ is orthogonal to both e_1 and e_2. So, the third vector of the sequence (8.11) can be taken as

$$e_3 = \frac{\tilde{v}_3}{\|\tilde{v}_3\|}. \qquad (8.22)$$

For the norm of $\tilde{v}_3$ in this expression we obtain,

$$\begin{aligned}
\|\tilde{v}_3\| &= (<\tilde{v}_3, \tilde{v}_3>)^{1/2} \\
&= \Big(<(v_3 - <v_3,e_2>\cdot e_2 - <v_3,e_1>\cdot e_1), \\
&\quad (v_3 - <v_3,e_2>\cdot e_2 - <v_2,e_1>\cdot e_1)>\Big)^{1/2}
\end{aligned}$$

$$\begin{aligned}
&= \Big(<v_3,v_3> - <(<v_3,e_2>\cdot e_2), v_3> \\
&\quad - <(<v_3,e_1>\cdot e_1), v_3> \\
&\quad - <v_3, <v_3,e_2>\cdot e_2> - <v_3, <v_3,e_1>\cdot e_1> \\
&\quad + <(<v_2,e_2>\cdot e_2), (<v_2,e_2>\cdot e_2)> \\
&\quad + <(<v_2,e_1>\cdot e_1), (<v_2,e_2>\cdot e_2)> \\
&\quad + <(<v_2,e_2>\cdot e_2), (<v_2,e_1>\cdot e_1)> \\
&\quad + <(<v_2,e_1>\cdot e_1), (<v_2,e_1>\cdot e_1)> \Big)^{1/2} \\
&= \Big(<v_3,v_3> - <v_3,e_2>\cdot <e_2,v_3> \\
&\quad - <v_3,e_1>\cdot <e_1,v_3> \\
&\quad - <v_3,e_2>^* \cdot <v_3,e_2> - <v_3,e_1>^* \cdot <v_3,e_1> \\
&\quad + <v_3,e_2><v_3,e_2>^* \cdot <e_2,e_2> \\
&\quad + <v_3,e_2><v_3,e_1>^* \cdot <e_2,e_1> \\
&\quad + <v_3,e_1><v_3,e_2>^* \cdot <e_1,e_2> \\
&\quad + <v_3,e_1><v_3,e_1>^* \cdot <e_1,e_1> \Big)^{1/2} \\
&= \Big(<v_3,v_3> - |<v_3,e_2>|^2 - |<v_3,e_1>|^2\Big)^{1/2}. \quad (8.23)
\end{aligned}$$

The linearly independent vectors e_1, e_2 and e_3 being linear combinations of vectors v_1, v_2, and v_2,

$$[\{e_1, e_2, e_3\}] = [\{v_1, v_2, v_3\}]. \quad (8.24)$$

Clearly, we can continue this process, involving more and more vectors of the sequence (8.10). As the result of this induction, we obtain the following expressions for the n-th element of the sequence (8.11):

$$e_n = \frac{\tilde{v}_n}{\|\tilde{v}_n\|}, \quad (8.25)$$

$$\tilde{v}_n = v_n - \sum_{i=1}^{n-1} <v_n, e_i> \cdot e_i, \quad (8.26)$$

$$\|\tilde{v}_n\| = \left(<v_n,v_n> - \sum_{i=1}^{n-1} |<v_n,e_i>|^2\right)^{1/2}, \quad (8.27)$$

with

$$[\{e_1, e_2, e_3, \ldots, e_n\}] = [\{v_1, v_2, v_3, \ldots, v_n\}]. \quad (8.28)$$

Thus, we have completed the proof of the Gram-Schmidt theorem, and, as a by-product, established the algorithm of the so-called Gram-Schmidt *orthonormalization*

176 ORTHONORMAL BASES AND FOURIER SERIES

procedure, which produces an *orthonormal basis* for each of the finite-dimensional subspaces generated by a finite number of the linearly independent vectors of the original sequence (8.10).

It is clear also that, being continued indefinitely, the Gram - Schmidt orthonormalization procedure should provide an orthonormal basis for the infinite-dimensional Hilbert space.

8.3 SAMPLE PROBLEM I

In the four-dimensional Hilbert space of four-component column vectors, consider the span of the following vectors;

$$\begin{pmatrix} 1 \\ 0 \\ 1 \\ 0 \end{pmatrix} ; \begin{pmatrix} 2 \\ 1 \\ 1 \\ 1 \end{pmatrix} ; \begin{pmatrix} 1 \\ 1 \\ 0 \\ 1 \end{pmatrix} ; \begin{pmatrix} 0 \\ 1 \\ -1 \\ 0 \end{pmatrix}.$$

Determine the dimension of this linear subspace and find its orthonormal basis.

Solution

The dimension of the linear subspace is equal to the number of linearly independent vectors among the given ones. The second vector is clearly the sum of the first and the third ones. To find how many vectors are independent out of the three remaining ones, we arrange these remaining vectors in a matrix. According to Section 5.6, the rank of this matrix is equal to the number of independent columns. To find the rank, we put the matrix to $\hat{U}$-form (see Sections 5.7 and 5.8):

$$\begin{pmatrix} 1 & 1 & 0 \\ 0 & 1 & 1 \\ 1 & 0 & -1 \\ 0 & 1 & 0 \end{pmatrix} \to \begin{pmatrix} 1 & 1 & 0 \\ 0 & 1 & 1 \\ 0 & -1 & -1 \\ 0 & 1 & 0 \end{pmatrix} \to \begin{pmatrix} 1 & 1 & 0 \\ 0 & 1 & 1 \\ 0 & 0 & 0 \\ 0 & 1 & 0 \end{pmatrix} \to$$

$$\begin{pmatrix} 1 & 1 & 0 \\ 0 & 1 & 1 \\ 0 & 0 & 0 \\ 0 & 0 & -1 \end{pmatrix} \to \begin{pmatrix} 1 & 1 & 0 \\ 0 & 1 & 1 \\ 0 & 0 & -1 \\ 0 & 0 & 0 \end{pmatrix}. \quad (8.29)$$

There are three independent rows here. Thus, the rank of the initial matrix is 3. Thus, all three remaining vectors are linearly independent and the dimension of the linear subspace is 3.

Orthonormalization

$$e_1 = \frac{v_1}{\|v_1\|} = \frac{1}{\sqrt{2}} \begin{pmatrix} 1 \\ 0 \\ 1 \\ 0 \end{pmatrix}. \quad (8.30)$$

$$<v_2, e_1> = \frac{1}{\sqrt{2}} \cdot 1. \tag{8.31}$$

$$\tilde{v}_2 = v_2 - <v_2, e_1> e_1 = \begin{pmatrix} 1 \\ 1 \\ 0 \\ 1 \end{pmatrix} - \frac{1}{2} \begin{pmatrix} 1 \\ 0 \\ 1 \\ 0 \end{pmatrix} = \begin{pmatrix} 1/2 \\ 1 \\ -1/2 \\ 1 \end{pmatrix}. \tag{8.32}$$

$$\|\tilde{v}_2\| = \sqrt{\frac{1}{4} + 2 + \frac{1}{4}} = \sqrt{\frac{5}{2}}. \tag{8.33}$$

$$e_2 = \sqrt{\frac{2}{5}} \begin{pmatrix} 1/2 \\ 1 \\ -1/2 \\ 1 \end{pmatrix}. \tag{8.34}$$

$$<v_3, e_1> = -\frac{1}{\sqrt{2}}. \tag{8.35}$$

$$\tilde{v}_3 = v_3 - <v_3, e_1> e_1 = \begin{pmatrix} 0 \\ 1 \\ -1 \\ 0 \end{pmatrix} + \frac{1}{2} \begin{pmatrix} 1 \\ 0 \\ 1 \\ 0 \end{pmatrix} = \begin{pmatrix} 1/2 \\ 1 \\ -1/2 \\ 0 \end{pmatrix}. \tag{8.36}$$

$$<\tilde{v}_3, e_2> = <v_3, e_2> = \sqrt{\frac{2}{5}} \cdot (1 + \frac{1}{2}) = \sqrt{\frac{2}{5}} \cdot \frac{3}{2}. \tag{8.37}$$

$$\tilde{\tilde{v}}_3 = \tilde{v}_3 - <\tilde{v}_3, e_2> e_2 = \begin{pmatrix} 1/2 \\ 1 \\ -1/2 \\ 0 \end{pmatrix} - \frac{3}{5} \begin{pmatrix} 1/2 \\ 1 \\ -1/2 \\ 1 \end{pmatrix} = \begin{pmatrix} 1/5 \\ 2/5 \\ -1/5 \\ -3/5 \end{pmatrix}. \tag{8.38}$$

$$\|\tilde{\tilde{v}}_3\| = \frac{1}{5} \cdot \sqrt{1 + 4 + 1 + 9} = \frac{\sqrt{15}}{5} = \sqrt{\frac{3}{5}}. \tag{8.39}$$

$$e_3 = \sqrt{\frac{5}{3}} \begin{pmatrix} 1/5 \\ 2/5 \\ -1/5 \\ -3/5 \end{pmatrix}. \tag{8.40}$$

8.4 SAMPLE PROBLEM II

Consider the span of the following vectors in the space of polynomials $p(x)$ on the interval $x \in [0, 1]$:

$$p_1(x) = 1; \quad p_2(x) = 1 - x; \quad p_3(x) = x;$$
$$p_4(x) = x^2; \quad p_5(x) = x^2 + 5x + 7. \tag{8.41}$$

Determine the dimension of this subspace and find its orthonormal basis.

Solution
It can be seen that:

$$p_2(x) = p_1(x) - p_3(x);$$
$$p_5(x) = p_4(x) + 5p_3(x) + 7p_1(x), \tag{8.42}$$

that is, p_2 and p_5 can be expressed through the other polynomials. These others are different powers of x, so, they are linearly independent. Thus, the dimension of this subspace is 3, and its possible basis is $\{p_1(x), p_3(x), p_4(x)\}$.

Orthonormalization:
We start with $p_1(x) = 1$.

$$\|p_1(x)\| = \left(\int_0^1 p_1^2 dx \right)^{\frac{1}{2}} = 1. \tag{8.43}$$

$$e_1 = \frac{p_1}{\|p_1\|} = 1. \tag{8.44}$$

$$\tilde{p}_3 = p_3 - <p_3, e_1> e_1. \tag{8.45}$$

$$<p_3, e_1> = \int_0^1 p_3 e_1 \, dx = \int_0^1 x \, dx = \frac{1}{2}. \tag{8.46}$$

$$\tilde{p}_3 = x - \frac{1}{2}. \tag{8.47}$$

$$\|\tilde{p}_3\| = \left(\int_0^1 (x - \frac{1}{2})^2 dx \right)^{\frac{1}{2}} = \left(\frac{2}{3} \left(\frac{1}{2} \right)^3 \right)^{\frac{1}{2}} = \frac{1}{2\sqrt{3}}. \tag{8.48}$$

$$e_3 = \frac{\tilde{p}_3}{\|\tilde{p}_3\|} = 2\sqrt{3}\left(x - \frac{1}{2}\right). \tag{8.49}$$

$$\tilde{p}_4 = p_4 - <p_4, e_1> e_1 - <p_4, e_3> e_3. \tag{8.50}$$

$$<p_4, e_1> = \int_0^1 x^2 \, dx = \frac{1}{3}. \tag{8.51}$$

$$<p_4, e_3> = \int_0^1 x^2 2\sqrt{3}\left(x - \frac{1}{2}\right) dx = 2\sqrt{3}\left(\frac{1}{4} - \frac{1}{6}\right)$$
$$= \frac{2\sqrt{3} \cdot 2}{4 \cdot 6} = \frac{1}{2\sqrt{3}}. \tag{8.52}$$

$$\tilde{p}_4 = x^2 - \frac{1}{3} - \frac{1}{2\sqrt{3}} 2\sqrt{3}\left(x - \frac{1}{2}\right) = x^2 - x + \frac{1}{6}. \tag{8.53}$$

$$\|\tilde{p}_4\| = \left(\int_0^1 \left(x^2 - x + \frac{1}{6}\right)^2 dx\right)^{\frac{1}{2}}$$
$$= \left(\int_0^1 \left(\left(x - \frac{1}{2}\right)^2 - \frac{1}{12}\right)^2 dx\right)^{\frac{1}{2}}$$
$$= \left(\int_0^1 \left(\left(x - \frac{1}{2}\right)^4 - \frac{1}{6}\left(x - \frac{1}{2}\right)^2 + \frac{1}{(12)^2}\right) dx\right)^{\frac{1}{2}}$$
$$= \left(2\left(\frac{1}{5}\left(\frac{1}{2}\right)^5\right) - \frac{1}{6}\frac{1}{3}\left(\frac{1}{2}\right)^3 + \frac{1}{(12)^2}\right)^{\frac{1}{2}}$$
$$= \frac{1}{2 \cdot 3\sqrt{5}}. \tag{8.54}$$

$$e_4 = \frac{\tilde{p}_4}{\|\tilde{p}_4\|} = 2 \cdot 3\sqrt{5}\left(x^2 - x + \frac{1}{6}\right). \tag{8.55}$$

So, the orthonormal basis is:

$$e_1 = 1; \quad e_3 = 2\sqrt{3}\left(x - \frac{1}{2}\right); \quad e_4 = 6\sqrt{5}\left(x^2 - x + \frac{1}{6}\right). \tag{8.56}$$

180 ORTHONORMAL BASES AND FOURIER SERIES

8.5 BESSEL'S INEQUALITY

To make proper use of the results of Gram-Schmidt orthonormalization in the infinite-dimensional case, we will establish Bessel's Inequality which is a fundamental property of the sequence (8.11). Dealing with decomposition over a basis in the infinite-dimensional case might appear to be a bleak perspective. However, in the case where the basis is an orthonormal sequence, the vectors of this sequence that have large numbers make negligible contribution to the decomposition: for any vector v of a Hilbert space V and an orthonormal sequence,

$$(e_1, e_2, e_3, \ldots, e_n, \ldots), \qquad (8.57)$$

in this space, any finite sum,

$$\sum_{i=1}^{n} |<v, e_i>|^2 \leq \|v\|^2. \qquad (8.58)$$

This fundamental property of orthonormal sequences is known as *Bessel's inequality*.

Proof
To prove Bessel's inequality, we start with obvious statement,

$$\|v - \sum_{i=1}^{n} <v, e_i> \cdot e_i\|^2 \geq 0, \qquad (8.59)$$

and transform the left-hand side of Eq. (8.59) as

$$\|v - \sum_{i=1}^{n} <v, e_i> \cdot e_i\|^2$$
$$= <(v - \sum_{i=1}^{n} <v, e_i> \cdot e_i), (v - \sum_{j=1}^{n} <v, e_j> \cdot e_j)>$$
$$= <v, v> - <v, \sum_{j=1}^{n} <v, e_j> \cdot e_j> - <\sum_{i=1}^{n} <v, e_i> \cdot e_i, v>$$
$$+ <\sum_{i=1}^{n} <v, e_i> \cdot e_i, \sum_{j=1}^{n} <v, e_j> \cdot e_j>$$
$$= <v, v> - \sum_{j=1}^{n} <v, e_j>^* <v, e_j> - \sum_{i=1}^{n} <v, e_i> <e_i, v>$$
$$+ \sum_{i=1}^{n} \sum_{j=1}^{n} <v, e_i> <v, e_j>^* <e_i, e_j> . \qquad (8.60)$$

Due to the orthogonality of the vectors of the sequence (8.57), in the double sum in (8.60) only the terms with $i = j$ survive. Then, we use i for the index in all remaining summations, and obtain

$$\|v - \sum_{i=1}^{n} <v,e_i> \cdot e_i\|^2$$

$$= <v,v> - \sum_{i=1}^{n} <v,e_i>^* <v,e_i>$$

$$- \sum_{i=1}^{n} <v,e_i><v,e_i>^* + \sum_{i=1}^{n} <v,e_i><v,e_i>^*$$

$$= <v,v> - \sum_{i=1}^{n} <v,e_i>^* <v,e_i>$$

$$= \|v\|^2 - \sum_{i=1}^{n} |<v,e_i>|^2. \tag{8.61}$$

Substituting this result in the left-hand side of Eq. (8.59), we get

$$\|v\|^2 - \sum_{i=1}^{n} |<v,e_i>|^2 \geq 0. \tag{8.62}$$

This is equivalent to the statement of Eq. (8.58), which is thus proven. Moreover, since (8.58) holds for any large number n, one can write its generalization as

$$\sum_{i=1}^{\infty} |<v,e_i>|^2 \leq \|v\|^2. \tag{8.63}$$

8.6 THE PROJECTION DECOMPOSITION

Now we are in a position to take full advantage of the system of orthonormal vectors we just constructed. The result we are looking for is really so useful that it alone completely justifies our efforts in constructing the orthonormal basis. Namely, we will use the orthonormal basis to obtain an explicit expression for the projection element m_0 which we introduced in the Projection Theorem.

The Theorem of Projection Decomposition
Let M be a complete subspace in a Hilbert space V, and let

$$(e_1, e_2, e_3, \ldots, e_n, \ldots) \tag{8.64}$$

be an orthonormal basis in M. Then the projection of an arbitrary vector v of the space V onto the subspace M is given by the formula

$$m_0 = \sum_{j=1}^{n} <v,e_j> \cdot e_j, \tag{8.65}$$

182 ORTHONORMAL BASES AND FOURIER SERIES

in the case when M is finite dimensional (n-dimensional), or

$$m_0 = \sum_{j=1}^{\infty} <v, e_j> \cdot e_j, \qquad (8.66)$$

if M is infinite-dimensional. (The formula (8.66) means that in the infinite-dimensional case the sum (8.65) converges to the projection m_0 as $n \to \infty$).

Proof
According to the Projection Theorem of Chapter 7, to prove the statement of Eq. (8.65), it is sufficient to show that the vector

$$v - m_0 = v - \sum_{i=1}^{n} <v, e_i> \cdot e_i \qquad (8.67)$$

is orthogonal to any vector of the subspace M.

a) The finite-dimensional case. Since the n-tuple of Eq. (8.64) is a basis in the space M, an arbitrary vector of this space can be represented as the decomposition

$$m = \sum_{j=1}^{n} \alpha_j \cdot e_j \qquad (8.68)$$

with arbitrary complex coefficients α_j. Then, the inner product of the vectors (8.67) and (8.68) is

$$<v - m_0, m>$$
$$= <v - \sum_{i=1}^{n} <v, e_i> \cdot e_i, \sum_{j=1}^{n} \alpha_j \cdot e_j>$$
$$= <v, \sum_{j=1}^{n} \alpha_j \cdot e_j>$$
$$- <\sum_{i=1}^{n} <v, e_i> \cdot e_i, \sum_{j=1}^{n} \alpha_j \cdot e_j>$$
$$= \sum_{j=1}^{n} \alpha_j^* <v, e_j> - \sum_{i=1}^{n}\sum_{j=1}^{n} <v, e_i> \alpha_j^* <e_i, e_j>. \qquad (8.69)$$

Then, as was the case in Bessel's inequality, in the latter double sum all the terms are zero, except for those with $i = j$. The expression of Eq. (8.69) is then reduced to

$$<v - m_0, m>$$
$$= \sum_{j=1}^{n} \alpha_j^* <v, e_j> - \sum_{j=1}^{n} <v, e_j> \alpha_j^*$$
$$= \sum_{j=1}^{n} (\alpha_j^* <v, e_j> - <v, e_j> \alpha_j^*) = 0. \qquad (8.70)$$

Thus, the vector m_0 of Eq. (8.65) indeed makes $v - m_0$ orthogonal to any vector of a finite subspace M. Thus, m_0 is the projection of the vector v onto this subspace.

b) The infinite-dimensional case. For any finite-dimensional space M the statement of Eq. (8.65) holds. Letting n infinitely grow in this expression, we come to the statement of Eq. (8.66) for the infinite-dimensional case. It is only necessary to check whether the sequence of *partial sums* (8.65) really converges to some element in the infinite-dimensional space M. So, we need to show the sequence

$$\left((<v,e_1>\cdot e_1), (<v,e_1>\cdot e_1 + <v,e_2>\cdot e_2), \right.$$
$$(<v,e_1>\cdot e_1 + <v,e_2>\cdot e_2 + <v,e_3>\cdot e_3), \ldots,$$
$$\left. \left(\sum_{j=1}^{n} <v,e_j>\cdot e_j \right), \ldots \right) \tag{8.71}$$

to be a Cauchy sequence. Bessel's inequality stipulates that the sums

$$\sum_{i=1}^{n} |<v,e_i>|^2 \tag{8.72}$$

are bounded by the value $\|v\|^2$. On the other hand, each term in these sums is a positive number, that is, the sum grows along with n growing. Thus, the sequence

$$\left(|<v,e_1>|^2, (|<v,e_1>|^2 + |<v,e_2>|^2), \right.$$
$$(|<v,e_1>|^2 + |<v,e_2>|^2 + |<v,e_3>|^2), \ldots,$$
$$\left. \left(\sum_{j=1}^{n} |<v,e_j>|^2 \right), \ldots \right) \tag{8.73}$$

is definitely a Cauchy sequence. Now, the Cauchy criterion for the sequence of Eq. (8.71) reads,

$$\left\| \sum_{j=1}^{n} <v,e_j>\cdot e_j - \sum_{j=1}^{l} <v,e_j>\cdot e_j \right\| \to 0, \tag{8.74}$$

as $n, l \to \infty$. Let $l < n$. We rewrite (8.74) as

$$\left\| \sum_{j=l}^{n} <v,e_j>\cdot e_j \right\|$$

$$= \left(< \left(\sum_{j=l}^{n} <v,e_j>\cdot e_j \right), \left(\sum_{i=l}^{n} <v,e_i>\cdot e_i \right) > \right)^{1/2}$$

$$= \left(\sum_{j=l}^{n} \sum_{i=l}^{n} <v,e_j><v,e_i>^*<e_j,e_i> \right)^{1/2}$$

$$= \left(\sum_{j=l}^{n} |<v,e_j>|^2 \right)^{1/2} \to 0. \qquad (8.75)$$

But

$$\sum_{j=l}^{n} |<v,e_j>|^2 = \left| \sum_{j=1}^{n} |<v,e_j>|^2 - \sum_{j=1}^{l} |<v,e_j>|^2 \right|, \qquad (8.76)$$

which is the metric distance between n-th and l-th elements of the sequence (8.73). The sequence (8.73) being a Cauchy sequence,

$$\sum_{j=l}^{n} |<v,e_j>|^2 \to 0, \qquad l, n \to \infty. \qquad (8.77)$$

This means that condition (8.74) is satisfied, and the sequence (8.71) is also a Cauchy sequence. The subspace M being complete, the limit of this latter Cauchy sequence, the element m_0 of Eq. (8.66), really exists and belongs to M.

This completes the proof.

The theorem we have just proven has numerous implications. The most obvious (and the most important) of them is the following. In the case when the subspace M coincides with the whole Hilbert space V, the formulas (8.65) and (8.66) of the Projection Decomposition Theorem give the decomposition of an arbitrary element v of a Hilbert space V over an orthonormal basis of this Hilbert space. In the case of an infinite-dimensional Hilbert space this decomposition (formula (8.66)) is known as the *Generalized Fourier Series*.

Thus, after quite a long path through the jungle of various mathematical spaces, we eventually completed our major goal put forward in Chapter 4. Now, in a Hilbert space, we are really able to explicitly decompose any element of the space over a special (orthonormal) basis.

8.7 SAMPLE PROBLEM III

In the Hilbert space of polynomials

$$p^{(2)}(x) = a + bx + cx^2, \qquad (8.78)$$

where $x \in [0, 1]$, find the projection of the vector

$$p_1(x) = 3 + 2x + x^2 \qquad (8.79)$$

onto the subspace spanned by the vectors

$$p_2(x) = 1 - 5x^2 \text{ and } p_3(x) = x. \tag{8.80}$$

Solution
The projection of vector $p_1 = 3 + 2x + x^2$ onto the subspace spanned by vectors $p_2 = 1 - 5x^2$ and $p_3 = x$ is given by the formula

$$(p_1)_p = <p_1, e_2> \cdot e_2 + <p_1, e_3> \cdot e_3, \tag{8.81}$$

where $\{e_2, e_3\}$ is an orthonormal basis in the subspace.
We proceed with the orthonormalization starting with p_3:

$$(\|p_3\|)^2 = <p_3, p_3> = \int_0^1 x^2 \, dx = \frac{1}{3}. \tag{8.82}$$

Thus,

$$e_3 = \frac{p_3}{\|p_3\|} = \sqrt{3}x. \tag{8.83}$$

Then,

$$\tilde{p}_2 = p_2 - <p_2, e_3> \cdot e_3; \tag{8.84}$$

$$<p_2, e_3> = \int_0^1 (1 - 5x^2)\sqrt{3}x \, dx$$

$$= \sqrt{3}\int_0^1 (x - 5x^3) \, dx = \sqrt{3}\left(\frac{1}{2} - \frac{5}{4}\right) = -\frac{3\sqrt{3}}{4}. \tag{8.85}$$

$$\tilde{p}_2 = 1 - 5x^2 + \frac{3 \cdot 3}{4}x. \tag{8.86}$$

Normalization of vector $\tilde{p}_2$:

$$(\|\tilde{p}_2\|)^2 = \int_0^1 \left(1 + \frac{9}{4}x - 5x^2\right)^2 dx$$

$$= \int_0^1 \left(1 + 25x^4 + \left(\frac{9}{4}\right)^2 x^2 + \frac{9}{2}x - 10x^2 - \frac{9 \cdot 5}{2}x^3\right) dx$$

$$= 1 + 5 + \left(\frac{9}{4}\right)^2 \cdot \frac{1}{3} + \frac{9}{4} - \frac{10}{3} - \frac{9 \cdot 5}{2 \cdot 4} = \frac{47}{48}. \tag{8.87}$$

$$e_2 = \frac{\tilde{p}_2}{\|\tilde{p}_2\|} = \sqrt{\frac{48}{47}} \cdot \left(1 + \frac{9}{4}x - 5x^2\right). \tag{8.88}$$

The coefficients of the decomposition:

$$\begin{aligned}
<p_1, e_3> &= \int_0^1 (3 + 2x + x^2)\sqrt{3}x\,dx \\
&= \sqrt{3}\int_0^1 (3x + 2x^2 + x^3)\,dx = \sqrt{3}\left(\frac{3}{2} + \frac{2}{3} + \frac{1}{4}\right) \\
&= \sqrt{3}\frac{21 + 8}{4 \cdot 3} = \frac{29}{4 \cdot \sqrt{3}}.
\end{aligned} \tag{8.89}$$

$$\begin{aligned}
<p_1, e_2> &= \sqrt{\frac{48}{47}} \int_0^1 (3 + 2x + x^2)\left(1 + \frac{9}{4}x - 5x^2\right)dx \\
&= \sqrt{\frac{48}{47}} \int_0^1 \left(3 + 2x + x^2 + \frac{9 \cdot 3}{4}x + \frac{9}{4}x^3\right. \\
&\quad \left. - 15x^2 - 10x^3 - 5x^4\right)dx \\
&= \sqrt{\frac{48}{47}}\left(3 + 1 + \frac{1}{3} + \frac{9 \cdot 3}{4 \cdot 2} + \frac{9}{2 \cdot 3} + \frac{9}{4 \cdot 4} - \frac{15}{3} - 10 \cdot \frac{1}{4} - \frac{5}{5}\right) \\
&= \sqrt{\frac{48}{47}}\left(\frac{63}{16} - 3 + \frac{1}{3}\right) = \sqrt{\frac{48}{47}} \cdot \frac{61}{48}.
\end{aligned} \tag{8.90}$$

We substitute the decomposition coefficients obtained in Eq. (8.89) and (8.90) in Eq. (8.81), and eventually get:

$$\begin{aligned}
(p_1)_p &= \frac{61}{47}\left(1 + \frac{9}{4}x - 5x^2\right) + \frac{29}{4}x \\
&= \frac{1}{47}(61 + 478x - 305x^2).
\end{aligned} \tag{8.91}$$

Thus, the projection of vector $p_1(x)$ onto the subspace is

$$(p_1)_p = \frac{1}{47}(61 + 478x - 305x^2). \tag{8.92}$$

8.8 COMPLETE ORTHONORMAL SEQUENCES

The illustrious results obtained earlier in this chapter raise, however, one important question. Suppose we are given an orthonormal sequence in a Hilbert space V. How can we be sure that this sequence is a basis in this space, that is, that the elements of the sequence are sufficient to decompose *any* element of the space? Note, that in the finite-dimensional case the problem was solved automatically: all we needed was to collect just as many linearly independent vectors as was the dimension of the space. Then the Gram-Schmidt orthonormalization provided us with a proper orthonormal basis to use in the decomposition (8.65). Now, in the infinite-dimensional case, the situation is much more complicated. Even having obtained an infinite number of linearly independent vectors, we are not sure that they constitute a true basis. For example, in the sequence of functions on the interval $[0, 1]$,

$$(x^2, x^4, x^6, \ldots, x^{2n}, \ldots) \tag{8.93}$$

all the elements are linearly independent. However, we definitely cannot express, say, the function x^3 through decomposition over the elements of the sequence (8.93). Thus, for a given sequence the problem is whether the sequence generates the entire space. If it does, the sequence is called *complete* (do not confuse this completeness with the notion of complete set !). Without going into terminological subtleties, a sequence being complete just means that it is a proper basis in the infinite-dimensional space.

In particular, a complete orthonormal sequence

$$(e_1, e_2, e_3, \ldots, e_n, \ldots) \tag{8.94}$$

allows any element of the space to be represented in the form

$$v = \sum_{i=1}^{\infty} <v, e_i> e_i, \tag{8.95}$$

where the infinite series in the right-hand side means

$$\lim_{m \to \infty} \|v - \sum_{i=1}^{m} <v, e_i> e_i\| = 0. \tag{8.96}$$

The following two theorems describing the basic properties of complete orthonormal sequences are almost obvious statements.

1) An orthonormal sequence (8.94) in a Hilbert space V is complete if (and only if) the only vector $v \in V$ orthogonal to all the elements of the sequence is the zero vector of the space V,

$$(\forall n (<v, e_n> = 0)) \leftrightarrow v = \theta. \tag{8.97}$$

Proof
a) Necessity. If the sequence is complete, then it is an orthonormal basis in the space V, that is, any element $v \in V$ can be represented through the infinite series (8.95). If,

according to (8.97), in this decomposition all the coefficients are zero, then it results in the zero vector.

b) *Sufficiency.* Let us take an arbitrary vector $v \in V$, compose the related series,

$$\tilde{v} = \sum_{i=1}^{\infty} <v, e_i> e_i, \tag{8.98}$$

and show that by virtue of (8.97) $\tilde{v} = v$. To this end, consider the inner product

$$<v - \tilde{v}, e_j> = <v, e_j> - <\tilde{v}, e_j>, \tag{8.99}$$

for an arbitrary element e_m of the sequence (8.94). This inner product equals

$$<v, e_j> - <\left(\sum_{i=1}^{\infty} <v, e_i> e_i\right), e_j>$$

$$= <v, e_j> - \sum_{i=1}^{\infty} <v, e_i><e_i, e_j>$$

$$= <v, e_j> - \sum_{i=1}^{\infty} <v, e_i> \delta_{ij}$$

$$= <v, e_j> - <v, e_j> = 0. \tag{8.100}$$

This implies that the vector

$$v - \tilde{v} = \theta. \tag{8.101}$$

Then, $v = \tilde{v}$, that is, the formula (8.98) is a proper decomposition of this arbitrary vector in the space V.

2) *Parseval's Identity*

An orthonormal sequence (8.94) in a Hilbert space V is complete, if and only if, for every element $v \in V$,

$$\|v\|^2 = \sum_{n=1}^{\infty} |<v, e_n>|^2. \tag{8.102}$$

Proof

In fact, this formula is just a particular case of Bessel's inequality (8.58) when the orthonormal sequence mentioned there is complete. Indeed, according to the formula (8.61), for any finite number n,

$$\|v - \sum_{i=1}^{n} <v, e_i> \cdot e_i\|^2 = \|v\|^2 - \sum_{i=1}^{n} |<v, e_i>|^2. \tag{8.103}$$

Now, if the sequence (8.94) is complete, then (8.96) holds, that is, the left-hand-side of Eq. (8.103) converges to zero as $n \to \infty$. Thus,

$$\lim_{n \to \infty} \left(\|v\|^2 - \sum_{i=1}^{n} |<v, e_i>|^2 \right)$$
$$= \|v\|^2 - \lim_{n \to \infty} \left(\sum_{i=1}^{n} |<v, e_i>|^2 \right)$$
$$= \|v\|^2 - \sum_{i=1}^{\infty} |<v, e_i>|^2 = 0, \qquad (8.104)$$

which proves the necessity of the theorem's statement.

To prove sufficiency, we just reverse the arguments. If the formula (8.102) holds, then, by virtue of (8.104), the right-hand-side of (8.103) converges to zero as $n \to \infty$, which means that the decomposition (8.95) is valid, that is, the sequence (8.94) is complete.

The two theorems we just proved demonstrate the most important properties of complete orthonormal sequences. Note, however, that neither of these theorems provides a working criterion to judge whether a given orthonormal sequence is complete or not. In fact, any time the question of completeness emerges, it becomes a serious problem to prove or disprove that a given sequence is complete. In this regard, we will only formulate without proof one mighty theorem which actually provides most of the known working examples of orthonormal sequences.

Theorem of Hermitian Operator
Let V be an infinite-dimensional Hilbert space, and let F be a linear Hermitian mapping of this space into itself, that is, a mapping satisfying condition (7.68). In analogy to the finite-dimensional case, Eq. (7.67), we formulate the eigenvalue problem with the characteristic equation,

$$f(v) = \lambda v. \qquad (8.105)$$

Then, if the mapping F is a surjection, and if there exists an infinite number of countable eigenvalues λ_k, then the corresponding eigenvectors v_k constitute a complete orthonormal sequence in the space V. (A slightly more general statement is known as the Hilbert-Schmidt Theorem).

This Theorem finally provides us with a useful tool to check the completeness of existing orthonormal sequences and to explicitly construct complete orthonormal sequences, that is, convenient bases in infinite-dimensional cases.

8.9 THE FOURIER SERIES

Consider the space V of complex-valued functions of the real variable x on the interval $[0, 1]$. Consider the sequence of functions in this space,

$$(1, e^{2\pi ix}, e^{-2\pi ix}, e^{4\pi ix}, e^{-4\pi ix}, e^{6\pi ix}, e^{-6\pi ix}, \ldots, e^{2\pi nix}, e^{-2\pi nix}, \ldots). \quad (8.106)$$

This sequence is an orthonormal basis in the space V.

It is easy to check that functions of the sequence (8.106) are orthonormal:

$$< e^{2\pi nix}, e^{2\pi mix} > = \int_0^1 e^{2\pi nix} \left(e^{2\pi imx}\right)^* dx$$

$$= \int_0^1 e^{2\pi inx} e^{-2\pi imx} dx = \int_0^1 e^{2\pi i(n-m)x} dx$$

$$= \left. \frac{e^{2\pi i(n-m)x}}{2\pi i(n-m)} \right|_0^1 = \frac{e^{2\pi i(n-m)} - 1}{2\pi i(n-m)}. \quad (8.107)$$

Since $e^{2\pi il} = 1$ for any integer number l, the expression (8.107) means

$$< e^{2\pi nix}, e^{2\pi mix} > = 0, \quad (8.108)$$

provided $n \neq m$. For $n = m$ this expression is undefined, as $0/0$. However, in this case we easily can calculate the integral,

$$< e^{2\pi nix}, e^{2\pi nix} > = \int_0^1 e^{2\pi nix} \left(e^{2\pi inx}\right)^* dx$$

$$= \int_0^1 e^{2\pi i(n-n)x} dx = \int_0^1 dx = 1. \quad (8.109)$$

Thus, using the symbol δ_{nm} of Section 8.1,

$$< e^{2\pi nix}, e^{2\pi mix} > = \delta_{nm}, \quad (8.110)$$

that is, the sequence of Eq. (8.106) is indeed orthonormal.

As for the completeness of this sequence, we mention only that the functions (8.106) are the solutions of the eigenvalue problem,

$$\frac{d^2 f}{dx^2} = \lambda f, \quad f(0) = f(1), \quad \left.\frac{df}{dx}\right|_{x=0} = \left.\frac{df}{dx}\right|_{x=1}, \quad (8.111)$$

that is, they are eigenvectors of some Hermitian operator.

The decomposition of functions of a real variable on the interval $[0, 1]$ over the orthonormal basis (8.106) is called a *Fourier decomposition*. In this case, the vectors e_n of the general theory just take the form

$$e_n = e^{2\pi i n x}, \tag{8.112}$$

provided $n \in \mathcal{I}$, that is, the summation in the formula (8.95) actually goes from $-\infty$ to $+\infty$. Thus, the formula (8.95) turns out to be

$$f(x) = \sum_{n=-\infty}^{\infty} C_n e^{2\pi i n x}, \tag{8.113}$$

where the *Fourier coefficients* C_n are given by the formula

$$C_n = <f(x), e_n> = \int_0^1 f(x) e^{-2\pi i n x} dx. \tag{8.114}$$

8.10 PROPERTIES OF THE FOURIER SERIES

Parseval's identity (8.102) in the case of the Fourier series reads

$$\|f(x)\|^2 = \sum_{n=-\infty}^{\infty} |C_n|^2. \tag{8.115}$$

Thus, obeying the general behavioral principles of orthonormal decompositions, the Fourier coefficients C_n vanish as $|n|$ goes to infinity. Moreover, in addition to this general statement, we can even say how fast they vanish. By using integration by parts in (8.114),

$$\begin{aligned} C_n &= -\frac{1}{2\pi n} \int_0^1 f(x) d\left(e^{-2\pi n x}\right) \\ &= -\frac{1}{2\pi n} f(x) \left(e^{-2\pi n x}\right)\Big|_0^1 + \frac{1}{2\pi n} \int_0^1 \frac{df(x)}{dx} \left(e^{-2\pi n x}\right) dx. \end{aligned} \tag{8.116}$$

Of course, this implies that the function $f(x)$ is at least piecewise differentiable (that is, both $f(x)$ and df/dx are piecewise continuous functions on the interval $[0, 1]$); in the case of only piecewise continuity both terms in the right-hand side of (8.116) should be replaced by sums of the corresponding terms for all the pieces.

Thus, as $|n| \to \infty$, the coefficients C_n go to zero *at least* as

$$|C_n| \propto \frac{1}{|n|}. \tag{8.117}$$

Note also, that for smooth functions the series (8.113) converges even faster: if $f(0) = f(1)$ and there are no discontinuities inside the interval, the first term in (8.116) equals zero, and $|C_n| \propto |n|^{-2}$ or faster. This means that the asymptotic behavior of the Fourier coefficients contains important qualitative information on the original function.

Another important (and useful) fact regarding Fourier decomposition is the relation between the Fourier coefficients of a *real* function. From the formula (8.114) we get for the complex conjugation of the coefficients

$$(C_n)^* = (<f(x), e_n>)^* = \left(\int_0^1 f(x) e^{-2\pi n x} dx\right)^*$$

$$= \int_0^1 (f(x) e^{-2\pi n x})^* dx = \int_0^1 f(x) \left(e^{-2\pi n x}\right)^* dx$$

$$= \int_0^1 f(x) e^{2\pi n x} dx = C_{-n}. \qquad (8.118)$$

Thus, for real functions,

$$C_{-n} = (C_n)^*. \qquad (8.119)$$

This relation can be used to check the correctness of the Fourier coefficient calculations since these calculations often lead to lengthy formulas.

8.11 SAMPLE PROBLEM IV

Find the Fourier series of the function

$$f(x) = \frac{3}{4} + x - x^2 \qquad (8.120)$$

determined on the interval [0,1].

Solution
We rewrite the function as

$$f(x) = 1 - (1/2 - x)^2. \qquad (8.121)$$

Then,

$$C_n = \int_0^1 e^{-2\pi i n x} \left(1 - \left(\frac{1}{2} - x\right)^2\right) dx$$

$$= \int_0^1 e^{-2\pi i n x} dx - \int_0^1 e^{-2\pi i n x} \left(\frac{1}{2} - x\right)^2 dx. \qquad (8.122)$$

According to (8.110), the first integral in (8.122),

$$\int_0^1 e^{-2\pi i n x} dx = \delta_{n,0}. \qquad (8.123)$$

In the second integral in (8.122), we change the dummy variable, $y = x - 1/2$, so that

$$\int_0^1 dx\, e^{-2\pi i n x} \left(\frac{1}{2} - x\right)^2 = \int_{-1/2}^{1/2} dy\, y^2 e^{-n\pi i} e^{-2\pi i n y}$$

$$= e^{-\pi i n} \int_{-1/2}^{1/2} dy\, y^2 e^{-2\pi i n y}$$

$$= e^{-\pi i n} \int_0^{1/2} dy\, y^2 \left(e^{-2\pi i n y} + e^{2\pi i n y}\right)$$

$$= 2e^{-\pi i n} \cdot \int_0^{1/2} dy\, y^2 \cos(2\pi n y). \qquad (8.124)$$

To evaluate the latter integral, we use integration by parts,

$$\int_0^{1/2} dy\, y^2 \cos(\alpha y) = \frac{1}{\alpha} \int y^2 d(\sin(\alpha y))$$

$$= \frac{1}{\alpha} y^2 \sin(\alpha y) - \frac{1}{\alpha} \int \sin(\alpha y) 2y\, dy$$

$$= \frac{1}{\alpha} y^2 \sin(\alpha y) + \frac{2}{\alpha^2} \int y\, d(\cos(\alpha y))$$

$$= \frac{1}{\alpha} y^2 \sin(\alpha y) + \frac{2}{\alpha^2} y \cos(\alpha y) - \frac{2}{\alpha^2} \int \cos(\alpha y) dy$$

$$= \frac{1}{\alpha} y^2 \sin(\alpha y) + \frac{2}{\alpha^2} y \cos(\alpha y) - \frac{2}{\alpha^3} \sin(\alpha y). \qquad (8.125)$$

Substituting this result in (8.124) with the parameter $\alpha = 2\pi n$, we get

$$\int_0^1 dx\, e^{-2\pi i n x} \left(\frac{1}{2} - x\right)^2$$

194 ORTHONORMAL BASES AND FOURIER SERIES

$$= 2e^{-\pi i n} \cdot \frac{1}{(2\pi n)^3} \Big((2\pi n)^2 y^2 \sin(2\pi n y)$$
$$+ 2(2\pi n) y \cos(2\pi n y) - 2\sin(2\pi n y) \Big) \Big|_0^{1/2}$$
$$2e^{-\pi i n} \cdot \frac{1}{(2\pi n)^3} \Big(\big((2\pi n)^2 \frac{1}{4} \sin(\pi n)$$
$$+ (2\pi n) \cos(\pi n) - 2\sin(\pi n) \big) - 0 \Big)$$
$$= 2(-1)^n \cdot \frac{1}{(2\pi n)^3} \big(0 + (2\pi n)(-1)^n - 0 \big) = \frac{2}{(2\pi n)^2}. \quad (8.126)$$

Definitely, for $n = 0$ this result is not valid, and the integral (8.124) requires separate consideration,

$$\int_0^1 dx\, e^{-2\pi i 0 x} (1/2 - x)^2 = \int_0^1 dx (1/2 - x)^2$$
$$= \int_{-1/2}^{1/2} dy (y)^2 = \frac{y^3}{3}\Big|_{-1/2}^{1/2} = \frac{1}{12}. \quad (8.127)$$

Combining the results of (8.123), (8.126), and (8.127) in one formula, the Fourier coefficients are:

$$C_n = \frac{11}{12}\delta_{n,0} - \frac{2}{(2\pi n)^2}(1 - \delta_{n,0}). \quad (8.128)$$

Note that $C_n \propto n^{-2}$. Thus, C_n becomes very small as $n \to \infty$ due to the original function (8.120) being continuous, with $f(0) = f(1)$.

8.12 SAMPLE PROBLEM V

Find the Fourier series of the function

$$f(x) = x^2 \sin\left(x + \frac{\pi}{3}\right),$$

determined on the interval [0,1].

Solution
We need to compute the integral

$$C_n = \int_0^1 dx\, e^{-2\pi i n x} x^2 \sin\left(x + \frac{\pi}{3}\right). \quad (8.129)$$

We represent the sine in exponential form,

$$\sin\left(x + \frac{\pi}{3}\right) = \frac{1}{2i}\left(e^{ix+i\frac{\pi}{3}} - e^{-ix-i\frac{\pi}{3}}\right), \tag{8.130}$$

and thus (8.129) becomes:

$$\begin{aligned} C_n &= \frac{1}{2i}\left[e^{i\frac{\pi}{3}}\int_0^1 dx\, x^2 e^{(i-2\pi in)x} - e^{-i\frac{\pi}{3}}\int_0^1 dx\, x^2 e^{(-i-2\pi in)x}\right] \\ &= \frac{1}{2i}\left[e^{i\frac{\pi}{3}} I(i - 2\pi in) - e^{-i\frac{\pi}{3}} I(-i - 2\pi in)\right]. \end{aligned} \tag{8.131}$$

Here, the generic integral,

$$I(\alpha) = \int_0^1 dx\, x^2 e^{\alpha x}, \tag{8.132}$$

can be calculated via integration by parts,

$$\begin{aligned} \int dx\, x^2 e^{\alpha x} &= \frac{1}{\alpha} x^2 e^{\alpha x} - \frac{2}{\alpha}\int dx\, x e^{\alpha x} \\ &= \frac{1}{\alpha} x^2 e^{\alpha x} - \frac{2}{\alpha^2} x e^{\alpha x} + \frac{2}{\alpha^2}\int dx\, e^{\alpha x} \\ &= \frac{1}{\alpha} x^2 e^{\alpha x} - \frac{2}{\alpha^2} x e^{\alpha x} + \frac{2}{\alpha^3} e^{\alpha x} \\ &= \frac{1}{\alpha^3}\left(\alpha^2 x^2 - 2\alpha x + 2\right) e^{\alpha x}. \end{aligned} \tag{8.133}$$

Thus,

$$I(\alpha) = \frac{1}{\alpha^3}\left((\alpha^2 - 2\alpha + 2) e^{\alpha} - 2\right). \tag{8.134}$$

Correspondingly, in Eq. (8.131) we get

$$\begin{aligned} C_n &= \frac{1}{2i}\Big[e^{i\frac{\pi}{3}}(i - 2\pi in)^{-3} \\ &\quad \times \left(((i - 2\pi in)^2 - 2(i - 2\pi in) + 2) e^{(i-2\pi in)} - 2\right) \\ &\quad - e^{-i\frac{\pi}{3}}(-i - 2\pi in)^{-3} \\ &\quad \times \left(((-i - 2\pi in)^2 - 2(-i - 2\pi in) + 2) e^{(-i-2\pi in)} - 2\right)\Big] \\ &= \frac{1}{2}\Big[-e^{i\frac{\pi}{3}}(1 - 2\pi n)^{-3} \\ &\quad \times \left((-(1 - 2\pi n)^2 - 2i(1 - 2\pi n) + 2) e^{i} - 2\right) \\ &\quad - e^{-i\frac{\pi}{3}}(1 + 2\pi n)^{-3} \end{aligned}$$

$$\times \left((-(1+2\pi n)^2 + 2i(1+2\pi n) + 2) e^{-i} - 2 \right)]$$
$$= e^{i\frac{\pi}{3}} \frac{1}{2(1-2\pi n)^3} \left(((1-2\pi n)^2 + 2i(1-2\pi n) - 2) e^i + 2 \right)$$
$$+ e^{-i\frac{\pi}{3}} \frac{1}{2(1+2\pi n)^3}$$
$$\times \left(((1+2\pi n)^2 - 2i(1+2\pi n) - 2) e^{-i} + 2 \right). \tag{8.135}$$

This expression could be further reduced to a complicated combination of sines and cosines. Note, that the general requirement for real functions, $C_{-n} = C_n^*$, is satisfied in this particular case.

8.13 SUMMARY OF CHAPTER 8

1) An orthogonal set of nonzero vectors is a linearly independent set.
2) In an infinite-dimensional Hilbert space, given a sequence of linearly independent vectors, there exist the sequence of orthonormal vectors such that for each number n the n-dimensional subspace generated by the first n vectors of the first sequence is the same as the n-dimensional subspace generate by the first n vectors of the second sequence. The orthonormal sequence is obtained from the given linearly independent sequence by means of *Gram-Schmidt orthonormalization*.
3) *The Projection Decomposition*: In a Hilbert space V, the projection of an element v onto a subspace M (finite- or infinite-dimensional) is given as a linear combination (finite or infinite, respectively) of the elements of an orthonormal basis in the subspace, with the coefficients being inner products of the vector v and the corresponding vector of the basis.
4) When M is the space V itself, and, correspondingly the projection of an element v is the element itself, the projection decomposition over an orthonormal basis gives a *generalized Fourier series*.
5) In the case of decomposition of a vector over an orthonormal sequence in an infinite-dimensional Hilbert space, Bessel's inequality ensures that the coefficients of those vectors of the sequence that have large numbers become negligibly small.
6) An orthonormal sequence in a Hilbert space is complete if and only if the only vector $v \in V$ orthogonal to all the elements of the sequence is the zero vector. For instance, eigenvectors of an Hermitian operator having an infinite number of countable eigenvalues constitute a complete orthonormal sequence.
7) *Parseval's Identity*: For any element of an infinite-dimensional Hilbert space, the square of its norm is equal to the sum of the squares of coefficients of its decomposition over a complete orthonormal sequence.
8) The set of functions $e_n(x) = e^{2\pi i n x}$ is an orthonormal basis in the Hilbert space of functions of the real variable x on the interval $[0, 1]$. The decomposition over this basis is called the *Fourier series*.
9) With $|n|$ growing, the coefficients in a Fourier series go to zero at least as $|n|^{-1}$. In the Fourier decomposition of a *real-valued* function, the coefficient of e_{-n} is the

complex conjugate coefficient of e_n.

8.14 PROBLEMS

I. Find an orthonormal basis of the linear space generated by the vectors

$$\begin{pmatrix} 1 \\ 2i \\ 2i \\ 1 \end{pmatrix} ; \begin{pmatrix} -3i \\ 1 \\ 1 \\ -3i \end{pmatrix} ; \begin{pmatrix} 1 \\ 3i \\ 3i \\ -1 \end{pmatrix} ; \begin{pmatrix} 2i \\ 1 \\ 1 \\ -2i \end{pmatrix}.$$

(This is the same space as in Problem III of Section 5.16).

II. In the Hilbert space of functions of a real variable x on the interval [-1, 1], find the projection of the element $\sin x$ onto the subspace generated by the elements e^x and e^{-x}.

III. Find the Fourier series of the function

$$f(x) = \begin{cases} x & x < \frac{1}{2} \\ 4x^3 & x > \frac{1}{2} \end{cases}$$

determined on the interval [0,1].

IV Find the Fourier series of the function

$$f(x) = x^2 e^{-\frac{x}{2}} \sin\left(\frac{\sqrt{3}}{2}x + \frac{\pi}{3}\right)$$

determined on the interval [0,1].

8.15 FURTHER READING

Introductory texts:

A. Naylor, G. R. Sell, *Linear Operator Theory in Engineering and Science*, Springer-Verlag, New York, 1982;

P. Beckman, *Orthogonal Polynomials for Engineers and Physicists*. The Golem Press, Boulder, Colorado, 1973;

T. S. Chihara, *An introduction to Orthogonal Polynomials*, Gordon & Breach, New York, 1978;

B. Osilenker, *Fourier Series in Orthogonal Polynomials*, World Scientific, Singapore, 1999;

C. Lanczos, *Discourse on Fourier Series*, Oliver & Boyd, Edinburgh, 1966.
Advanced Texts:

P. E. Edwards, *Fourier Series: a Modern Introduction*, Holt, Rinehart, and Winston, New York, 1967;

B. Szokefalvi-Nagy, *Introduction to Real Functions and Orthogonal Expansions*, Oxford University Press, New York, 1965;

G. Freud, *Orthogonal Polynomials*, Pergamon Press, Oxford, 1971.
Applications:

H. F. Harmuth, *Transmission of Information by Orthogonal Functions*, 2nd ed., Springer-Verlag, New York, 1972;

H. Jeffris, B. S. Jeffris, *Methods of Mathematical Physics*, Cambridge University Press, Cambridge, 1966.

9

Operator Equations

This chapter begins the second part of this course, in which we will consider various applications of the general ideas, principles, concepts, and notions, elaborated in the first part of the course. In the practice-oriented chapters which follow we will continue to implement our deductive approach. That is, we first consider the problem at the most abstract level possible and obtain general results which are applicable to many situations. Then we consider particular applications either as examples or as problems at the ends of the chapters.

Adhering to this strategy, in this chapter we consider the general and generic problem of solving an *operator equation* or finding (exactly or approximately) the *inverse mapping* with respect to a given direct mapping. To do this, we carefully investigate the conditions under which this generic problem has a solution, and when this solution is unique. We also introduce two basic approaches to approximate solution of the problem and discuss their comparative advantages and shortcomings. We apply these general results to the very important case of *differential equations*. Finally, we discuss the method of *least-square* approximation most appropriate in analysis of experimental data.

9.1 INVERSE MAPPINGS AND OPERATOR EQUATIONS

We often encounter situations where the result of some mapping is known, and we are required to find the original element which produced the result, that is, to perform the inverse mapping. Let the mapping under consideration be a mapping F from set

V into set W, so that the known value of the unknown element $v \in V$ is $w \in W$:

$$f(v) = w. \tag{9.1}$$

Here, $f(v)$ is often called the *operator* performing the mapping F. Correspondingly, equation (9.1) is called an *operator equation*. Its formal solution is given by the inverse mapping,

$$v = f^{-1}(w). \tag{9.2}$$

This is, however, easier to say than to do. Cases where it is possible to find the exact solution of such a problem are very rare. Thus, in most cases, only an approximate solution is possible. Whether exactly or approximately, solving the operator equation in any realistic situation is a task requiring much time and effort. Therefore, understanding the properties of the solution is highly desirable, and the main questions leading to this understanding are:
i) whether the solution exists at all;
ii) whether the solution is unique;
iii) what technique is most suitable to approach the solution approximately and what will be the error introduced by this approximation.

In what follows we will assume the sets V and W to be Banach spaces. The reason for this is that treating approximations will imply metric distances between approximate and exact solutions and considering limit transitions from approximate solutions to the exact solution.

9.2 THREE GENERIC PROBLEMS AND THEIR EQUIVALENCE

There are three typical forms of operator equations relating to mappings of a Banach space V into itself:
a) The *root-finding problem* :

$$f(v) = 0; \tag{9.3}$$

solutions of this equation are called *roots* of the operator f.
b) Finding the *fixed point* of a mapping, that is, finding an element $v \in V$ which is unaffected by operator f,

$$f(v) = v, \tag{9.4}$$

c) The *eigenvalue and eigenvector* problem: finding a *nonzero* vector $v \in V$ and a complex number λ which satisfy the equation

$$f(v) = \lambda v. \tag{9.5}$$

The vector V which is a solution of this equation, is called the *eigenvector* of operator f, and the number $\lambda = \lambda(v)$ is called the *eigenvalue* of this operator. Note that the requirement for v to be a nonzero vector is necessary for $\lambda(v)$ to be unique. Note also, that we already encountered this problem formulated for *matrices* in Chapters 5 and 7 and for the more general case of a *linear* mapping in Chapter 8.

In fact, the three problems (9.3) – (9.5) are equivalent in the sense that each of them can be reformulated in terms of any of the others. In the next section, we choose the fixed-point problem as the generic one. This form of operator equation will be very convenient for qualitative analysis and approximate solution of non-linear algebraic equations, ordinary differential equations, and integral equations in various applications.

The other typical forms of operator equations can be reduced to fixed-point problems in the following ways:

a) *The root-finding problem* of Eq. (9.3): consider an auxiliary operator

$$\tilde{f}(v) = v - f(v). \tag{9.6}$$

It is apparent that any vector v being a solution of Eq. (9.3) satisfies the equation

$$\tilde{f}(v) = v, \tag{9.7}$$

that is, a vector v being a root of an operator f is equivalent to its being a fixed point of the operator $\tilde{f}$.

b) *The eigenvalue problem.* Here the reduction is less straightforward. Let the vector v be an eigenvector of operator f, that is, a solution of Eq. (9.5). To keep track of the v being a nonzero vector (say, unit-norm vector), we consider the cross-product set $V \times \mathcal{C}$, where V is the original Banach space, and $\mathcal{C}$ is the set of complex numbers. The set V being a Banach space implies a norm mapping

$$V \to \mathcal{R} \subset \mathcal{C}, \tag{9.8}$$

with the norm $\|v\| \in \mathcal{R}$ for each vector $v \in V$. Now, we construct a "combined" mapping S of the kind

$$V \times \mathcal{C} \to V \times \mathcal{C}, \tag{9.9}$$

which transforms a pair (v, λ) into the pair $(f(v) - \lambda \cdot v, \|v\| - 1)$:

$$s((v, \lambda)) = (f(v) - \lambda \cdot v, \|v\| - 1). \tag{9.10}$$

Comparing Eq. (9.10) and Eq. (9.5), we easily see that a pair (v, λ) being the root of operator s is just an eigenvector and corresponding eigenvalue pair of the original eigenvalue problem. Thus, we have succeeded in reducing the eigenvalue problem to the root-finding one. Then, going to case a), we can reduce this root-finding problem to the fixed-point one.

9.3 FIXED-POINT PROBLEM. THE CONTRACTION MAPPING THEOREM

In part, the fixed-point problem is so popular because in this case there exists a well-defined criterion for the operator equation to have a unique solution, and this criterion

OPERATOR EQUATIONS

is easy to deal with in applications. The criterion is formulated in the following theorem, which is so fundamental and important that it even has two names.

Contraction Mapping Theorem (Banach Fixed-Point Theorem)
Let V be a complete metric space, and let F be a *contraction* mapping of the space V into itself, $V \to V$. Then F has one and only one fixed point, that is, the equation $f(v) = v$ has a solution, and this solution is unique.

As a handy means to prove this Theorem, we use a special construction.

The Picard sequence
For an arbitrary element of the space V, v_0, the infinite sequence is constructed as follows:

$$(v_0, v_1 = f(v_0), v_2 = f(v_1), \ldots, v_{n+1} = f(v_n) \ldots), \tag{9.11}$$

that is, as the sequence of consecutive values under the mapping F. The sequence (9.11) is called the *Picard sequence* for the given operator f. Correspondingly, the algorithm used to produce it is called the *Picard algorithm*; the n-th element of the sequence (9.11), v_n, is often referred to as the nth *Picard iterate*.

The most attractive and the most important feature of the Picard sequence of a contraction mapping is that whatever its initial point $v_0 \in V$, this sequence converges, and its limit is exactly the fixed point of operator f.

Convergence proof for the Picard sequence
According to the definition of a contraction mapping we introduced in Chapter 6 (Section 6.11), the mapping F being a contraction means that for any two elements of the set V, $\tilde{v}_1$ and $\tilde{v}_2$, the metric distance

$$\rho(f(\tilde{v}_1), f(\tilde{v}_2)) \leq \mu \rho(\tilde{v}_1, \tilde{v}_2), \tag{9.12}$$

where the Lipschitz constant $\mu < 1$. Let the distance between the first and the second elements of the Picard sequence,

$$\rho(v_0, f(v_0)) = d. \tag{9.13}$$

Then, the distance between the second and the third elements of the sequence,

$$\rho(v_1, v_2) = \rho(f(v_0), f(v_1)) \leq \mu \rho(v_0, v_1) = \mu d. \tag{9.14}$$

Applying this inequality for further and further pairs of consecutive elements of the sequence, we get

$$\rho(v_n, v_{n+1}) \leq \mu \rho(v_{n-1}, v_n) \leq \mu^n \rho(v_0, v_1) = \mu^n d. \tag{9.15}$$

Now, we choose large numbers m and n and consider two elements of the sequence (9.11), v_n and v_m (to be certain, let $m > n$). According to the triangle axiom of metric distance, (6.4),

$$\begin{aligned} \rho(x_m, x_n) &\leq \rho(x_m, x_{m-1}) + \rho(x_{m-1}, x_n) \\ &\leq \rho(x_m, x_{m-1}) + \rho(x_{m-1}, x_{m-2}) + \ldots + \rho(x_{n+1}, x_n) \\ &\leq (\mu^{m-1} + \mu^{m-2} + \ldots + \mu^n) \cdot d \\ &\leq \mu^n \cdot (1 + \mu + \ldots + \mu^{m-n-1}) \cdot d. \end{aligned} \quad (9.16)$$

The sum in parentheses is less than the sum of the infinite geometric series, which converges since $\mu < 1$,

$$1 + \mu + \mu^2 + \ldots + \mu^{m-n-1} \leq \sum_{k=0}^{\infty} \mu^k = \frac{1}{1-\mu}. \quad (9.17)$$

Substituting (9.17) in (9.16), for $m, n > N$, we obtain the estimate:

$$\rho(v_m, v_n) \leq \frac{d \cdot \mu^n}{1-\mu} \leq \frac{d\mu^N}{1-\mu}. \quad (9.18)$$

Then, we put

$$\frac{d\mu^N}{1-\mu} = \varepsilon, \quad (9.19)$$

and solve this equation for N:

$$N = \frac{\ln(\varepsilon(1-\mu)/d)}{\ln(\mu)}. \quad (9.20)$$

This means that for any small positive number ε we explicitly find $N = N(\varepsilon)$, such that for $n, m > N$ the distance $\rho(v_m, v_n) < \varepsilon$. Thus, the Picard sequence (9.11) is a Cauchy sequence. Therefore, it converges to some limiting element $v^\star$. The space V being complete, the element $v^\star$ necessarily belongs to V.

Now, it is not hard to show that $v^\star$ is the fixed point of the mapping F. By virtue of being a Lipschitz mapping, the mapping F is *continuous*. Therefore,

$$f(v^\star) = f(\lim_{n \to \infty} v_n) = \lim_{n \to \infty} (f(v_n)) = \lim_{n \to \infty} v_{n+1} = v^\star. \quad (9.21)$$

This just means that the element $v^\star$ is a fixed point of the operator f, $f(v^\star) \equiv v^\star$. Thus, we have proven that a Picard sequence starting at an arbitrary element $v_0 \in V$ converges to a limiting element $v^\star$ being the fixed point of the mapping F.

The only thing left to prove is that $v^\star$ is the only fixed point of the operator f. Applying our favorite method of *reductio ad absurdum*, assume the opposite. Let there be two different fixed points of the operator f, $v^\star$ and $v^{\star\star}$. According to Eq. (9.12), the metric distance between these two fixed points,

$$\rho(v^\star, v^{\star\star}) = \rho(f(v^\star), f(v^{\star\star})) \leq \mu \rho(v^\star, v^{\star\star}). \quad (9.22)$$

Consequently,
$$\rho(v^\star, v^{\star\star}) - \mu\rho(v^\star, v^{\star\star}) = \rho(v^\star, v^{\star\star})(1 - \mu) \leq 0. \tag{9.23}$$

The second factor in (9.23) is definitely positive, because $\mu < 1$. Thus, $\rho(v^\star, v^{\star\star})$ must be either negative or zero. Since by definition it cannot be negative,
$$\rho(v^\star, v^{\star\star}) = 0. \tag{9.24}$$

According to the properties of metric distance, this implies $v^\star = v^{\star\star}$, contradicting our assumption. Thus, the assumption was wrong, and the fixed point is unique.

This completes the proof of the Contraction Mapping Theorem.

9.4 SUCCESSIVE APPROXIMATIONS

In fact, in the preceding section we have advanced far beyond our initial goal of proving the existence and uniqueness of the solution of the fixed-point problem for a contraction mapping F. We have obtained, in the form of the Picard sequence, a powerful tool for approaching the solution approximately. Indeed, since the Picard sequence converges its elements lie progressively closer to the limiting point, i. e. to the fixed point of the operator f. Thus, having started at an arbitrary element $v_0 \in V$ and having proceeded a sufficient number of iterations, we can get however close to the exact solution we desire.

This method of approaching the solution by iterations is in fact a manifestation of the general approach of *successive approximations*. Correspondingly, the points of the Picard sequence are successive approximations to the fixed point $v^\star$. Based on the results of the previous section, we can estimate the accuracy of the n-th approximation in terms of this approximation and the preceding one. The error of the n-th approximation, that is, its distance to the exact solution, is estimated as

$$\rho(x_n, x^\star) \leq \rho(x_{n-1}, x_n) \frac{\mu}{1 - \mu}. \tag{9.25}$$

Clearly, the Picard Sequence converges quite fast and the error of the n-th approximation decreases rapidly when the constant μ is small. In contrast, in unpleasant cases when the value of μ is close to 1, obtaining an accurate approximation can take very many iterations.

Example
Consider once more a contraction mapping of the interval $[0, 1]$ into itself,
$$f(x) = \cos x. \tag{9.26}$$

In Section 6.11, we have obtained that in this case the Lipschitz constant
$$\mu = \sin(1) < 1. \tag{9.27}$$

Consider now the related fixed-point problem,

$$\cos x = x. \tag{9.28}$$

According to the Contraction Mapping Theorem, this problem has a unique solution, and the Picard sequence of successive approximations to this solution is given by the recursion

$$x_{n+1} = \cos(x_n). \tag{9.29}$$

If, for instance, we choose the initial approximation $x_0 = \pi/3$, then $x_1 = \cos(\pi/3) = 1/2$, $x_2 = \cos(1/2)$, and so on.

Since $\mu \approx 0.84$ is quite close to 1, this is a case when obtaining accurate approximation should take many steps. For instance, for the second Picard iterate, x_2, the approximation error is estimated as

$$\rho(x_2, x^\star) \leq \left| \frac{1}{2} - \cos\left(\frac{1}{2}\right) \right| \frac{\sin(1)}{1 - \sin(1)} \approx 2 \quad (!). \tag{9.30}$$

In this case we can easily visualize the Picard sequence and the process of its convergence. We represent mappings $\mathcal{R} \to \mathcal{R}$ as sets of points in the cartesian plane (x, y). Then, the graphical solution of the fixed-point equation (9.28) corresponds to the abscissa of the point of intersection of the curve

$$y = \cos x \tag{9.31}$$

and the straight line

$$y = x \tag{9.32}$$

approaching this solution by the Picard sequence is represented in Fig. 9.1. The sequence starts with an arbitrary point x_0 of the interval $[0, 1]$. The next Picard iterate,

$$x_1 = \cos x_0, \tag{9.33}$$

is found as the *ordinate* of the intersection of the vertical line drawn from point x_0 and the curve $y = \cos x$. To continue the iterations, we need to transform this ordinate into an abscissa. Doing this graphically, we draw a horizontal line from point $y = x_1$ to the intersection with the straight line $y = x$. Then we draw a vertical line from this intersection point to the x-axis, and get the desired point $x = x_1$. Then the whole procedure is repeated.

As you might notice, this procedure includes redundant moves to axes x and y and back to the points of intersection with the lines of Eq. (9.31) and Eq. (9.32). Upon excluding these redundancies, the process of Picard iterations is reduced to a rectangular helix converging to the point whose abscissa is the solution of Eq. (9.28). The abscissas of the angles of this helix just constitute the Picard sequence.

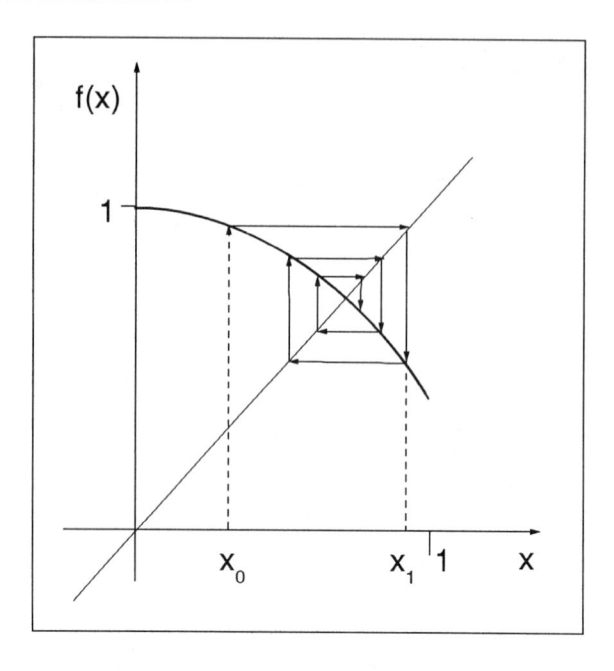

Fig. 9.1 Convergence of the Picard sequence.

Fig. 9.1 also clearly illustrates the significance of the mapping being a contraction. In the case under consideration, we recall the mean-value theorem of differential calculus, which states that for an arbitrary differentiable function $f(x)$,

$$f(x_1) - f(x_2) = \left.\frac{df}{dx}\right|_{\tilde{x}} (x_1 - x_2), \tag{9.34}$$

where $\tilde{x}$ is some point in the interval $[x_1, x_2]$. This means that the condition for the mapping represented by such function to be a contraction is

$$\max_{x \in [0,1]} \left|\frac{df}{dx}\right| < 1. \tag{9.35}$$

Now, let us take a fixed-point problem with a mapping which does not meet this condition,

$$x = 1 - x^2. \tag{9.36}$$

As is shown in Fig. 9.2, this fixed-point equation still has a quite definite graphical solution. However, the rectangular helix of the Picard sequence *diverges* away from the point of solution!

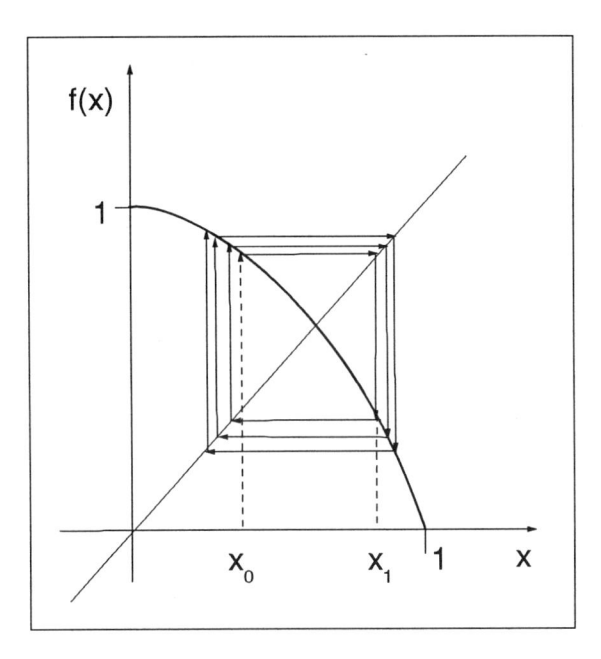

Fig. 9.2 Example of a divergent Picard sequence.

9.5 APPLICATION TO DIFFERENTIAL EQUATIONS

One of the most important applications of the Picard sequence to both theoretical analysis and approximate solution of problems is ordinary differential equations. Historically, this was the field where the method first made an appearance.

Basic form of an ordinary differential equation
In its general form, an ordinary differential equation of n-th order for function $y(t)$ of variable t defined on some interval of real numbers is

$$f\left(t, y, \frac{dy}{dt}, \frac{d^2y}{dt^2}, \frac{d^3y}{dt^3}, \ldots, \frac{d^ny}{dt^n}\right) = 0. \tag{9.37}$$

We, however, will always pull out the highest derivative, and consider equations of the form

$$\frac{d^ny}{dt^n} = v\left(t, y, \frac{dy}{dt}, \frac{d^2y}{dt^2}, \frac{d^3y}{dt^3}, \ldots, \frac{d^{n-1}y}{dt^{n-1}}\right). \tag{9.38}$$

This equation, is equivalent to a first-order *vector* differential equation. We can easily construct this equivalent first-order equation, if we consider the column vector with components

$$\vec{y} = \begin{pmatrix} y_1 \\ y_2 \\ y_3 \\ y_4 \\ \vdots \\ y_n \end{pmatrix} = \begin{pmatrix} y \\ \dfrac{dy}{dt} \\ \dfrac{d^2 y}{dt^2} \\ \dfrac{d^3 y}{dt^3} \\ \vdots \\ \dfrac{d^{n-1} y}{dt^{n-1}} \end{pmatrix}. \tag{9.39}$$

Then,

$$\frac{d\vec{y}}{dt} = \begin{pmatrix} \dfrac{dy}{dt} \\ \dfrac{d^2 y}{dt^2} \\ \dfrac{d^3 y}{dt^3} \\ \dfrac{d^4 y}{dt^4} \\ \vdots \\ \dfrac{d^n y}{dt^n} \end{pmatrix} = \begin{pmatrix} y_2 \\ y_3 \\ y_4 \\ y_5 \\ \vdots \\ v(t, y_1, y_1, y_2, y_3, y_4, \ldots, y_n) \end{pmatrix}, \tag{9.40}$$

which is just a first-order equation in the vector-function $\vec{y}(t)$:

$$\frac{d\vec{y}}{dt} = \vec{v}(t, \vec{y}). \tag{9.41}$$

Thus, in our theoretical considerations, we can (and will) consider simple scalar first-order equations,

$$\frac{dy}{dt} = v(t, y), \tag{9.42}$$

APPLICATION TO DIFFERENTIAL EQUATIONS

keeping in mind that the results can be easily generalized to the case of the vector equation, and, then, to the scalar equation of arbitrary order.

Formulation of the related fixed-point problem

The basic equation (9.42) does not look like a fixed-point problem. Thus, before applying the Picard-sequence method, we need to reformulate the problem, i. e. to reduce it to the fixed-point one.

Without loss of generality, we take the variable t to be in some definite interval of the real numbers, say, our favorite closed interval $[0, 1]$. Then, when we deal with the first-order differential equation, we need one boundary (or initial) condition, which we will put at the point $t = 0$. Therefore, the complete definition of the problem we will consider is as follows:

$$\frac{dy}{dt} = v(t, y), \quad t \in [0, 1], \quad y(0) = y_0. \tag{9.43}$$

Thus, in this case the Banach space V is the space of *differentiable* functions of the real variable t on the interval $[0, 1]$; the elements of this space are functions, $v = y(t)$.

Now, to reformulate the problem as a fixed-point one, it is sufficient to formally integrate both sides of the differential equation from the initial point of the interval, $t = 0$, to the current value of the variable t:

$$\int_0^t dt' \frac{dy}{dt'} = \int_0^t dt' v(t', y(t')). \tag{9.44}$$

This integration gives

$$y(t) = y_0 + \int_0^t dt' v(t', y(t')). \tag{9.45}$$

Equation (9.45) is already in the form of a fixed-point problem,

$$y(t) = \hat{f}(y(t)), \tag{9.46}$$

for a mapping from the Banach space V into itself performed by the *integral operator*

$$\hat{f}(y(t)) = y_0 + \int_0^t dt' v(t', y(t')). \tag{9.47}$$

In fact, Eq. (9.45) is the *integral equation* equivalent to the original differential equation (9.43).

The contraction criterion

Having reformulated the initial differential equation as a fixed-point problem, we can investigate the existence and uniqueness of its solution. To do so, we have to

210   OPERATOR EQUATIONS

understand which properties of the original differential equation (9.43), that is, of the function $v(t, y)$, determine whether the mapping (9.47) is or is not a contraction. We will find this contraction criterion by means of the following heuristic consideration. The metric distance

$$\rho(\hat{f}(y_1(t)) - \hat{f}(y_2(t))) = \|\hat{f}(y_1(t)) - \hat{f}(y_2(t))\|$$

$$= \int_0^1 dt \left| \hat{f}(y_1(t)) - \hat{f}(y_2(t)) \right|$$

$$= \int_0^1 dt \left| y_0 + \int_0^t dt' v(t', y_1(t')) - y_0 - \int_0^t dt' v(t', y_2(t')) \right|$$

$$= \int_0^1 dt \left| \int_0^t dt' \left(v(t', y_1(t')) - v(t', y_2(t')) \right) \right|$$

$$\leq \int_0^1 dt \int_0^t dt' \left| v(t', y_1(t')) - v(t', y_2(t')) \right|. \tag{9.48}$$

In order to obtain something like a Lipschitz condition, we need to express this in terms of $\|y_1(t') - y_2(t')\|$. To do so, we consider for a moment the function $v(t', y_1(t'))$ as a function of two *independent* variables, t' and y. From this standpoint, the integrand in the last integral of Eq. (9.48) is just the difference of the values of the function $v(t', y)$ taken at y_1 and y_2. Then, we can apply to this difference the mean value theorem which you should know from calculus,

$$v(t', y_1) - v(t', v_2) = \left. \frac{\partial v}{\partial y} \right|_{\tilde{y}} (y_1 - y_2), \tag{9.49}$$

where $\tilde{y}$ lies between y_1 and y_2, provided $v(t', y)$ is a differentiable function of the variable y. Based on this relation, we write

$$|v(t', y_1(t')) - v(t', y_2(t'))| = \left| \left. \frac{\partial v}{\partial y} \right|_{\tilde{y}} (y_1(t') - y_2(t')) \right|$$

$$= \left| \left. \frac{\partial v}{\partial y} \right|_{\tilde{y}} \right| |y_1(t') - y_2(t')|, \tag{9.50}$$

and estimate

$$\left| \left. \frac{\partial v}{\partial y} \right|_{\tilde{y}} \right| \leq \max_{t' \in [0,1]} \left| \frac{\partial v}{\partial y} \right|. \tag{9.51}$$

The latter maximum being just a positive number, we get in Eq. (9.48):

$$\|\hat{f}(y_1(t)) \quad - \quad \hat{f}(y_2(t))\|$$

$$\leq \max_{t\in[0,1]}\left|\frac{\partial v}{\partial y}\right|\int_0^1 dt\int_0^t dt'|y_1(t')-y_2(t')|. \qquad (9.52)$$

The integral over t' is an integral of a definitely positive function. So, its (positive) value can only increase with the extension of the interval of integration to the entire interval $[0, 1]$:

$$\int_0^t dt'|y_1(t')-y_2(t')| \leq \int_0^1 dt'|y_1(t')-y_2(t')| = \|y_1(t)-y_2(t)\|. \qquad (9.53)$$

Substituting this in Eq. (9.52), we obtain

$$\|\hat{f}(y_1(t))-\hat{f}(y_2(t))\|$$

$$\leq \max_{t\in[0,1]}\left|\frac{\partial v}{\partial y}\right|\int_0^1 dt\|y_1(t)-y_2(t)\|$$

$$= \max_{t\in[0,1]}\left|\frac{\partial v}{\partial y}\right|\|y_1(t)-y_2(t)\|\int_0^1 dt$$

$$= \max_{t\in[0,1]}\left|\frac{\partial v}{\partial y}\right|\|y_1(t)-y_2(t)\|. \qquad (9.54)$$

Thus,

$$\|\hat{f}(y_1(t))-\hat{f}(y_2(t))\| \leq \max_{t\in[0,1]}\left|\frac{\partial v}{\partial y}\right|\|y_1(t)-y_2(t)\|. \qquad (9.55)$$

Therefore, the mapping of Eq. (9.47) is a Lipschitz mapping, with the Lipschitz constant

$$\mu = \max_{t\in[0,1]}\left|\frac{\partial v}{\partial y}\right|, \qquad (9.56)$$

provided, of course, that this maximum exists, that is, that the function $v(t, y)$ does not diverge as a function of y. Then, the desired criterion for the mapping to be a contraction, $\mu < 1$, is

$$\max_{t'\in[0,1]}\left|\frac{\partial v}{\partial y}\right| < 1. \qquad (9.57)$$

The only question which remains is what to do when this maximum value is greater than 1. Does this mean that the method is completely inapplicable? No, this only shows that we cannot construct the convergent sequence of the Picard iterates on the entire interval $[0, 1]$ at once. Indeed, at the final step, in Eq. (9.54), we used

$$\int_0^1 dt = 1. \qquad (9.58)$$

If, instead of the interval $[0, 1]$, we had an interval $[0, \delta]$, $\delta < 1$, we would get

$$\int_0^\delta dt = \delta, \tag{9.59}$$

all the other details of the above consideration are the same. Thus, in this case, with a smaller interval, the Lipschitz constant would be

$$\mu = \delta \max_{t' \in [0,1]} \left| \frac{\partial v}{\partial y} \right|. \tag{9.60}$$

We see that for any *finite* value of

$$L = \max_{t' \in [0,1]} \left| \frac{\partial v}{\partial y} \right|, \tag{9.61}$$

we always can take a sufficiently small δ for the mapping in question to be a contraction on the interval $[0, \delta]$. Then, if we are still interested in the solution on the entire interval $[0, 1]$, we should use the result obtained for the interval $[0, \delta]$ to calculate $y(\delta)$, take this value as the initial condition for interval $[\delta, 2\delta]$, and so on.

The conclusions of our considerations can be rigorously formulated as the following theorem, which we give here without proof.

The Picard Theorem
If $v(t, y(t))$ as a function of y is continuously differentiable in a neighborhood of the initial point $(0, y_0)$, then there is a neighborhood of this point, where the solution $y(t), y(0) = y_0$ is defined and unique. This solution depends continuously on y_0 and is given by the limit of the Picard approximations,

$$y_n(t) = y_0 + \int_0^t v(\tau, y_{n-1}(\tau)) d\tau. \tag{9.62}$$

Generally speaking, one should not confuse the initial Picard iterate $y_0(t)$ with the initial condition $y(0) = y_0$ in Eq. (9.43). However, there is no harm in this similarity of notations, because the natural choice of the initial iterate is just this very constant $y_0(t) = y_0$.

9.6 SAMPLE PROBLEM I

Consider the differential equation

$$\frac{dy}{dt} = t \cos^3 y; \quad t \in \left[0, \frac{3}{5}\right], \quad y(0) = 0. \tag{9.63}$$

Reformulate it as a fixed-point problem, check the convergence of the Picard sequence on the whole interval of t, and find the first three Picard iterates, $y_0(t)$, $y_1(t)$, and $y_2(t)$.

Solution
1) Since the original differential equation has the standard form

$$\frac{dy}{dt} = f(t, y), \tag{9.64}$$

the equivalent integral equation of the fixed point problem takes the form:

$$y(t) = y(0) + \int_0^t dt' f(t', y(t'))$$

For the particular right-hand side and initial condition of equation (9.63) this reads:

$$y(t) = \int_0^t dt' t' \cos^3(y(t')) \tag{9.65}$$

2) *Checking convergence of the Picard sequence*
The convergence criterion requires:

$$\left| \frac{\partial f(t, y)}{\partial y} \right| < 1 \tag{9.66}$$

on the entire interval $t \in [0, \frac{3}{5}]$. For the right-hand side of Eq. (9.63) we have:

$$\left| \frac{\partial f}{\partial y} \right| = \left| t(-3 \cos^2 y \sin y) \right| = \left| 3t \sin y (1 - \sin^2 y) \right|, \tag{9.67}$$

which is definitely a bounded function. Its extremum with respect to y is determined by the equation:

$$\frac{\partial}{\partial y} \left(\left| \frac{\partial f}{\partial y} \right| \right) \bigg|_{y_m} = 0, \tag{9.68}$$

that is,

$$3t \cos y_m (1 - 3 \sin^2 y_m) = 0. \tag{9.69}$$

The first solution of this equation,

$$\cos y_{m_1} = 0, \tag{9.70}$$

implies

$$\sin y_{m_1} = 1 \tag{9.71}$$

and

$$\left(\left| \frac{\partial f}{\partial y} \right| \right) \bigg|_{y_{m_1}} = 0. \tag{9.72}$$

214 OPERATOR EQUATIONS

Thus, this point corresponds to *minimum* and is not the point of interest. The other solution,

$$1 - 3\sin^2 y_{m_2} = 0, \tag{9.73}$$

gives

$$\sin y_{m_2} = \pm \frac{1}{\sqrt{3}}, \tag{9.74}$$

so that

$$\left(\left|\frac{\partial f}{\partial y}\right|\right)\bigg|_{y_{m_2}} = 3t \frac{1}{\sqrt{3}} \left(1 - \frac{1}{3}\right) = \frac{2t}{\sqrt{3}}. \tag{9.75}$$

Thus, on all interval $t \in [0, \frac{3}{5}]$

$$\left|\frac{\partial f}{\partial y}\right| < \frac{2}{\sqrt{3}} \frac{3}{5} = \frac{2\sqrt{3}}{5} < 1 \tag{9.76}$$

($\sqrt{3} \approx 1.71 < 2$). Thus, the Picard sequence converges.

3) *The First Picard Iterates*

We naturally choose $y_0(t) = y(0) = 0$.
The following iterates are determined by the relation:

$$y_n(t) = \int_0^t dt' t' \cos^3(y_{n-1}(t')). \tag{9.77}$$

Correspondingly, the first iterate is found as

$$y_1(t) = \int_0^t dt' t' \cos^3 0 = \int_0^t dt' t' = \frac{t^2}{2}. \tag{9.78}$$

The second iterate,

$$y_2(t) = \int_0^t dt' t' \cos^3\left(\frac{t'^2}{2}\right) = \int_0^{\frac{t^2}{2}} d\tau \cos^3 \tau, \tag{9.79}$$

where $\tau = \frac{t^2}{2}$. The integral

$$\int d\tau \cos^3 \tau = \int d\tau \cos\tau (1 - \sin^2 \tau)$$
$$= \int d(\sin\tau)(1 - \sin^2\tau) = \sin\tau - \frac{1}{3}\sin^3\tau, \tag{9.80}$$

which gives

$$y_2(t) = \left(\sin\tau - \frac{1}{3}\sin^3\tau\right)\bigg|_0^{\frac{t^2}{2}} = \sin\left(\frac{t^2}{2}\right) - \frac{1}{3}\sin^3\left(\frac{t^2}{2}\right). \tag{9.81}$$

Thus, the first three Picard iterates are obtained as

$$y_0(t) = 0;$$
$$y_1(t) = \frac{t^2}{2};$$
$$y_2(t) = \sin\left(\frac{t^2}{2}\right) - \frac{1}{3}\sin^3\left(\frac{t^2}{2}\right). \tag{9.82}$$

9.7 RESIDUAL PRINCIPLE AND RESIDUAL APPROXIMATION

Our construction of the Picard sequence in the previous sections was a typical example of the successive approximation approach. There exist more effective (and more tricky) sequences of iterates defined by recursive relations, but the main idea of a step-by-step approximation approaching the unknown solution is the same.

The appeal of such an approach is that by a finite number of iterations it automatically produces an approximate solution with a prescribed accuracy. However, if the Lipschitz constant μ is not sufficiently small, this "finite number" may be quite large. In such cases we seek another approach which would be more appropriate. The basic idea of this complementary approach is to intentionally construct an approximate solution in a form which can be optimized, and then to optimize this solution by minimizing its approximation error. In practice this approach proceeds as follows.

Let us be given an operator equation in the general form,

$$f(v) = w, \tag{9.83}$$

where $v \in V$, $w \in W$, V and W are Banach spaces. Let then $V_1 \subset V$ be a Banach subspace (in practice, V_1 is usually chosen as the subspace of elements being especially easy to operate with). Our goal is to find an element $\tilde{v} \in V_1$ which is the best approximation to the solution of Eq. (9.83) among all the elements of the subspace V_1.

Pursuing this goal, we reformulate Eq. (9.83) as a *root-finding* problem:

$$f(v) - w = \theta. \tag{9.84}$$

Equation (9.84) says that if we substitute in the expression in the left-hand side the *exact* solution, $\bar{v}$, we obtain the zero element. Of course, substitution of any other element v_s in this left-hand side produces a non-zero result:

$$f(v_s) - w = r(v_s). \tag{9.85}$$

This result, $r(v_s)$, is called the *residual* corresponding to the *substitution* v_s. In particular, all elements of the subspace V_1 yield some residuals. Then, V_1 being a complete set, there definitely exists a $\tilde{v} \in V_1$ whose residual has the least norm among

216 OPERATOR EQUATIONS

all these residuals. This element $\tilde{v}$ is called the *minimal residual approximation*, and this is precisely the best approximation we wanted to find. In rigorous terms, $\tilde{v}$ is the minimal residual approximation, if

$$\forall v_1 \in V_1 \; (\|f(v_1) - w\| \geq \|f(\tilde{v}) - w\|). \tag{9.86}$$

Note, however, that while setting the minimal residual approximation as the best approximation to the unknown solution, we tacitly started using a new criterion of accuracy, different from that we exploited in the previous sections. Indeed, the fact that the element $\tilde{v}$ minimizes the norm of the *residual*, $\| r(v_1) \|$, does not necessarily mean that this $\tilde{v}$ minimizes the *error*, $\| v_1 - \bar{v} \|$.

A direct correspondence between these two criteria can be established only under certain (and rather strict) conditions, which are considered in the following important theorem.

Residual Theorem
Let the mapping $f(v)$ of Eq. (9.83) be an injection, that is, a one-to-one mapping implying the existence of the inverse operator $f^{-1}(w)$. Let then this inverse mapping be a Lipschitz mapping with constant μ. Then the approximation error for a substitution v_s is bounded by

$$\| v_s - \bar{v} \| \leq \mu \, \| r(v_s) \|. \tag{9.87}$$

Proof
Since the inverse mapping f^{-1} exists, we can formally write the exact solution of Eq. (9.83) as

$$\bar{v} = f^{-1}(w). \tag{9.88}$$

On the other hand, we can formally express v_s from Eq. (9.85) as

$$v_s = f^{-1}(w + r(v_s)). \tag{9.89}$$

Then, the approximation error of the substitution v_s is written as

$$\| v_s - \bar{v} \| = \| f^{-1}(w + r(v_s)) - f^{-1}(w) \|. \tag{9.90}$$

The mapping f^{-1} being a Lipschitz mapping, the expression in the right-hand side of Eq. (9.90) is bounded as

$$\| f^{-1}(w + r(v_s)) - f^{-1}(w) \| \leq \mu \, \| (r(v_s) + w) - w \| = \mu \, \| r(v_s) \|. \tag{9.91}$$

Thus, we get in Eq. (9.90):

$$\| v_s - \bar{v} \| \leq \mu \, \| r(v_s) \|, \tag{9.92}$$

the inequality which had to be proven.

The trick with using the inverse mapping f^{-1} here is that we do not know this inverse mapping explicitly. (If we knew it, we would solve Eq. (9.83) exactly and would not need any approximations!). There is, however, a catch here. Since we have no idea of what f^{-1} is, it is almost impossible to check whether f^{-1} is a Lipschitz mapping. To comprehend the difficulties, we consider the following example.

Example

Consider the ordinary differential equation,

$$\frac{dx}{dt} + tx = y(t), \quad t \in [0, 1], \quad x(0) = x_0, \tag{9.93}$$

$y(t)$ is a given function. In this case, the operator equation (9.1) is written for a linear differential operator,

$$\hat{f}(x(t)) = \left(\frac{d}{dt} + t\right) x(t), \tag{9.94}$$

while both the domain and the codomain of the mapping performed by this operator are the space of functions on the interval [0, 1]. We need to check whether the inverse mapping $\hat{f}^{-1}$ is a Lipschitz mapping. To do so, we should compare the metric distance $\|x_1(t) - x_2(t)\|$ between the solutions of the equation (9.93) for different functions $y_1(t)$ and $y_2(t)$ in its right-hand side and the metric distance $\|y_1(t) - y_2(t)\|$ between these different right-hand-side functions. We approach this in an oblique manner. First, we rewrite the equation (9.93) in integral form (cf. Eq. (9.45):

$$x(t) = x_0 + \int_0^t dt' y(t') - \int_0^t dt' t' x(t'). \tag{9.95}$$

Then, we write the equalities for the unknown solutions,

$$x_1(t) = x_0 + \int_0^t dt' y_1(t') - \int_0^t dt' t' x_1(t'). \tag{9.96}$$

$$x_2(t) = x_0 + \int_0^t dt' y_2(t') - \int_0^t dt' t' x_2(t'). \tag{9.97}$$

We subtract the second equality from the first to obtain

$$x_1(t) - x_2(t) = \int_0^t dt' (y_1(t') - y_2(t')) - \int_0^t dt' t' (x_1(t') - x_2(t')). \tag{9.98}$$

218   OPERATOR EQUATIONS

In the following analysis, it is convenient to use the second definition of metric distance in the space of functions, that of Eq. (6.19). From Eq. (9.98), we obtain the metric distance between $x_1(t)$ and $x_2(t)$ as

$$\|x_1(t) - x_2(t)\| = \int_0^1 dt \, |x_1(t) - x_2(t)|$$

$$= \int_0^1 dt \left| \int_0^t dt'(y_1(t') - y_2(t')) - \int_0^t dt' \, t' \, (x_1(t') - x_2(t')) \right|. \quad (9.99)$$

Now, from the triangle axiom for the norm we get

$$\left| \int_0^t dt'(y_1(t') - y_2(t')) - \int_0^t dt' \, t' \, (x_1(t') - x_2(t')) \right|$$

$$\leq \left| \int_0^t dt'(y_1(t') - y_2(t')) \right| + \left| \int_0^t dt' \, t' \, (x_1(t') - x_2(t')) \right|, \quad (9.100)$$

which gives in Eq. (9.99):

$$\|x_1(t) - x_2(t)\|$$

$$\leq \int_0^1 dt \left(\left| \int_0^t dt'(y_1(t') - y_2(t')) \right| \right.$$

$$\left. + \left| \int_0^t dt' \, t' \, (x_1(t') - x_2(t')) \right| \right)$$

$$= \int_0^1 dt \, (I_1(t) + I_2(t)), \quad (9.101)$$

where

$$I_1(t) = \left| \int_0^t dt'(y_1(t') - y_2(t')) \right| \leq \int_0^t dt' \, |y_1(t') - y_2(t')|; \quad (9.102)$$

$$I_2(t) = \left| \int_0^t dt' \, t' \, (x_1(t') - x_2(t')) \right| \leq \int_0^t dt' \, t' \, |x_1(t') - x_2(t')|. \quad (9.103)$$

To estimate the integral I_1, we note that its integrand is definitely a positive function. Thus, the expansion of the integration to all the interval $[0, 1]$ can only increase the result:

$$I_1(t) \leq \int_0^1 dt' \, |y_1(t') - y_2(t')| \equiv \|y_1(t) - y_2(t)\|; \tag{9.104}$$

Overcoming the integral I_2 requires a more subtle strategy: we first substitute for the factor t' in the integrand the largest value it can take, t:

$$I_2(t) \leq \int_0^t dt' \, t \, |x_1(t') - x_2(t')| = t \int_0^t dt' \, |x_1(t') - x_2(t')|$$

$$\leq t \int_0^1 dt' \, |x_1(t') - x_2(t')| \equiv t\|x_1(t) - x_2(t)\|. \tag{9.105}$$

The results of Eqs. (9.104) and (9.105) enable us to write the estimation:

$$\|x_1(t) - x_2(t)\| \leq \int_0^1 dt \, (I_1(t) + I_2(t))$$

$$\leq \int_0^1 dt \|y_1(t) - y_2(t)\| + \int_0^1 dt \, t \, \|x_1(t) - x_2(t)\|$$

$$= \|y_1(t) - y_2(t)\| \int_0^1 dt + \|x_1(t) - x_2(t)\| \int_0^1 t \, dt$$

$$= \|y_1(t) - y_2(t)\| + \frac{1}{2}\|x_1(t) - x_2(t)\|. \tag{9.106}$$

Thus,

$$\|x_1(t) - x_2(t)\| \leq \|y_1(t) - y_2(t)\| + \frac{1}{2}\|x_1(t) - x_2(t)\|. \tag{9.107}$$

So, the bottom line is:

$$\|x_1(t) - x_2(t)\| \leq 2\|y_1(t) - y_2(t)\|. \tag{9.108}$$

This means that the inverse mapping $\hat{f}^{-1}$ is a Lipschitz mapping with the Lipschitz constant $\mu = 2$.

Approximate solution of Eq. (9.93)
Consider now, how the residual approximation works in practice. We continue treating Eq. (9.93), but to be certain (and for the sake of simplicity) we set the function $y(t)$ to be, say,

$$y(t) = t. \tag{9.109}$$

220 OPERATOR EQUATIONS

Moreover, we impose the simplest initial condition:

$$x(0) = x_0 = 0. \tag{9.110}$$

Then, let us look for the best approximation on the set of polynomial functions,

$$\tilde{x}_n(t) = \alpha + \beta t + \gamma t^2 + \ldots + \omega t^n, \tag{9.111}$$

and restrict our quest to polynomials of order 1. Thus, we take

$$\tilde{x}(t) = \alpha + \beta t. \tag{9.112}$$

Now, our task is to find the best values of the parameters α and β, that is, those values which provide the least norm of the residual.

The initial condition dictates at once that α should be zero. So, in fact, our *trial* function $\tilde{x}(t)$ contains only one adjustable parameter, β.

We substitute $\tilde{x}(t)$ of Eq. (9.112) into Eq. (9.93), and obtain the residual $r(t)$ in the form:

$$\beta + t^2\beta - t = r(t, \beta). \tag{9.113}$$

Depending on β, the expression in the left-hand side can change its sign while t runs the interval $[0, 1]$. This makes it too complicated to use the norm of Eq. (6.19) for our practical purposes. Thus, we return to the regular norm of Eq. (6.18),

$$\|r\| = \sqrt{\int_0^1 dt(\beta + t^2\beta - t)^2}. \tag{9.114}$$

However, it is easier to minimize the square of this norm,

$$\begin{aligned}
\|r\|^2 &= \int_0^1 dt(\beta^2 t^4 + \beta^2 + t^2 + 2\beta^2 t^2 - 2\beta t - 2\beta t^3) \\
&= \frac{\beta^2}{5} + \beta^2 + \frac{1}{3} + 2\frac{\beta^2}{3} - 2\frac{\beta}{2} - 2\frac{\beta}{4} \\
&= \beta^2\left(\frac{1}{5} + 1 + \frac{2}{3}\right) - \beta\left(1 + \frac{1}{2}\right) + \frac{1}{3} \\
&= \frac{28}{15}\beta^2 - \frac{3}{2}\beta + \frac{1}{3}.
\end{aligned} \tag{9.115}$$

The minimum of $\|r\|^2$ as a function of the parameter β is determined by the condition:

$$\frac{\partial}{\partial \beta}(\|r\|^2) = 0, \tag{9.116}$$

which gives us the equation for the parameter β,

$$\frac{56}{15}\beta - \frac{3}{2} = 0. \tag{9.117}$$

The solution of this equation,

$$\beta = \frac{45}{112}, \qquad (9.118)$$

corresponds to the minimal norm of the residual. Thus, the minimal residual solution of Eq. (9.93) on the subspace of linear functions of t is found as

$$\tilde{x}(t) = \frac{45}{112} t. \qquad (9.119)$$

Unfortunately, we cannot combine this result with $\mu = 2$ in formula (9.87) to estimate the error of our approximation, because that value of the Lipschitz constant was obtained with another definition of norm. However, if we wished to improve the accuracy, we would need to consider a trial function containing more free parameters suitable for minimization, say, a polynomial of second order. It is also interesting to note that Eq. (9.93) taken on the entire interval $[0, 1]$ does not satisfy the contraction criterion of Section 9.5. This means that it is not straightforward to apply the alternative method of successive approximations to this equation: the interval first should be broken into smaller portions, as we have outlined at the end of Section 9.5.

9.8 INCOMPLETELY SPECIFIED EQUATIONS. LEAST-SQUARE APPROXIMATION

An interesting and important feature of the residual principle is that in the process of minimizing the residual we, in fact, do not refer to the exact solution. This feature provides an extension of the residual principle to situations where the problem as formulated has no exact solution at all. A typical example of such situation is the problem of measurements.

Example — Multiple measurements
A common situation in the life of an engineer is when it it necessary to find a value of some pertinent quantity by means of measuring this quantity or properties related to it. For example, to know the mass of an apple, we put it on scale. To know the mass of a molecule, we use a mass-spectrometer. To know the mass of a subatomic particle, we photograph its track in a bubble chamber and do some calculations...

But then, do we obtain the mass of the apple as a result of our measurement? By no means. If we repeat the measurement five minutes later, we will get another result (if even we did not bite the apple). This is because of *measurement errors* . Having weighted the apple several times, the results will differ from one another, and none of them will give us the true value. In our abstract mathematical parlance this means that the measuring device (and the measurement procedure) reveals some relation between the set of the true values of whatever quantity we are interested in and the results of measurements. Due to the mentioned measurement errors involved, this relation is not a mapping because one element in the domain corresponds to several values in the codomain. Thus, the would-be problem of inverse mapping is unsolvable and even meaningless. The question is how to cope with this situation?

In the general framework of residual ideology, we can try to find a value which is in the best correspondence to all the measurements. Note, that in this case we cannot even say, "the best approximation", because the exact solution does not exist. In mathematical language, this means that instead of the operator equation

$$f(v) = w \qquad (9.120)$$

we have a set of equations:

$$\begin{cases} f(v) & = & w + \epsilon_1 \\ f(v) & = & w + \epsilon_2 \\ & \vdots & \\ f(v) & = & w + \epsilon_n \end{cases}, \qquad (9.121)$$

where $\epsilon_1, \epsilon_2, \ldots, \epsilon_n$ are the measurement errors. Of course, we do not know these errors and cannot separate them explicitly in the right-hand sides of Eqs. (9.121). Thus, the system looks like:

$$\begin{cases} f(v) & = & w_1 \\ f(v) & = & w_2 \\ & \vdots & \\ f(v) & = & w_n \end{cases}. \qquad (9.122)$$

Definitely, this system does not have a solution. Of course, in this particular instance the question is rhetorical. We know that we need to take the *mean* value. This is, however, just common sense without any solid mathematical basis.

Now we seek to establish the solid basis in terms of a generalized situation. We consider a linear mapping from an m-dimensional Hilbert space V into an n-dimensional Hilbert space W, $(n > m)$. Thus, we represent an element of V by an m-component column vector of complex numbers, and the elements of W by an n-component column vector. The mapping itself is then represented by an $m \times n$ matrix $\hat{A}$:

$$\hat{A}\vec{v} = \vec{w}. \qquad (9.123)$$

This linear system of equations on the components of vector v is *overspecified* — it has more equations then unknowns. In the general case the system is inconsistent and does not have a solution.

We, however, set our goal to be that of finding the vector $\vec{v}$ minimizing the norm of the residual,

$$\|r\| = \|\hat{A}\vec{v} - \vec{w}\|, \qquad (9.124)$$

that is, we seek a vector $\vec{v}$ such that

$$\forall \vec{v} \in V : \|w - \hat{A}\vec{v}\| \geq \|w - \hat{A}\vec{\tilde{v}}\|. \qquad (9.125)$$

Since we defined the norm of a column vector as

$$\|\vec{r}\| = \sqrt{r_1 \cdot r_1^* + r_2 \cdot r_2^* + \ldots + r_n \cdot r_n^*}, \qquad (9.126)$$

we need to find the vector producing the minimal

$$(\|\vec{r}\|)^2 = <\vec{w} - \hat{A}\vec{v}, \ \hat{w} - \hat{A}\vec{v}>. \qquad (9.127)$$

This is why this solution is called the *least square* approximation. The algorithm for finding the vector $\vec{\tilde{v}}$ happens to be unexpectedly simple. Namely, it is given by the following theorem.

Least-Square Theorem
For equation (9.123), the *least-square solution*, $\tilde{v}$, always exists, and this $\tilde{v}$ satisfies the linear matrix equation (called the *normal equation*),

$$(\hat{A}^T)^* \hat{A}\tilde{v} = (\hat{A}^T)^* w, \qquad (9.128)$$

where $(\hat{A}^T)^*$ is the Hermitian conjugate of matrix $\hat{A}$ that we introduced and discussed in Chapter 7 (Section 7.8).

Proof
For an arbitrary element of the space V, a value of the mapping performed by the matrix $\hat{A}$, the element $\hat{A}v \in W$, belongs to the range space of the matrix $\hat{A}$, $R(\hat{A})$. According to the theorem of Section 7.4, this finite-dimensional subspace of the Hilbert space W is a closed subset. Thus, we can apply the *Projection Theorem* (Section 7.13) which states that for any element w in the Hilbert space W there exists an element r_0 of the given closed Hilbert subspace $R(\hat{A})$ such that
1) $\|w - r_0\| \le \|w - r\| \ \forall r \in R(\hat{A})$;
2) $(w - r_0) \perp R(\hat{A})$.
The latter condition means that for any element $r \in R(\hat{A})$

$$< (w - r_0), r >= 0. \qquad (9.129)$$

Substituting here $r_0 = \hat{A}\tilde{v}$ and $r = \hat{A}v$, we get

$$< (w - \hat{A}\tilde{v}), \hat{A}v >= 0. \qquad (9.130)$$

Then, according to the relation (7.57) (Section 7.8),

$$< (\hat{A}^T)^*(w - \hat{A}\tilde{v}), v >= 0. \qquad (9.131)$$

The equality of Eq. (9.131) holds for *any* element $v \in V$. In particular, this means, that

$$< (\hat{A}^T)^*(w - \hat{A}\tilde{v}), (\hat{A}^T)^*(w - \hat{A}\tilde{v}) >= 0. \qquad (9.132)$$

But we know that the only element of the space V which has zero inner product with itself is the zero element, θ. Thus,

$$(\hat{A}^T)^*(w - \hat{A}\tilde{v}) = \theta, \qquad (9.133)$$

224 OPERATOR EQUATIONS

or
$$(\hat{A}^T)^*w - (\hat{A}^T)^*(\hat{A}\tilde{v}) = 0. \tag{9.134}$$
Thus,
$$((\hat{A}^T)^*\hat{A})\tilde{v} = (\hat{A}^T)^*w, \tag{9.135}$$
and the theorem is proven.

Examples of least-square approximation
a) Let us return to the problem of *multiple measurements* and consider this problem from the standpoint of least-square approximation. Suppose, we have measured some quantity n times and obtain the results $y_1, y_2, y_3, \ldots, y_n$. This means that we obtain a system of n equations for the unknown quantity x:

$$\begin{cases} x = y_1 \\ x = y_2 \\ x = y_3 \\ \vdots \\ x = y_n \end{cases} \tag{9.136}$$

In matrix notation, this system looks like

$$\begin{pmatrix} 1 \\ 1 \\ 1 \\ \vdots \\ 1 \end{pmatrix} x = \begin{pmatrix} y_1 \\ y_2 \\ y_3 \\ \vdots \\ y_n \end{pmatrix}. \tag{9.137}$$

Thus, in this case the matrix $\hat{A}$ degenerates into a column vector,

$$\hat{A} = \begin{pmatrix} 1 \\ 1 \\ 1 \\ \vdots \\ 1 \end{pmatrix}. \tag{9.138}$$

Equation (9.137) certainly has no *exact* solution, the system is apparently inconsistent. However, we easily obtain the least-square solution. The matrix
$$(\hat{A}^T)^* = (\hat{A}^T) = (1\ 1\ 1\ \ldots\ 1) \tag{9.139}$$
is a single-row matrix. Thus,

$$\hat{A}^T\hat{A} = (1\ 1\ 1\ \ldots\ 1) \begin{pmatrix} 1 \\ 1 \\ 1 \\ \vdots \\ 1 \end{pmatrix} = n, \tag{9.140}$$

just a number;

$$\hat{A}^T \begin{pmatrix} y_1 \\ y_2 \\ y_3 \\ \vdots \\ y_n \end{pmatrix} = (1\ 1\ 1\ \ldots\ 1) \begin{pmatrix} y_1 \\ y_2 \\ y_3 \\ \vdots \\ y_n \end{pmatrix} = y_1 + y_2 + y_3 + \ldots + y_n. \quad (9.141)$$

The normal equation turns out to be

$$nx = y_1 + y_2 + y_3 + \ldots + y_n, \quad (9.142)$$

and its obvious solution is

$$x = \frac{y_1 + y_2 + y_3 + \ldots + y_n}{n}, \quad (9.143)$$

that is, the mean value of the results of our measurements. We see that in this case the mathematical approach remarkably agrees with the common-sense approach. We have obtained, however, something more that this nice feeling. Namely, now we know that the mean value of multiple measurements is not only aesthetically attractive, but it also minimizes the *mean square deviation*,

$$\sqrt{(x - y_1)^2 + (x - y_2)^2 + (x - y_3)^2 + \ldots + (x - y_n)^2}. \quad (9.144)$$

b) Consider a more complicated situation. Suppose, we know that some entity z depends on two other entities x and y. We have measured their values, (x, y, z) three times and obtained the data: $(1, 2, 1)$, $(3, 4, 1)$, and $(5, 6, 2)$. We suspect z to be a *linear* function of x and y,

$$z = \alpha x + \beta y, \quad (9.145)$$

and want to find the coefficients α and β. From our experimental data, we obtain the system of equations:

$$\begin{cases} \alpha \cdot 1 + \beta \cdot 2 = 1 \\ \alpha \cdot 3 + \beta \cdot 4 = 1 \\ \alpha \cdot 5 + \beta \cdot 6 = 2 \end{cases}. \quad (9.146)$$

Thus, the two-component column vector of the coefficients satisfies the matrix equation:

$$\begin{pmatrix} 1 & 2 \\ 3 & 4 \\ 5 & 6 \end{pmatrix} \begin{pmatrix} \alpha \\ \beta \end{pmatrix} = \begin{pmatrix} 1 \\ 1 \\ 2 \end{pmatrix}. \quad (9.147)$$

This is again a least-square problem, because the system (9.147) is inconsistent and has no exact solution. All the components of matrix

$$\hat{A} = \begin{pmatrix} 1 & 2 \\ 3 & 4 \\ 5 & 6 \end{pmatrix} \quad (9.148)$$

being real numbers,

$$(\hat{A}^T)^* = (\hat{A}^T) = \begin{pmatrix} 1 & 3 & 5 \\ 2 & 4 & 6 \end{pmatrix}. \tag{9.149}$$

Then, the right-hand side of the normal equation is obtained as

$$\begin{pmatrix} 1 & 3 & 5 \\ 2 & 4 & 6 \end{pmatrix} \begin{pmatrix} 1 \\ 1 \\ 2 \end{pmatrix} = \begin{pmatrix} 14 \\ 18 \end{pmatrix}. \tag{9.150}$$

$$\hat{A}^T \hat{A} = \begin{pmatrix} 1 & 3 & 5 \\ 2 & 4 & 6 \end{pmatrix} \begin{pmatrix} 1 & 2 \\ 3 & 4 \\ 5 & 6 \end{pmatrix} = \begin{pmatrix} 35 & 44 \\ 44 & 56 \end{pmatrix}. \tag{9.151}$$

Thus, the normal equation takes the form:

$$\begin{pmatrix} 35 & 44 \\ 44 & 56 \end{pmatrix} \begin{pmatrix} \alpha \\ \beta \end{pmatrix} = \begin{pmatrix} 14 \\ 18 \end{pmatrix}. \tag{9.152}$$

Its solution gives the desired coefficients in the relation (9.145) as

$$\begin{pmatrix} \alpha \\ \beta \end{pmatrix} = \begin{pmatrix} -\frac{1}{3} \\ \frac{7}{12} \end{pmatrix} = \frac{1}{12} \begin{pmatrix} -4 \\ 7 \end{pmatrix}. \tag{9.153}$$

9.9 PROBLEM OF UNIQUENESS. PSEUDO-INVERSE MAPPING

In the normal equation (9.135), we can consider both the right-hand side and the left-hand side as some linear combinations of the columns of the matrix $(\hat{A}^T)^*$. This means that the rank of the matrix of this linear system, $(\hat{A}^T)^*\hat{A}$, is the same as the rank of the *extended* matrix, that is, the matrix with added column $\hat{A}^T w$, $((\hat{A}^T)^*\hat{A}|\hat{A}^T w)$; this rank is equal to the number of linearly independent columns of the matrix $(\hat{A}^T)^*$. From the standpoint of the systems of linear algebraic equations, this means that the normal equation always has a solution. There is no guarantee, however, that this solution is unique. Moreover, in the general case the least-square solution $\tilde{v}$ is definitely not unique, because we can always add to it any vector from the nullspace $\mathcal{N}(\hat{A})$ (cf. Section 5.6). Thus, we may suspect the criterion for the uniqueness of the least-square solution to be the nullity $\nu(\hat{A}) = 0$, or the rank $\rho(\hat{A}) = n$, the number of the columns in this matrix. More rigorously, this criterion is formulated as the following theorem.

Theorem

The nullspace of the matrix $\hat{A}$ being zero-dimensional, $\mathcal{N}(\hat{A}) = \theta$, is equivalent to the matrix $(\hat{A}^T)^*\hat{A}$ being non-singular. In this case, the solution of the least-square problem of Eq. (9.123) is unique and given by the formula:

$$\tilde{v} = ((\hat{A}^T)^*\hat{A})^{-1}(\hat{A}^T)^*w. \tag{9.154}$$

(This was the case in both examples considered in the previous section.)

Proof

Let us first prove the *necessity*, that is, show that if the nullspace of the matrix $\hat{A}$ is zero-dimensional, then the matrix $(\hat{A}^T)^*\hat{A}$ is nonsingular.

Let v be such an element of the space V that

$$(\hat{A}^T)^*\hat{A}v = \theta, \tag{9.155}$$

the zero element of the same space. We take the inner product,

$$< (\hat{A}^T)^*\hat{A}v, v > = < \theta, v > = 0, \tag{9.156}$$

and transform it (according to formula (8.47)) to

$$< \hat{A}v, \hat{A}v > = 0, \tag{9.157}$$

the inner product of elements of the space W. The only vector having zero inner product with itself is the zero vector, so

$$\hat{A}v = \theta, \tag{9.158}$$

the zero vector of the W-space. Thus, the vector v belongs to the null-space of the matrix $\hat{A}$. But this space is zero-dimensional and consisting of the only zero vector. Thus, the only vector v satisfying equation (9.155) is the zero vector of the V-space. This means that the null-space of the matrix $(\hat{A}^T)^*\hat{A}$ is also zero-dimensional, that is, $\nu((\hat{A}^T)^*\hat{A}) = 0$. This implies that $\rho((\hat{A}^T)^*\hat{A}) = n$. Thus, the matrix $(\hat{A}^T)^*\hat{A}$ is nonsingular. Thus, it is invertible, and the solution of equation (9.123) is obtained using formula (9.154).

Now, we prove the *sufficiency* of the statement of the theorem, that is, we show that if the matrix $(\hat{A}^T)^*\hat{A}$ is nonsingular, then the nullspace of the matrix $\hat{A}$ is zero-dimensional, $\mathcal{N}(\hat{A}) = \theta$.

The matrix $\hat{A}$ being nonsingular implies $\rho((\hat{A}^T)^*\hat{A}) = n$ and $\nu((\hat{A}^T)^*\hat{A}) = 0$. Now, let the vector v' of the space V satisfy equation

$$\hat{A}v' = \theta, \tag{9.159}$$

which means that this vector belongs to the null-space of the matrix $\hat{A}$. We multiply both sides of this equation by the matrix $(\hat{A}^T)^*$:

$$(\hat{A}^T)^*\hat{A}v' = (\hat{A}^T)^*\theta = \theta. \tag{9.160}$$

228 OPERATOR EQUATIONS

This shows that the vector v' also belongs to the nullspace of the matrix $(\hat{A}^T)^* \hat{A}$. But this nullspace is zero-dimensional, so, the only vector satisfying equation (9.159) is the zero vector of the space V. Thus, the nullspace of the matrix $\hat{A}$ is also zero-dimensional, $\mathcal{N}(\hat{A}) = \theta$.

A practical conclusion from this consideration is that we can tell from the very beginning whether the least square solution of the equation (9.123) is unique. We only need to find the rank of the matrix $\hat{A}$, for instance, by reducing the matrix to echelon form. If $\rho(\hat{A}) = n$, the solution is unique, if $\rho(\hat{A}) < n$, it is not.

Pseudo-inverse operator
The problem still remains, what to do in the case where $\rho(\hat{A}) < n$. In this situation, a solution of the normal equation will contain free parameters. To make the solution of the initial problem unique, we need to impose additional conditions. The accepted convention is to take the vector $\tilde{v}_0$ which has minimum norm among all the vectors $\tilde{v}$ being solutions of the normal equation. Since the set of vectors $\tilde{v}$ is closed (each such vector can be produced from one particular solution of the normal equation by adding some vector of a closed subspace, the nullspace of the matrix $\hat{A}$), the minimum-norm vector always exists and is unique.

The mapping of the space W into space V that transforms the element w from the right-hand side of equation (9.123) into vector $\tilde{v}_0$ is called *pseudo-inverse* mapping and denoted as $A^+(w) = \tilde{v}_0$. The operator A^+ is often referred to as the pseudo-inverse operator with respect to the matrix $\hat{A}$. Of course, in the case $\rho(\hat{A}) = n$ the pseudo-inverse mapping just produces the least square solution, so that the pseudo-inverse operator is expressed in the matrix form,

$$\hat{A}^+ = ((\hat{A}^T)^* \hat{A})^{-1} (\hat{A}^T)^*. \tag{9.161}$$

Note, however, that in the general case a simple formula expressing the pseudo-inverse operator in terms of matrix $\hat{A}$ does not exist.

Finally, we just list here, without proof, the most fundamental properties of the pseudo-inverse operator, A^+.

Properties of A^+
a) A^+ is linear.
b) $(A^+)^+ = A$.
c) $AA^+A = A$.

9.10 SAMPLE PROBLEM II

Find the least-square solution of the following inconsistent system:

$$\begin{pmatrix} 1 & 0 & 0 \\ 1 & 1 & 0 \\ 0 & 1 & 1 \\ 0 & 0 & 1 \end{pmatrix} \begin{pmatrix} x_1 \\ x_2 \\ x_3 \end{pmatrix} = \begin{pmatrix} 1 \\ 1 \\ 1 \\ -1 \end{pmatrix}.$$

Solution
The equation has the standard form,

$$\hat{A}\vec{x} = \vec{b}, \tag{9.162}$$

with the rectangular matrix $\hat{A}$.

1) *Uniqueness.* The three columns of the matrix $\hat{A}$, $\vec{a}_1$, $\vec{a}_2$, and $\vec{a}_3$ are evidently independent: in a trivial linear combination,

$$\alpha_1 \cdot \vec{a}_1 + \alpha_2 \cdot \vec{a}_2 + \alpha_3 \cdot \vec{a}_3 = \theta, \tag{9.163}$$

we must take $\alpha_1 = \alpha_3 = 0$ to provide zeros for the first and the last components of the resulting vector; then, α_2 also must be zero because $\vec{a}_2 \neq \theta$. So, the rank of matrix $\hat{A}$ is 3. Thus, there exists a unique least-square solution.

Another (regular) approach: by the standard procedure of elimination (Section 5.7) we can reduce $\hat{A}$ to an echelon matrix, which in this particular case happens to be of especially simple form:

$$\hat{U} = \begin{pmatrix} 1 & 0 & 0 \\ 0 & 1 & 0 \\ 0 & 0 & 1 \\ 0 & 0 & 0 \end{pmatrix}. \tag{9.164}$$

The rank of this matrix, and, thus, the rank of the original matrix $\hat{A}$ is 3.

2) *Normal equation.* The transpose matrix,

$$\hat{A}^T = \begin{pmatrix} 1 & 1 & 0 & 0 \\ 0 & 1 & 1 & 0 \\ 0 & 0 & 1 & 1 \end{pmatrix}. \tag{9.165}$$

The matrix of the left-hand side of the normal equation;

$$\hat{A}^T \hat{A} = \begin{pmatrix} 1 & 1 & 0 & 0 \\ 0 & 1 & 1 & 0 \\ 0 & 0 & 1 & 1 \end{pmatrix} \begin{pmatrix} 1 & 0 & 0 \\ 1 & 1 & 0 \\ 0 & 1 & 1 \\ 0 & 0 & 1 \end{pmatrix} = \begin{pmatrix} 2 & 1 & 0 \\ 1 & 2 & 1 \\ 0 & 1 & 2 \end{pmatrix}. \tag{9.166}$$

The right-hand side of the normal equation:

$$\hat{A}^T \vec{b} = \begin{pmatrix} 1 & 1 & 0 & 0 \\ 0 & 1 & 1 & 0 \\ 0 & 0 & 1 & 1 \end{pmatrix} \begin{pmatrix} 1 \\ 1 \\ 1 \\ -1 \end{pmatrix} = \begin{pmatrix} 2 \\ 2 \\ 0 \end{pmatrix}. \tag{9.167}$$

So, the normal equation takes the form:

$$\begin{pmatrix} 2 & 1 & 0 \\ 1 & 2 & 1 \\ 0 & 1 & 2 \end{pmatrix} \begin{pmatrix} x_1 \\ x_2 \\ x_3 \end{pmatrix} = \begin{pmatrix} 2 \\ 2 \\ 0 \end{pmatrix}. \quad (9.168)$$

3) *Least-square solution.* The matrix of the normal equation containing many zeros, it is easy to find the solution in a special way. From the last equation of the system,

$$x_2 + 2x_3 = 0, \quad (9.169)$$

we obtain

$$x_3 = -\frac{x_2}{2}. \quad (9.170)$$

Then, from the first equation of the system,

$$2x_1 + x_2 = 2, \quad (9.171)$$

we get

$$x_1 = 1 - \frac{x_2}{2}. \quad (9.172)$$

We substitute these results in the second equation of the system, and obtain equation for x_2:

$$\left(1 - \frac{x_2}{2}\right) + 2x_2 - \frac{x_2}{2} = 2, \quad (9.173)$$

which gives

$$x_2 = 1. \quad (9.174)$$

Eventually, we get the least-square solution $\vec{x}$ as:

$$\vec{x} = \begin{pmatrix} x_1 \\ x_2 \\ x_3 \end{pmatrix} = \begin{pmatrix} 1/2 \\ 1 \\ -1/2 \end{pmatrix}. \quad (9.175)$$

9.11 SUMMARY OF CHAPTER 9

1) Operator equations deal with the problem of inverse mapping, that is, how to find an element of the domain which has been transformed into a given element of the codomain.

2) There are three generic problems involving operator equations which are most common in applications. They are: (1) the root-finding problem, (2) the eigenvector problem, and (3) the fixed-point problem.

3) Provided the conditions of the Banach Fixed-Point Theorem are met, the fixed-point problem has a unique solution, which is the limit of the Picard sequence.

4) The Picard iterates are successive approximations to the solution of a fixed-point problem. Taking a sufficient number of the iterates, one can approximate the exact solution with an *a priori* prescribed accuracy.

5) The most popular use of successive approximations is for analyzing and solving ordinary differential equations.

6) The residual approximation approach sets its goal as that of finding the best approximation on a given closed subset of the elements of the domain. This best approximation is achieved by minimizing the norm of the residual.

7) An important particular case of the residual approximation approach is the least-square solution of problems where the exact solution does not exist, for instance, in the analysis of experimental data.

8) In the case of a linear mapping into a space of increased dimension, the least-square solution always exists and satisfies the square-matrix normal equation.

9) If the least-square solution is not unique, the uniqueness can be imposed by the additional requirement for the solution to have minimal norm. The resulting two-step procedure is called pseudo-inverse mapping.

10) There are no general simple formula for the pseudo-inverse mapping. However, this mapping possesses many properties of the usual inverse mapping.

9.12 PROBLEMS

I. Consider the differential equation,

$$\frac{dy}{dt} = \frac{\pi}{4}\sin^2 y; \quad y(0) = \frac{\pi}{6}.$$

Reformulate it as a fixed-point problem and find the first three Picard iterates, $y_0(t)$, $y_1(t)$, and $y_2(t)$.

II. Consider the differential equation,

$$2\frac{dy}{dt} = \frac{ty}{y^2+1}; \quad t \in [0,1], \quad y(0) = 1.$$

Reformulate it as a fixed-point problem, check the convergence of the Picard sequence, and find the first three Picard iterates, $y_0(t)$, $y_1(t)$, and $y_2(t)$.

III. Consider the differential equation,

$$\frac{dy}{dt} = \frac{2t^2}{1+e^y}; \quad t \in [0,1], \quad y(0) = 0.$$

Reformulate it as a fixed-point problem, check the convergence of the Picard sequence on the whole interval of t, and find the first three Picard iterates, $y_0(t)$, $y_1(t)$, and $y_2(t)$.

IV. Find the least-square solution of the following inconsistent system:

$$\begin{pmatrix} 1 & 0 & 1 \\ 0 & 1 & 0 \\ 1 & 0 & 2 \\ 0 & 1 & 0 \end{pmatrix} \begin{pmatrix} x_1 \\ x_2 \\ x_3 \end{pmatrix} = \begin{pmatrix} 2 \\ 1 \\ 1 \\ 2 \end{pmatrix}.$$

V. Consider the incompletely specified matrix equation,

$$\begin{pmatrix} 1 & 0 & 2 \\ 0 & 3 & 0 \\ 4 & 0 & 5 \\ 0 & 6 & 0 \end{pmatrix} \begin{pmatrix} x_1 \\ x_2 \\ x_3 \end{pmatrix} = \begin{pmatrix} 1 \\ 2 \\ 3 \\ 4 \end{pmatrix}.$$

Find the pseudo-inverse solution.

VI. Find the least-square or pseudo-inverse solution (whichever is applicable) of the following inconsistent system:

$$\begin{pmatrix} 1 & 2 & -1 \\ 2 & 1 & 1 \\ 1 & -1 & 2 \\ -1 & 3 & 0 \end{pmatrix} \begin{pmatrix} x_1 \\ x_2 \\ x_3 \end{pmatrix} = \begin{pmatrix} 1 \\ 2 \\ 3 \\ 5 \end{pmatrix}.$$

9.13 FURTHER READING

General issues in operator equations:
 A. N. Kolmogorov, S. V. Fomin, *Elements of the Theory of Functions and Functional Analysis*, v. I, Graylock Press, Rochester, NY, 1957;
 A. Friedman, *Foundations of Modern Analysis*, Holt, Rinehart, and Winston, New York, 1970;
 A. Wouk, *A Course of Applied Functional Analysis*, Wiley, New York, 1979;
 S. Zaidman, *Topics in Abstract Differential Equations*, Longman, Harlow, Essex, U.K., v. 1, 1994, v. 2, 1995;
 V. Lakshmikantham, S. Leela, *Nonlinear Differential Equations in Abstract Spaces*, Pergamon, Oxford, U.K., 1981;
 R. H. Martin, *Nonlinear Operators and Differential Equations in Banach Spaces*, Wiley, New York, 1976.
 Applications to solving differential equations:

V. I. Arnold, *Ordinary Differential Equations*, Springer-Verlag, Berlin, 1992;

F. Bauer, J. A. Nohel, *The Qualitative Theory of Ordinary Differential Equations*, W. A. Benjamin, New York, 1969;

J. L. Brenner, *Problems in Differential Equations*, W. H. Freeman, San Francisco, 1963;

P. Biler, T. Nadzieja, *Problems and Examples in Differential Equations*, Marcel Dekker, New York, 1992.

The method of successive approximations and its variants:

W. M. Patterson, *Iterative Methods for the Solution of a Linear Operator Equation in Hilbert Space*, Springer-Verlag, Berlin, 1974;

V. K. Dzyadyk, *Approximation Methods for Solutions of Differential and Integral Equations*, VSP, Utrecht, The Netherlands, 1995;

C. Den Heijer, *The numerical solution of nonlinear operator equations*, Mathematisch Centrum, Amsterdam, 1979.

Minimal residual approximations and related topics:

C. W. Groetsch, *Generalized Inverses of Linear Operators: Representation and Approximation*, Marcel Dekker, New York, 1977;

V. P. Tanana, *Methods for Solution of Nonlinear Operator Equations*, VSP, Utrecht, The Netherlands, 1997;

D. G. Luenberger, *Optimization by Vector Space Methods*, Wiley, New York, 1969.

Least-square and pseudo-inverse methods:

A. Ben-Israel, T. N. E. Greville, *Generalised Inverses: Theory and Applications*, Wiley, New York, 1974;

B. Noble, J. W. Daniel, *Applied Linear Algebra*, Prentice-Hall, Englewood Cliffs, NJ, 1977;

G. H. Golub, C. F. Van Loan, *Matrix Computations*, Johns Hopkins University Press, Baltimore, 1996;

C. L. Lawson, R. J. Hanson, *Solving Least Square Problems*, Prentice-Hall, Englewood Cliffs, NJ, 1981.

10
Fourier and Laplace Transforms

10.1 FOURIER INTEGRAL

Fourier series — reminiscence
In Chapter 8, we have introduced an orthonormal basis in the infinite-dimensional Hilbert space of functions of the real variable t on the interval $[0, 1]$. This infinite basis is given by the functions

$$e_n(t) = e^{2\pi i n t}, \tag{10.1}$$

where n is arbitrary integer. Correspondingly, any function $f(t)$ which is an element of the space is represented by means of the *Fourier series*,

$$f(t) = \sum_{n=-\infty}^{\infty} C_n e_n(t), \tag{10.2}$$

with coefficients

$$C_n = <f(t), e_n(t)> = \int_0^1 dt f(t) e^{-2\pi i n t}. \tag{10.3}$$

This representation is just the decomposition of the function over the given basis (10.1).

236   FOURIER AND LAPLACE TRANSFORMS

The formulas (10.2), (10.3) can also be applied to functions defined on the entire real axis, but periodic with period 1. In this case, it is sufficient to write the decomposition for the interval $[0, 1]$ and then to extent it periodically on the entire axis t. In this chapter, we address a more challenging question, whether it is possible to write an analog of the formulas (10.2), (10.3) for a *non-periodic* function $f(t)$ defined on the entire axis. We attack the problem in two steps. First, we will modify the formulas (10.2), (10.3) to fit an arbitrary interval of the real numbers. Then, we will extend this arbitrary interval to the entire axis and see the drastic change in the form and meaning of the formulas when passing to this limit.

Arbitrary symmetric interval

Let us write the Fourier decomposition of a function $f(t)$ defined on an arbitrary *symmetric* interval, $t \in [-\frac{T}{2}, \frac{T}{2}]$. To do this, we just use formulas (10.2) and (10.3), and change the variable in them in such a way that when the old variable takes the values from 0 to 1, the new variable takes the values from $-T/2$ to $T/2$. Namely, let $t = T(s - \frac{1}{2})$, $s \in [0, 1]$. Then, the arbitrary function on the interval $[-\frac{T}{2}, \frac{T}{2}]$,

$$f(t) = f\left(T\left(s - \frac{1}{2}\right)\right) = \tilde{f}(s), \tag{10.4}$$

is a function of s. The Fourier coefficients of this function $\tilde{f}(s)$ are

$$C_n = \int_0^1 ds\, \tilde{f}(s) e^{-2\pi i n s} = \frac{1}{T}\int_{-\frac{T}{2}}^{\frac{T}{2}} dt\, f(t) e^{-2\pi i n(\frac{t}{T} + \frac{1}{2})} = e^{-\pi i n} \tilde{C}_n, \tag{10.5}$$

where

$$\tilde{C}_n = \frac{1}{T}\int_{-\frac{T}{2}}^{\frac{T}{2}} dt\, f(t) e^{-2\pi i n \frac{t}{T}}. \tag{10.6}$$

Now, for the expression of the function $f(t)$ through its Fourier coefficients we have:

$$\begin{aligned}f(t) &= \sum_{n=-\infty}^{\infty} C_n e^{2\pi i n s} = \sum_{n=-\infty}^{\infty} \tilde{C}_n e^{-\pi i n} e^{2\pi i n(\frac{t}{T}+\frac{1}{2})} \\ &= \sum_{n=-\infty}^{\infty} \tilde{C}_n e^{-\pi i n} e^{2\pi i n \frac{t}{T}} e^{\pi i n} = \sum_{n=-\infty}^{\infty} \tilde{C}_n e^{2\pi i n \frac{t}{T}}.\end{aligned} \tag{10.7}$$

Infinite interval

Let us see now what happens when $T \to \infty$, that is, when we consider the function $f(t)$ defined for all t. In trying to do so straightforwardly, however, we face at once some difficulties. On the one hand, we have to demand that the function $f(t)$ be such that the integral in expression (10.6) remains finite as $T \to \infty$. On the other hand, this integral being finite, we immediately get $\tilde{C}_n \to 0$ at $T \to \infty$. Thus, representation

(10.2) becomes an infinite sum of vanishing terms. To avoid this ambiguity, let us redefine the Fourier coefficients as:

$$\bar{C}_n = T\tilde{C}_n = \int_{-\frac{T}{2}}^{\frac{T}{2}} dt f(t) e^{-2\pi i \frac{n}{T} t}, \tag{10.8}$$

so that formula (10.2) becomes

$$f(t) = \sum_{n=-\infty}^{\infty} \bar{C}_n \frac{1}{T} e^{2\pi i \frac{n}{T} t}. \tag{10.9}$$

This seemingly insignificant transfer of the factor $1/T$ from one formula to the other radically improves the situation. Now, the meaning of formula (10.9) at $T \to \infty$ is that instead of the discrete sum of formula (10.2) we have an integral sum, the terms of which vary vanishingly little with changing n. To show this explicitly, we denote:

$$\frac{n}{T} = \omega_n; \quad \frac{1}{T} \to \Delta\omega. \tag{10.10}$$

Then, the formula (10.9) is transformed into

$$f(t) = \sum_{n=-\infty}^{\infty} \bar{C}_{\omega_n} \Delta\omega e^{2\pi i \omega_n t}. \tag{10.11}$$

At this point, we are able to proceed to the limit, $\Delta\omega \to 0$. Then, ω_n is transformed into a *continuous* variable ω, and the sum in (10.11) is transformed into an integral representation:

$$f(t) = \int_{-\infty}^{\infty} d\omega \bar{C}_\omega e^{2\pi i \omega t}, \tag{10.12}$$

with $\bar{C}_\omega$ obtained from Eq. (10.8) as

$$\bar{C}_\omega = \int_{-\infty}^{\infty} dt f(t) e^{-2\pi i \omega t}. \tag{10.13}$$

Formulas (10.12) and (10.13) are often referred as a *Fourier pair*, or as the direct (10.13) and inverse (10.12) *Fourier transforms.*

10.2 THE FOURIER TRANSFORM AS A MAPPING

Note, that the limit transition we have performed in the previous section cardinally changed the meaning of the formulas. Originally, in formulas (10.2) and (10.3), we had the decomposition of function $f(t)$ over the infinite orthonormal basis $e_n(t)$. Now, in formulas (10.12) and (10.13), we have mutual transformations of two functions of continuous variables, $f(t)$ and $\bar{C}_\omega$. This means that the formulas (10.12) and (10.13) perform mappings between a set of functions of the real variable t on an infinite interval and the set of functions of real variable ω on an infinite interval. (We assume, of course, that functions $f(t)$ and $\bar{C}_\omega$ are functions that provide convergence of the integrals in both (10.12) and (10.13).)

Both the domain and codomain of the mapping (10.13) are linear spaces. Moreover, we can show that the mapping (10.13) is a linear mapping. Let us take two Fourier-transformable functions, $f_1(t)$ and $f_2(t)$. Their Fourier transforms are $\bar{C}_{\omega 1}$ and $\bar{C}_{\omega 2}$, respectively. Then, for a linear combination, $\alpha_1 \cdot f_1(t) + \alpha_2 \cdot f_2(t)$, the Fourier transform is obtained as

$$\begin{aligned} \bar{C}_\omega &= \int_{-\infty}^{\infty} dt (\alpha_1 \cdot f_1(t) + \alpha_2 \cdot f_2(t)) e^{-2\pi i \omega t} \\ &= \alpha_1 \int_{-\infty}^{\infty} dt\, f_1(t) e^{-2\pi i \omega t} + \alpha_2 \int_{-\infty}^{\infty} dt\, f_2(t) e^{-2\pi i \omega t} \\ &= \alpha_1 \bar{C}_{\omega 1} + \alpha_2 \bar{C}_{\omega 2}. \end{aligned} \quad (10.14)$$

This Fourier transform is the linear combination of the Fourier transforms of the functions $f_1(t)$ and $f_2(t)$ with the same coefficients. Thus, the Fourier transform is a linear mapping, according to the definition of Section 4.5.

We can also show that the Fourier transform is an injection, that is, that the elements of a Fourier pair uniquely correspond to each other. On the one hand, the unique correspondence of $\bar{C}_\omega$ and $f(t)$ is clear from $\bar{C}_\omega$ being developed from the unique coefficients of the decomposition $\tilde{C}_n$. On the other hand, we did this development is a somewhat heuristic way. So, it is worthwhile to check the fact directly by restoring the original function $f(t)$ through the inverse transform of formula (10.12), having taken $\bar{C}_\omega$ from formula (10.13):

$$\begin{aligned} f(t) &= \int_{-\infty}^{\infty} d\omega\, e^{2\pi i \omega t} \int_{-\infty}^{\infty} dt'\, f(t') e^{-2\pi i \omega t'} \\ &= \int_{-\infty}^{\infty} dt'\, f(t') \int_{-\infty}^{\infty} d\omega\, e^{2\pi i \omega (t-t')} \\ &= \int_{-\infty}^{\infty} dt'\, f(t') \delta(t-t'), \end{aligned} \quad (10.15)$$

where we introduced a new function,

$$\delta(x) = \int_{-\infty}^{\infty} d\omega\, e^{2\pi i \omega x}. \quad (10.16)$$

10.3 DIGRESSION. PROPERTIES OF THE δ-FUNCTION

It is hard to say something definite about the δ-function at first glance, because the integral in formula (10.16) seems to be undefined. Indeed, the integration gives

$$\int_{-\infty}^{\infty} d\omega e^{2\pi i \omega x} = \frac{1}{2\pi i x} e^{2\pi i \omega x} \bigg|_{\omega=-\infty}^{\omega=\infty}$$

$$= \frac{1}{2\pi i x} \left(e^{2\pi i x \infty} - e^{-2\pi i x \infty} \right) = \frac{\sin(2\pi x \infty)}{\pi x}. \quad (10.17)$$

But the sine endlessly oscillates between $+1$ and -1 while its argument goes to infinity. Thus, we cannot ascribe any value to this expression.

We can, however, treat the integral in (10.16) as a result of some passage to the limit. The idea is the following: let us make the integral in (10.16) convergent, by adding to the integrand a factor $\exp{-\gamma|\omega|}$ which vanishes at $|\omega| \to \infty$ and impose that the integrand vanishes as well:

$$\delta_\gamma(x) = \int_{-\infty}^{\infty} d\omega e^{2\pi i \omega x - \gamma|\omega|}. \quad (10.18)$$

Here γ is a positive number, which can be arbitrarily small. Upon evaluating the integral in (10.18), we will let $\gamma \to 0$. Indeed, the integral in (10.18) is now easy to evaluate,

$$\begin{aligned}\delta_\gamma(x) &= \int_{-\infty}^{0} d\omega e^{2\pi i \omega x - \gamma|\omega|} + \int_{0}^{\infty} d\omega e^{2\pi i \omega x - \gamma|\omega|} \\ &= \frac{e^{2\pi i \omega x + \gamma \omega}}{2\pi i x + \gamma} \bigg|_{-\infty}^{0} + \frac{e^{2\pi i \omega x - \gamma \omega}}{2\pi i x - \gamma} \bigg|_{0}^{\infty} \\ &= \frac{1}{2\pi i x + \gamma} - \frac{1}{2\pi i x - \gamma} = \frac{-2\gamma}{(2\pi i x)^2 - (\gamma)^2} \\ &= \frac{2\gamma}{(2\pi x)^2 + (\gamma)^2}. \end{aligned} \quad (10.19)$$

This function is displayed in Fig. 10.1 for several values of the parameter γ. Thus,

$$\delta(x) = \lim_{\gamma \to 0} (\delta_\gamma(x)) = \lim_{\gamma \to 0} \left(\frac{2\gamma}{(2\pi x)^2 + (\gamma)^2} \right). \quad (10.20)$$

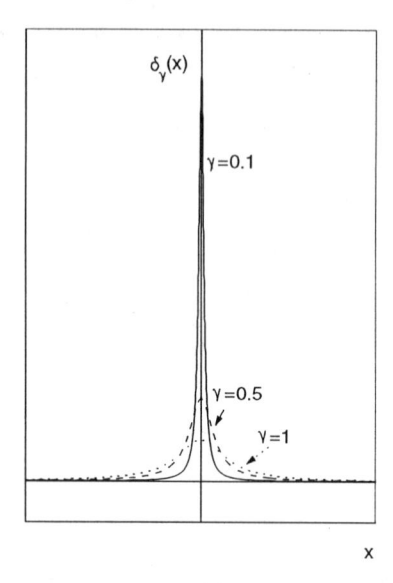

Fig. 10.1 The representation of the δ-function.

This formula, however, seems to be of a little help, since it defines a very strange function: at any point $x \neq 0$ the function is zero, while at $x = 0$ it is infinitely large! The remarkable feature of the function defined by (10.20) and its predecessor of Eq. (10.19) is that at any value of γ some invariant holds, namely,

$$\int_{-\infty}^{\infty} \delta_\gamma(x) dx = \int_{-\infty}^{\infty} dx \frac{2\gamma}{(2\pi x)^2 + (\gamma)^2}$$

$$= \frac{1}{\pi} \int_{-\infty}^{\infty} dy \frac{1}{y^2 + 1} = \frac{1}{\pi} \tan^{-1} y \bigg|_{-\infty}^{\infty} = 1, \qquad (10.21)$$

where we used the substitution

$$y = \frac{2\pi x}{\gamma}. \qquad (10.22)$$

Now, since $\delta(x) = 0$ at all the points but $x = 0$, the result (10.21) provides a very peculiar (and very useful) feature of the δ-function: for any continuous function $f(x)$,

$$\int_{-\infty}^{\infty} f(x)\delta(x) dx = f(0). \qquad (10.23)$$

As we see, the δ-function can not really be perceived as a function in the regular meaning of the word. In fact, it is not a function in *any* interpretation. It makes sense only in integral formulas like (10.23). This is the first and the typical example of the so-called *singularity functions*, alias *generalized functions*, alias *distributions*.

10.4 BACK TO FOURIER TRANSFORM MAPPING

Now we can make use of the main property of the δ-function, expressed in formula (10.23), and obtain in Eq. (10.15):

$$\int_{-\infty}^{\infty} dt' f(t')\delta(t-t') = f(t). \tag{10.24}$$

This result finally proves that the Fourier transform is a *one-to-one* mapping. Then, the very symmetry of the pair (10.12) and (10.13) shows that the Fourier Transform is a *bijection* between the space of complex-valued absolutely integrable functions of real variable t and the space of complex-valued absolutely integrable functions of real variable ω.

Notation
To underline the symmetry we just mentioned, we will denote the Fourier transform of a function $f(t)$ as F_ω, and, when convenient, write this transform in operator form:

$$F_\omega = \hat{F}(f(t)) = \int_{-\infty}^{\infty} dt\, f(t) e^{-2\pi i \omega t}; \tag{10.25}$$

$$f(t) = \hat{F}^{-1}(F_\omega) = \int_{-\infty}^{\infty} d\omega\, F_\omega e^{2\pi i \omega t}. \tag{10.26}$$

10.5 PROPERTIES OF THE FOURIER TRANSFORM

In this section we consider those basic properties of the Fourier transform that make this mapping so useful in applications.

The direct and inverse Fourier transforms are linear mappings
This property simply follows from the definitions (10.12) and (10.13). We have already shown the linearity of the direct Fourier transform in Eq. (10.14). Now we merely rewrite the property in the new notation:

$$\hat{F}(\alpha_1 f_1(t) + \alpha_2 f_2(t)) = \alpha_1 \hat{F}(f_1(t)) + \alpha_2 \hat{F}(f_2(t)), \tag{10.27}$$

for any Fourier-transformable functions, $f_1(t)$ and $f_2(t)$, and arbitrary complex numbers, α_1 and α_2. The reader is invited to show the linearity of the inverse Fourier

242 FOURIER AND LAPLACE TRANSFORMS

transform, following the same approach.

Inner-product conservation
Both the "t-space" and the "ω-space" of functions are Hilbert spaces with the inner products properly defined, say, by formula (8.33). The question is whether these two inner products are related, and, if so, what is this relation. To give the answer, we consider the inner product of two functions, $f_1(t)$ and $f_2(t)$:

$$<f_1, f_2> = \int_{-\infty}^{\infty} dt \, f_1(t) f_2^*(t). \tag{10.28}$$

In this formula, we substitute for f_1 and f_2 their expressions in terms of $F_1(\omega)$ and $F_2(\omega)$ through the inverse Fourier transform:

$$<f_1, f_2> = \int_{-\infty}^{\infty} dt \left(\int_{-\infty}^{\infty} d\omega \, F_1(\omega) e^{2\pi i \omega t} \right) \left(\int_{-\infty}^{\infty} d\omega' \, F_2^*(\omega') e^{-2\pi i \omega' t} \right). \tag{10.29}$$

(We use different notations for the dummy variables in the inverse-transform integrals). In this expression all the integrals duly converge, so we can change the order of integrations. We note that the factors depending on t do not relate to the functions $F_1(\omega)$ and $F_2(\omega)$ whatsoever. Thus, it is natural to collect these functions containing t in one universal multiplier:

$$<f_1, f_2> = \int_{-\infty}^{\infty} d\omega \, F_1(\omega) \int_{-\infty}^{\infty} d\omega' \, F_2^*(\omega') \int_{-\infty}^{\infty} dt \, e^{2\pi i \omega t} e^{-2\pi i \omega' t}$$

$$= \int_{-\infty}^{\infty} d\omega \, F_1(\omega) \int_{-\infty}^{\infty} d\omega' \, F_2^*(\omega') \int_{-\infty}^{\infty} dt \, e^{2\pi i t (\omega - \omega')}. \tag{10.30}$$

Then, we joyfully recognize in the last integral our good friend, $\delta(\omega - \omega')$. As usual, the presence of the δ-function allows us to easily compute one of the remaining integrals:

$$<f_1, f_2> = \int_{-\infty}^{\infty} d\omega \, F_1(\omega) \int_{-\infty}^{\infty} d\omega' \, F_2^*(\omega') \delta(\omega - \omega')$$

$$= \int_{-\infty}^{\infty} d\omega \, F_1(\omega) F_2^*(\omega). \tag{10.31}$$

But this is just the expression for $<F_1, F_2>$. So, we get to a remarkable result: the two inner products not only are related, but they are equal:

$$<f_1(t), f_2(t)> = <F_1(\omega), F_2(\omega)>. \tag{10.32}$$

This important relation once more emphasizes the fundamental equality of the t- and ω-spaces.

The relation of Eq. (10.32) has at least two important corollaries.

a) *The Fourier transform preserves the norm of a function.*
Indeed, the norm being the square root of the inner product of a function and itself, we obtain from Eq. (10.32):

$$< f(t), f(t) > = < F(\omega), F(\omega) >, \qquad (10.33)$$

hence,

$$\|f(t)\| = \|F(\omega)\|. \qquad (10.34)$$

b) *The Fourier transform preserves the equality of functions.*
We come to this conclusion in the following way; if two functions are equal in the t-space,

$$f_1(t) = f_2(t), \qquad (10.35)$$

then

$$f_1(t) - f_2(t) = 0. \qquad (10.36)$$

This implies

$$< (f_1(t) - f_2(t)), (f_1(t) - f_2(t)) > = 0. \qquad (10.37)$$

By virtue of Eq. (10.32) this means

$$< (F_1(\omega) - F_2(\omega)), (F_1(\omega) - F_2(\omega)) > = 0. \qquad (10.38)$$

Then, according to the second axiom of the inner product, Eq. (10.38) is tantamount to

$$F_1(\omega) - F_2(\omega) = 0 \qquad (10.39)$$

and, eventually, to

$$F_1(\omega) = F_2(\omega). \qquad (10.40)$$

The statement of this corollary seems natural and even trivial, but it has far-reaching consequences. It means that any equation written in t-space can be transformed to (and solved in) ω-space. Based on this property, solving differential and integral equations is a traditional (and a major) application of the Fourier transform.

Change of time scale

Let us see what happens with the Fourier transform of a function $f(t)$ when we change the scale of variable t to αt, that is, stretch (if $\alpha > 1$) or compress (if $\alpha < 1$) this variable. We apply formula (10.12) to the function of the modified variable, $f(\alpha t)$:

$$\hat{F}(f(\alpha t)) = \int_{-\infty}^{\infty} dt\, f(\alpha t) e^{-2\pi i \omega t}. \tag{10.41}$$

Our goal is to express this in terms of F_ω. To do so, we change the dummy variable of integration:

$$\hat{F}(f(\alpha t)) = \frac{1}{\alpha} \int_{-\infty}^{\infty} d(\alpha t) f(\alpha t) e^{-2\pi i \alpha t (\omega/\alpha)}$$

$$= \frac{1}{\alpha} \int_{-\infty}^{\infty} dt'\, f(t') e^{-2\pi i t' (\omega/\alpha)}, \tag{10.42}$$

where $t' = \alpha t$. Comparing Eq. (10.42) with Eq. (10.12), we see that

$$\hat{F}(f(\alpha t)) = \frac{1}{\alpha} F_{\omega/\alpha}. \tag{10.43}$$

Thus, stretching of the time scale leads to a reciprocal compressing of the frequency scale, and vice versa.

Time shifting
Let us see now what happens with the Fourier transform if the argument t is shifted, $t \to t - t_0$. We again straightforwardly apply formula (10.12), this time to the function with shifted argument, $f(t - t_0)$:

$$\hat{F}(f(t - t_0)) = \int_{-\infty}^{\infty} dt\, f(t - t_0) e^{-2\pi i \omega t}. \tag{10.44}$$

Then we change the dummy variable,

$$\hat{F}(f(t - t_0)) = \int_{-\infty}^{\infty} d(t - t_0) f(t - t_0) e^{-2\pi i \omega (t - t_0 + t_0)}$$

$$= \int_{-\infty}^{\infty} dt'\, f(t') e^{-2\pi i \omega (t' + t_0)}$$

$$= e^{-2\pi i \omega t_0} \int_{-\infty}^{\infty} dt'\, f(t') e^{-2\pi i \omega t'} = e^{-2\pi i \omega t_0} F_\omega. \tag{10.45}$$

PROPERTIES OF THE FOURIER TRANSFORM 245

Thus, the translation of a function by t_0 results in its Fourier transform acquiring the phase factor, $e^{-2\pi i \omega t_0}$:

$$\hat{F}(f(t-t_0)) = e^{-2\pi i \omega t_0} F_\omega. \tag{10.46}$$

Of course, Eq. (10.46) can be read in the inverse direction, too: if a Fourier transform has a constant phase factor, this results in the argument shifting in the original function.

Frequency shifting
This property is the counterpart to the previous one. We are interested in what happens with the Fourier transform if the original function gets multiplied by a *phasor*,

$$f(t) \longrightarrow e^{2\pi i \omega_0 t} f(t). \tag{10.47}$$

The Fourier transform of the modified function is found as

$$\begin{aligned}
\hat{F}\left(e^{2\pi i \omega_0 t} f(t)\right) &= \int_{-\infty}^{\infty} dt\, e^{2\pi i \omega_0 t} f(t) e^{-2\pi i \omega t} \\
&= \int_{-\infty}^{\infty} dt\, f(t) e^{2\pi i \omega_0 t - 2\pi i \omega t} \\
&= \int_{-\infty}^{\infty} dt\, f(t) e^{-2\pi i (\omega - \omega_0) t} = F_{\omega - \omega_0}. \tag{10.48}
\end{aligned}$$

The multiplication of a function by a phasor causes the Fourier transform frequency to be shifted by the frequency of the phasor.

The Fourier transform of derivatives
The following property might be counted as the most important, or, at least, most often used in applications. We intend to express the Fourier transform of the first derivative of a function in terms of the Fourier transform of the function itself:

$$\hat{F}\left(\frac{df(t)}{dt}\right) = \int_{-\infty}^{\infty} dt\, \frac{df(t)}{dt} e^{-2\pi i \omega t}. \tag{10.49}$$

To get from here to something similar to formula (10.12), we use integration by parts:

$$\int_{-\infty}^{\infty} dt\, \frac{df(t)}{dt} e^{-2\pi i \omega t}$$

$$= f(t) e^{-2\pi i \omega t} \Big|_{t=-\infty}^{t=\infty} - \int_{-\infty}^{\infty} dt\, f(t) \frac{d}{dt}\left(e^{-2\pi i \omega t}\right)$$

$$= f(t)e^{-2\pi i\omega t}\Big|_{t=-\infty}^{t=\infty} - \int_{-\infty}^{\infty} dt f(t)(-2\pi i\omega)e^{-2\pi i\omega t}$$

$$= f(t)e^{-2\pi i\omega t}\Big|_{t=-\infty}^{t=\infty} + (2\pi i\omega)\int_{-\infty}^{\infty} dt f(t)e^{-2\pi i\omega t}. \quad (10.50)$$

The non-integral term is equal to zero, because the Fourier-transformable function $f(t)$ definitely vanishes when t goes to ∞ or $-\infty$. The integral just gives the Fourier transform of the function $f(t)$. Thus, we get to the important formula:

$$\hat{F}\left(\frac{df(t)}{dt}\right) = (2\pi i\omega)\hat{F}(f(t)). \quad (10.51)$$

The Fourier transform of the derivative just equals the Fourier transform of the function times $2\pi i\omega$.

Now, it is quite understandable why this formula is so useful. Indeed, it can transform, as in a fairy tale, an ordinary differential equation of a high order into a simple algebraic equation! To emphasize this use (and usefulness), we generalize the formula (10.51) to the case of n-th derivative:

$$\hat{F}\left(\frac{d^n f(t)}{dt^n}\right) = (2\pi i\omega)\hat{F}\left(\frac{d^{n-1} f(t)}{dt^{n-1}}\right)$$

$$= (2\pi i\omega)^2 \hat{F}\left(\frac{d^{n-2} f(t)}{dt^{n-2}}\right) = \ldots = (2\pi i\omega)^n \hat{F}(f(t)). \quad (10.52)$$

These basic properties of the Fourier transform constitute a handy set of tools which allow us to compute Fourier transforms of complicated functions and to use the results in various applications.

10.6 APPLICATION TO LINEAR SYSTEMS. THE CONVOLUTION PRODUCT

One of the major fields where Fourier transforms are applied is in signal processing for linear systems. In many areas of engineering, special systems are designed to manipulate various signals, that is, to transform an input function of time, $f_{in}(t)$ into an output function of time, $f_{out}(t)$ (see Fig. 10.2). From our mapping-oriented viewpoint, such a transformation is a mapping L of a linear space of functions on the interval $(-\infty, \infty)$ into itself:

$$f_{out}(t) = l(f_{in}(t)). \quad (10.53)$$

It is often the case that the system under consideration is a *linear system*, that is, the mapping of Eq. (10.53) is a linear mapping:

$$l(\alpha_1 f_{in1}(t) + \alpha_2 f_{in2}(t)) = \alpha_1 l(f_{in1}(t)) + \alpha_2 l(f_{in2}(t)), \quad (10.54)$$

APPLICATION TO LINEAR SYSTEMS. THE CONVOLUTION PRODUCT   247

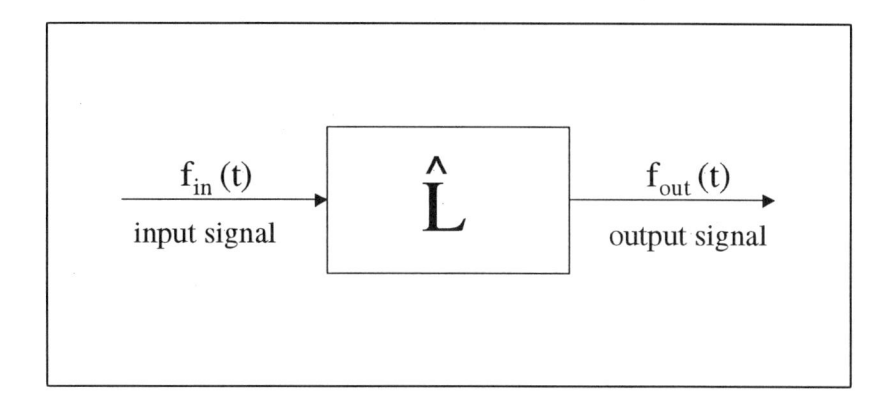

Fig. 10.2   Signal transformation by a linear system.

where α_1 and α_2 are arbitrary constants. Then, an action of a linear system on an input function $f_{in}(t)$ can be written in some general form. To obtain this form, assume that we are dealing with functions determined on a finite interval, say, $t \in [0, 1]$, and use the same heuristic approach as we did in Chapter 6 when deriving the expression of metric distance in the space of functions.

We suppose that both of the functions $f_{in}(t)$ and $f_{out}(t)$ are *sampled*, that is, recorded at discrete times $t_n = n/N$, where $n = 0, 1, 2, 3, \ldots, N$. Then these functions turn out to be N-component vectors, and the system under consideration just maps one of these vectors into another. But we know quite well (from Chapter 5) that such a linear mapping between vectors can be expressed as the action of an $N \times N$ matrix:

$$\begin{pmatrix} f_{out}(0) \\ f_{out}(1/N) \\ f_{out}(2/N) \\ \vdots \\ f_{out}((N-1)/N) \\ f_{out}(1) \end{pmatrix} = \begin{pmatrix} l_{11} & l_{12} & l_{13} & \cdots & l_{1,N-1} & l_{1N} \\ l_{21} & l_{22} & l_{23} & \cdots & l_{2,N-1} & l_{2N} \\ l_{31} & l_{32} & l_{33} & \cdots & l_{3,N-1} & l_{3N} \\ \vdots & \vdots & \vdots & \ddots & \cdots & \vdots \\ l_{N-1,1} & l_{N-1,2} & \cdots & \cdots & \cdots & l_{1N} \\ l_{N1} & l_{N2} & \cdots & \cdots & \cdots & l_{NN} \end{pmatrix} \times \begin{pmatrix} f_{in}(0) \\ f_{in}(1/N) \\ f_{in}(2/N) \\ \vdots \\ f_{in}((N-1)/N) \\ f_{in}(1) \end{pmatrix}. \quad (10.55)$$

This means that at any of the moments t_n,

$$f_{out}(t_n) = \sum_{n'=0}^{N} l_{nn'} f_{in}(t_{n'}). \tag{10.56}$$

As the limit of this expression at $N \to \infty$ we obtain

$$f_{out}(t) = \int_0^1 dt' L(t, t') f_{in}(t'), \tag{10.57}$$

where $L(t, t')$, a function of two variables, is called the *kernel*.

The above consideration has not used any specific features of the interval $[0, 1]$. Thus, the result (10.57) can be extended to any interval, just by putting in the corresponding limits on the integral. Considering wider and wider intervals, we eventually come to the formula for the entire axis of real numbers,

$$f_{out}(t) = \int_{-\infty}^{\infty} dt' L(t, t') f_{in}(t'). \tag{10.58}$$

This is the most general form of a *linear operator* describing the action of a linear system in the space of functions.

An important modification can be made in formula (10.58) in the case of *time-invariant* systems, that is, systems whose properties do not change with time. If the same input signal passes through such a system at two different times, the same output signal appears at both times. Expressed mathematically, this means

$$f_{out}(t + T) = \int_{-\infty}^{\infty} dt' L(t, t') f_{in}(t' + T), \tag{10.59}$$

where T is an arbitrary time shift. In particular, putting $t = 0$ in Eq. (10.59), we have

$$f_{out}(T) = \int_{-\infty}^{\infty} dt' L(0, t') f_{in}(t' + T). \tag{10.60}$$

Note now, that in this latter formula the meaning of the shift T is also slightly changed. Here, $f_{out}(T)$ is just function at an arbitrary moment T, that is, T has taken the role of the time variable t,

$$f_{out}(t) = \int_{-\infty}^{\infty} dt' L(0, t') f_{in}(t' + t). \tag{10.61}$$

The considerable improvement, compared to Eq. (10.58), is that now the kernel is effectively a function of *one* variable, $L(0, t') = L(t')$. Finally, to rewrite (10.60) in the conventional form, we change the variable of integration, $t'' = t + t'$:

$$f_{out}(t) = \int_{-\infty}^{\infty} dt'' \, K(t - t'') f_{in}(t''), \tag{10.62}$$

where the kernel is

$$K(t - t'') = L(0, t'' - t). \tag{10.63}$$

The expression (10.62) is the general form of input–output relation for a time-invariant system. The final touch is to mention an important property of the kernel $K(t - t'')$ called *causality*: the fact that a system cannot physically affect an input signal at future moments causes $K(t - t'') = 0$ at $t < t''$.

The integral in Eq. (10.62) is encountered so often in signal processing and other fields that it has gained a special name. It is called the *convolution product* of the functions $K(t)$ and $f_{in}(t)$.

Now, let us see, what use we can make of the Fourier transform in this regard. In Eq. (10.62) we express the functions $K(t - t'')$ and $f_{in}(t'')$ in terms of their Fourier transforms,

$$K(t - t'') = \int_{-\infty}^{\infty} d\omega \, e^{2\pi i \omega(t-t'')} K_\omega;$$

$$f_{in}(t'') = \int_{-\infty}^{\infty} d\omega' \, e^{2\pi i \omega' t''} F_{\omega' \, in}. \tag{10.64}$$

Substituting these expressions in (10.62), we have

$$f_{out}(t) = \int_{-\infty}^{\infty} dt'' \int_{-\infty}^{\infty} d\omega \, e^{2\pi i \omega(t-t'')} K_\omega \int_{-\infty}^{\infty} d\omega' \, e^{2\pi i \omega' t''} F_{\omega' \, in}. \tag{10.65}$$

Then we change the order of integration:

$$\begin{aligned}
f_{out}(t) &= \int_{-\infty}^{\infty} d\omega \, K_\omega \int_{-\infty}^{\infty} d\omega' \, F_{\omega' \, in} \int_{-\infty}^{\infty} dt'' \, e^{2\pi i \omega(t-t'')} e^{2\pi i \omega' t''} \\
&= \int_{-\infty}^{\infty} d\omega \, K_\omega e^{2\pi i \omega t} \int_{-\infty}^{\infty} d\omega' \, F_{\omega' \, in} \int_{-\infty}^{\infty} dt'' \, e^{2\pi i (\omega' - \omega) t''} \\
&= \int_{-\infty}^{\infty} d\omega \, K_\omega e^{2\pi i \omega t} \int_{-\infty}^{\infty} d\omega' \, F_{\omega' \, in} \delta(\omega' - \omega) \\
&= \int_{-\infty}^{\infty} d\omega \, K_\omega e^{2\pi i \omega t} F_{\omega \, in}
\end{aligned}$$

$$= \int_{-\infty}^{\infty} d\omega e^{2\pi i \omega t} K_\omega F_{\omega\ in}. \tag{10.66}$$

Thus, the expression for the output signal $f_{out}(t)$ takes the form of the inverse Fourier transform of the product $K_\omega F_{\omega\ in}$. Therefore, the Fourier transform of this output signal is just the product itself:

$$F_{\omega\ out} = K_\omega F_{\omega\ in}. \tag{10.67}$$

The Fourier transform of the convolution product of two functions is equal to the product of their Fourier transforms.

Formula (10.67) is the basis and essence of all linear signal processing. Its meaning is that a linear system acts on a signal as a *frequency filter*: by means of its *window function*, K_ω, it changes both the amplitude and phase of the signal frequency components, so that some frequencies are suppressed while the others are enhanced.

Later, in Chapter 11, we will introduce other important implications of the remarkable formula (10.67).

10.7 SAMPLE PROBLEM I: SMOOTHING OPERATOR

Consider a linear system whose action on a signal is to produce a running average of the signal over a window of size T:

$$f_{out} = \frac{1}{T} \int_{t-T/2}^{t+T/2} dt'\, f_{in}(t'). \tag{10.68}$$

Write the kernel of this transformation, $K(t)$, and find the corresponding frequency filter, K_ω.

Solution
We rewrite Eq. (10.68) in the form of Eq. (10.62):

$$\begin{aligned}
f_{out} &= \frac{1}{T} \int_{t-T/2}^{t+T/2} dt'\, f_{in}(t') \\
&= \frac{1}{T} \int_{-\infty}^{\infty} dt'\, u\left(t + \frac{T}{2} - t'\right) u\left(t' - \left(t - \frac{T}{2}\right)\right) f_{in}(t') \\
&= \frac{1}{T} \int_{-\infty}^{\infty} dt'\, u\left(\frac{T}{2} + (t - t')\right) u\left(\frac{T}{2} - (t - t')\right) f_{in}(t')
\end{aligned}$$

$$= \int_{-\infty}^{\infty} dt' \frac{1}{T} u\left(\frac{T}{2} + (t - t')\right) u\left(\frac{T}{2} - (t - t')\right) f_{in}(t'), \quad (10.69)$$

using the unit-step function, defined as

$$u(t) = \begin{cases} 1 & t \geq 0 \\ 0 & t < 0 \end{cases}. \quad (10.70)$$

Thus,

$$K(t - t') = \frac{1}{T} u\left(\frac{T}{2} + (t - t')\right) u\left(\frac{T}{2} - (t - t')\right). \quad (10.71)$$

The frequency filter is

$$\begin{aligned} K_\omega &= \int_{-\infty}^{\infty} dt\, e^{2\pi i \omega t} K(t) \\ &= \int_{-\infty}^{\infty} dt\, e^{2\pi i \omega t} \frac{1}{T} u\left(\frac{T}{2} + t\right) u\left(\frac{T}{2} - t\right) \\ &= \frac{1}{T} \int_{-T/2}^{T/2} dt\, e^{2\pi i \omega t} = \frac{1}{2\pi i \omega T} \left(e^{\pi i \omega T} - e^{-\pi i \omega T}\right) \\ &= \frac{\sin(\pi \omega T)}{\pi \omega T}. \end{aligned} \quad (10.72)$$

Thus,

$$K_\omega = \frac{\sin(\pi \omega T)}{\pi \omega T}. \quad (10.73)$$

The kernel $K(t)$ and the corresponding filter K_ω are presented in Figs. 10.3 and 10.4.

10.8 RESIDUE THEORY

Both direct and inverse Fourier transforms involve evaluation of some integrals taken over the entire axis of real numbers. In this section, we consider one general and elegant method of handling such integrals, which results from the theory of functions of a complex variable. Here, we will use this method and the underlying concepts for the practical purpose of evaluating definite integrals. Quite shortly, however, we will use these concepts again, to develop and elaborate another celebrated transform, the Laplace transform. Of course, our brief exposition cannot and does not pretend to

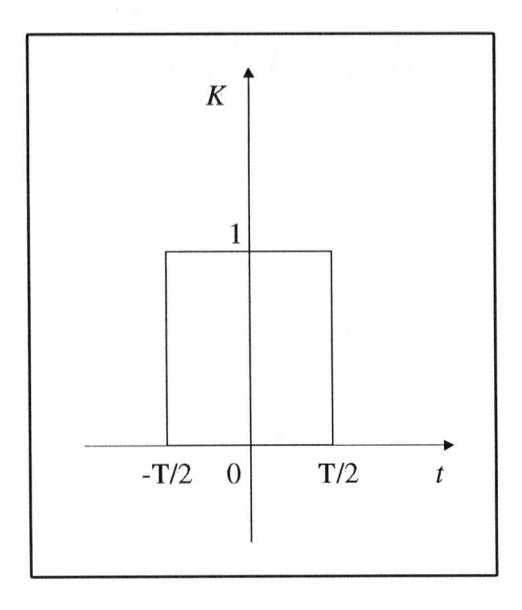

Fig. 10.3 The kernel $K(t)$.

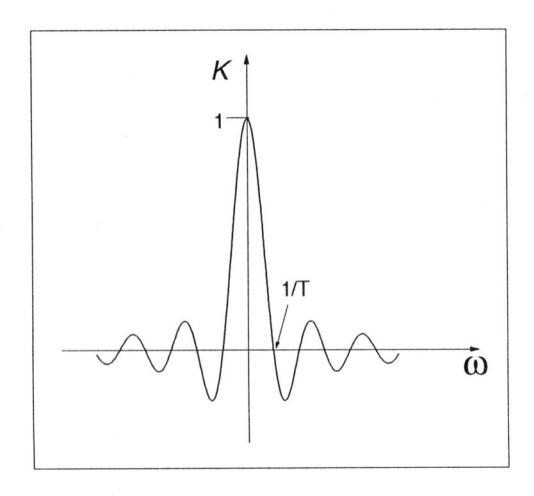

Fig. 10.4 The frequency filter K_ω.

offer an in-depth understanding of functions of a complex variable. Its mere purpose is to refresh the material which is prone to be forgotten when not used.

We start with some general results which we take here for granted. We consider complex-valued functions $f(z)$ of a *complex variable* z, that is, mappings from the complex plane z into the complex plane f. Most functions, and certainly all functions we will deal with here, are regular, so called analytic functions of the variable z. We do not present a rigorous definition of an analytic function nor criteria for a function to be analytic. We merely mention, that any analytic function in the neighborhood of a point z_0 of the complex plane z can be represented as *Laurent expansion*:

$$f(z) = \sum_{n=-\infty}^{\infty} a_n(z_0) \cdot (z - z_0)^n. \tag{10.74}$$

Note that the sum in this expression contains both positive and negative powers of $(z - z_0)$. However, not every point z_0 yields the Laurent expansion containing negative powers; in fact, most points do not. If the negative powers participate in the Laurent expansion near a point z_0, such a point is called a *singular point* of the function $f(z)$. Then, for a given singular point, the negative powers usually do not go as far as $-\infty$ but stop at some N. If this is the case, the singular point is called a *pole* of the function $f(z)$ of order N. In what follows, we will deal with functions whose only singular points are isolated poles, since these represent most functions of interest in engineering applications.

The fundamental *Residue Theorem* reads:
The integral of an analytic function $f(z)$ over a closed path (or loop), C, is expressed through the coefficients of the Laurent expansions of the function $f(z)$ at all the poles lying *inside* the path, $z_1, z_2, z_3, \ldots, z_m$, as

$$\int_C f(z)dz = 2\pi i(a_{-1}(z_1) + a_{-1}(z_2) + a_{-1}(z_3) + \ldots + a_{-1}(z_m)), \tag{10.75}$$

where the integration is assumed to run around the path *counterclockwise*. This is a really remarkable formula: it gives the value of the integral through just a few Laurent coefficients, one per pole! The important coefficient $a_{-1}(z_j)$ is called the Residue of the function f(z) at the pole z_j, and denoted as

$$a_{-1}(z_j) = \operatorname*{Res}_{z=z_j} (f(z)). \tag{10.76}$$

With these notations, the residue theorem reads:

$$\int_C f(z)dz = 2\pi i \sum_{j=1}^{m} \operatorname*{Res}_{z=z_j} (f(z)), \tag{10.77}$$

the integral over a closed loop is equal to $2\pi i$ times the sum of residues of the integrand at all the singular points inside the loop.

Now, the question is, what can we gain from this remarkable theorem? In fact, it is directly applicable to calculating integrals over the real axis, which are typical in direct and inverse Fourier transforms. Consider the integral of a function $f(x)$,

$$\int_{-\infty}^{\infty} f(x)dx. \tag{10.78}$$

In this expression, we can treat the function $f(x)$ as a function of a complex variable, $f(z)$, taken on the real axis. This function certainly vanishes when $|x| \to \infty$, otherwise the integral would not exist. Let this vanishing continue to also take place for infinitely large z, say, in the upper half-plane of the complex plane z, that is, for the semi-circle $|z| \to \infty$. Then, we can close the path of integration in (10.78) with this infinite semi-circle, without changing the value of the integral: the integration of the zero integrand over the semicircle results in zero contribution. Thus, the integral of Eq. (10.78) turns out to be

$$\int_C f(z)dz, \tag{10.79}$$

with the loop C depicted in Fig. 10.5. We can then apply the residue theorem along the path C. This produces the result:

$$\int_{-\infty}^{\infty} f(x)dx = 2\pi i \sum_{z_j} \operatorname*{Res}_{z=z_j} (f(z)), \tag{10.80}$$

the sum going over all the poles of the function $f(z)$ lying in the upper half-plane. (As we will see in examples, sometimes we have to close the loop through the lower half-plane, which adds the factor (-1) in the formula (10.80).)

The last, technical, question that remains in this regard is how to calculate the residues. In many cases the form of the function $f(z)$ allows one to find the values of the residues by inspection. In other, more complicated cases, the general formula is needed. It is also not too hard to obtain, if we look attentively at the Laurent expansion at a pole of N-th order,

$$f(z) = \sum_{n=-N}^{\infty} a_n(z_0) \cdot (z - z_0)^n, \tag{10.81}$$

with the intention to extricate the coefficient $a_{-1}(z_0)$ from the whole sum. An ample means of nullifying the coefficients of greater negative powers is to multiply $f(z)$ by the corresponding *positive* powers of $(z - z_0)$ and differentiate the result. For an N-th-order pole, this results in

$$\operatorname*{Res}_{z=z_j} (f(z)) = \frac{1}{(n-1)!} \lim_{z \to z_j} \left(\frac{d^{N-1}}{dz^{N-1}} \left((z - z_j)^N f(z) \right) \right). \tag{10.82}$$

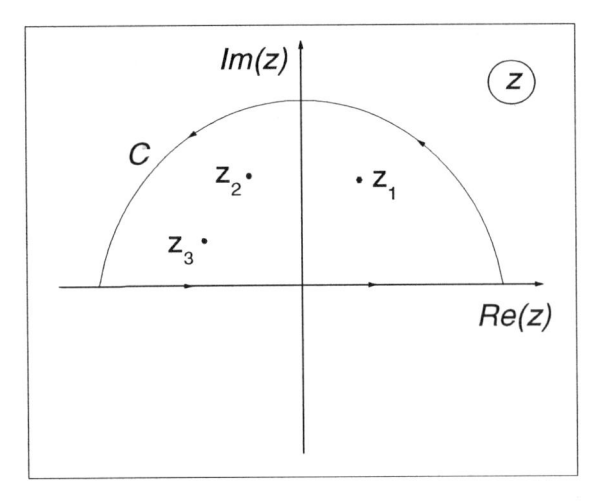

Fig. 10.5 The integration path closing in the upper half-plane.

Example: integral of $\delta_\gamma(x)$
As an example of the application of the Residue Theorem, consider the integral of the function

$$\delta_\gamma(x) = \frac{2\gamma}{(2\pi x)^2 + (\gamma)^2}, \tag{10.83}$$

which we introduced in Section 10.3 as a precursor to the δ-function:

$$\int_{-\infty}^{\infty} \delta_\gamma(x)\,dx = \int_{-\infty}^{\infty} dx \frac{2\gamma}{(2\pi x)^2 + (\gamma)^2}. \tag{10.84}$$

The corresponding function of the complex variable,

$$\delta_\gamma(z) = \frac{2\gamma}{(2\pi z)^2 + (\gamma)^2}, \tag{10.85}$$

vanishes as $|z| \to \infty$ in whatever direction. Thus, we can close the loop of integration, say, in the upper half-plane, and obtain the value of the integral (10.84) as the sum of the residues of the function $\delta_\gamma(x)$ at the poles in the upper half-plane. Then, we rewrite (10.85) as

$$\delta_\gamma(z) = \frac{2\gamma}{(2\pi)^2} \frac{1}{(z + i\gamma/(2\pi))(z - i\gamma/(2\pi))}, \tag{10.86}$$

and see that this function has only one pole (of first order) in the upper half-plane, at $z = i\gamma/2\pi$. Comparing expression (10.86) with the Laurent expansion, Eq. (10.74), we see the easy way to get a_{-1} from (10.86): it is sufficient to drop $(z - i\gamma/2\pi)$ from the denominator, and to put $z = i\gamma/2\pi$ in the rest of the expression. Thus,

$$\underset{z=i\gamma/2\pi}{\text{Res}} \ (\delta_\gamma(z)) = \frac{2\gamma}{(2\pi)^2} \frac{1}{(i\gamma/(2\pi) + i\gamma/(2\pi))} = \frac{1}{2\pi i}. \tag{10.87}$$

Accordingly,

$$\int_{-\infty}^{\infty} \delta_\gamma(x) dx = 2\pi i \ \underset{z=i\gamma/2\pi}{\text{Res}} \ (\delta_\gamma(z)) = 2\pi i \frac{1}{2\pi i} = 1, \tag{10.88}$$

which coincides with the result we obtained earlier, in Eq. (10.21).

10.9 FROM FOURIER TRANSFORM TO LAPLACE TRANSFORM

The requirement for a function $f(t)$ of a real variable t to go to zero fast enough for its Fourier transform,

$$F(\omega) = \int_{-\infty}^{+\infty} e^{-2\pi i \omega t} f(t) dt \tag{10.89}$$

to exist, is quite restrictive. Definitely, this requirement cannot be met in all applications. It is, then, a natural desire to modify somehow the Fourier transform for it to be applicable for functions which are not so eager to vanish as $t \to \infty$, and, may be, for functions which do not vanish at all. It happens that such a modified transform really exists, but, of course, one has to pay some price for this modification. This price is that the new transform works only with functions defined on the half-axis,

$$f(t) = \begin{cases} f(t) & t \geq 0 \\ 0 & t < 0 \end{cases}. \tag{10.90}$$

Consider the simplest case of such a function: this is the so-called unit step function (or Heaviside function),

$$u(t) = \begin{cases} 1 & t \geq 0 \\ 0 & t < 0 \end{cases}. \tag{10.91}$$

Let us try to Fourier transform this function and do all the necessary modifications for this transform to be possible. We write formally:

$$U(\omega) = \int_0^{\infty} e^{-2\pi i \omega t} dt \ \to \ ?, \tag{10.92}$$

FROM FOURIER TRANSFORM TO LAPLACE TRANSFORM 257

the integral is not defined. The possible remedy is to slightly modify the function $u(t)$. Let us take $\tilde{u}(t)$,

$$\tilde{u}(t) = \begin{cases} e^{-\delta t} & t \geq 0 \\ 0 & t < 0 \end{cases}, \tag{10.93}$$

where δ is an auxiliary parameter which we hope to put to zero in the end of the calculations. Then, we have the Fourier transform of this modified function as

$$\tilde{U}(\omega) = \int_0^\infty e^{-2\pi i \omega t - \delta t} dt = \frac{1}{\delta + 2\pi i \omega}. \tag{10.94}$$

Letting δ go to zero, we obtain the transform of the initial function:

$$U(\omega) = \frac{1}{2\pi i \omega}. \tag{10.95}$$

However, with $\tilde{U}(\omega)$ we obtained additional knowledge that we cannot throw away. Namely, this is how to set the integration loop in the ω plane for the inverse transform:

$$\begin{aligned} u(t) &= \int_{-\infty}^{+\infty} d\omega \, e^{2\pi i \omega t} \cdot \frac{1}{\delta + 2\pi i \omega} \\ &= \frac{1}{2\pi i} \int_{-\infty}^{+\infty} d\omega \, \frac{e^{2\pi i \omega t}}{\omega - (i\delta)/(2\pi)}. \end{aligned} \tag{10.96}$$

(Note that if we pretended to know nothing about this δ, we would have a real problem with the pole obnoxiously sitting on the real axis...).

In Eq. (10.96), we have one pole at $\omega_0 = (i\delta)/(2\pi)$. For $t > 0$, the contour is set in the upper half-plane, as depicted, and the integral gives:

$$2\pi i \left(\frac{1}{2\pi i} e^{2\pi i t (i\delta)/(2\pi)} \right) = e^{-\delta t}. \tag{10.97}$$

For $t < 0$, we close the integral loop in the lower half-plane, and we have no pole inside the loop. So, in this case the integral is equal to zero. Thus, we just have restored the modified function of Eq. (10.93),

$$\tilde{u}(t) = \begin{cases} e^{-\delta t} & t \geq 0 \\ 0 & t < 0 \end{cases}. \tag{10.98}$$

Note now, that the exponential is Eq. (10.93) a very "strong" function, whose rapid decrease to zero can guarantee the existence of the Fourier integral not only for the step-function, but also for many other functions, even for functions which grow when $t \to \infty$. The main conclusion (or mental leap) we can draw from this example is that it is reasonable to use the exponential not as an additional auxiliary factor, but to include it in the transform *by definition*. Thus, we come to the following transform.

The Laplace transform

$$\hat{\mathcal{L}}(f(t)) = \mathcal{F}(p) = \int_0^\infty dt\, f(t) \cdot e^{-pt}; \qquad (10.99)$$

$$f(t) = \frac{1}{2\pi i} \int_{\delta-i\infty}^{\delta+i\infty} dp\, e^{pt} \cdot \mathcal{F}(p). \qquad (10.100)$$

In these equations, we put $\delta + 2\pi i\omega = p - $ complex variable. The formulas (10.99) and (10.100) are referred to as the *direct* and *inverse Laplace transforms*, respectively.

We suppose that δ in the last formula is large enough, so that it is greater than the real part of all the singular points of the function $\mathcal{F}(p)$. Taken in this way, the vertical line $(\delta - i\infty,\ \delta + i\infty)$ in the complex p-plane is often referred as the *Bromwich path* of integration and denoted as B.

Inversion Theorem

We take the following statement for granted. If $e^{-pt} \cdot f(t)$ is absolutely integrable for some complex $p = p_0$, that is,

$$\int_0^\infty dt\, |e^{-pt} f(t)| < \infty, \qquad (10.101)$$

then there exists a function $\mathcal{F}(p)$ defined by formula (10.99), such that formula (10.100) produces the original $f(t)$ for any Bromwich path of integration.

10.10 PROPERTIES OF THE LAPLACE TRANSFORM

a) *Convergence abscissa.* Let $f(t)e^{-pt}$ be absolutely integrable for some $p = p_0$. Then for other values of p we can write:

$$\begin{aligned}
|f(t) \cdot e^{-pt}| &= |f(t)| \cdot |e^{-pt}| = |f(t)|e^{-Re(p)t} \\
&= |f(t)|e^{-Re(p_0)t} e^{-(Re(p)-Re(p_0))t} \\
&\leq |f(t)e^{-p_0 t}| e^{-(Re(p)-Re(p_0))t}. \qquad (10.102)
\end{aligned}$$

This means that $f(t) \cdot e^{-pt}$ is absolutely integrable for any p with $Re(p) > Re(p_0)$. But in this case, there must exist a minimum value of the real part of p which still provides absolute convergence. This real number is denoted as R_0 and called the *abscissa of absolute convergence*.

b) *Linearity.* For arbitrary constants α_1, α_2, the linear combination

$$f(t) = \alpha_1 f_1(t) + \alpha_2 f_2(t) \tag{10.103}$$

is Laplace transformable,

$$\hat{L}(f(t)) = \alpha_1 \hat{L}(f_1(t)) + \alpha_2 \hat{L}(f_2(t)), \tag{10.104}$$

and the abscissa of absolute convergence of the combination f will be

$$R = \max\{R_1, R_2\}, \tag{10.105}$$

where R_1 and R_2 are the abscissas of absolute convergence of the functions f_1 and f_2, respectively.

This property is based on the following estimates:

$$|\alpha_1 f_1(t) + \alpha_2 f_2(t)| \leq |\alpha_1| \cdot |f_1(t)| + |\alpha_2| \cdot |f_2(t)|, \tag{10.106}$$

so that

$$\int_0^\infty dt |f(t) e^{-pt}| \leq \int_0^\infty dt |e^{-pt}| (|\alpha_1| \cdot |f_1(t)| + |\alpha_2| \cdot |f_2(t)|)$$

$$= |\alpha_1| \cdot \int_0^\infty dt \, e^{-Re(p)t} |f_1(t)|$$

$$+ |\alpha_2| \cdot \int_0^\infty dt \, e^{-Re(p)t} |f_2(t)|, \tag{10.107}$$

where both integrals converge if $Re(p) \geq \max\{R_1, R_2\}$.

c) *Differentiability.* The Laplace transform $F(p)$ is an infinitely differentiable function of p. In (10.99) we can formally differentiate the integral with respect to p however many times, and this will not affect the absolute convergence.

d) *Vanishing upon infinitely growing argument.* $F(p) \to 0$ if $|p| \to \infty$, $Re(p) > R_0$. We rewrite formula (10.99) as

$$F(p) = \int_0^\infty dt (e^{-Re(p)\cdot t} \cdot f(t)) \cdot e^{-i \cdot Im(p) \cdot t}. \tag{10.108}$$

Obviously, $F(p) \to 0$ if $Re(p) \to \infty$, because the integrand vanishes. If, on the contrary, $Im(p) \to \infty$, $(Re(p) > R_0)$, we just treat (10.108) as the *Fourier transform* of the function $e^{-Re(p)\cdot t} \cdot f(t) \cdot u(t)$ with $\omega = -Im(p)/2\pi$ and employ the property that the Fourier transform vanishes with $\omega \to \infty$ (see Eq.(8.117)).

e) *The Laplace transform being zero corresponds to the original function being zero.*

$$f(t) = 0 \Leftrightarrow F(p) = 0. \tag{10.109}$$

This property, so important in applications, is merely a corollary of the Inversion Theorem.

f) *Laplace transforms of derivatives and integrals.* These transforms can be expressed in terms of the Laplace transform of the function itself in the following way.

$$\begin{aligned}
\hat{L}\left(\frac{df}{dt}\right) &= \int_0^\infty dt\, e^{-pt} \cdot \frac{df}{dt} \\
&= f(t) \cdot e^{-pt}\Big|_{t=0}^{t=\infty} - \int_0^\infty dt\, f(t)\frac{d(e^{-pt})}{dt} \\
&= -f(0) + p\int_0^\infty dt\, f(t)e^{-pt} = -f(0) + pF(p). \tag{10.110}
\end{aligned}$$

Making use of (10.110), we easily obtain the formula for the second derivative:

$$\begin{aligned}
\hat{L}\left(\frac{d^2 f}{dt^2}\right) &= \hat{L}\left(\frac{d}{dt}\left(\frac{df}{dt}\right)\right) = \frac{df}{dt}\Big|_{t=0} + p\hat{L}\left(\frac{df}{dt}\right) \\
&= \frac{df}{dt}\Big|_{t=0} + p(-f(0) + pF(p)) \\
&= p^2 F(p) - pf(0) - \frac{df}{dt}\Big|_{t=0}. \tag{10.111}
\end{aligned}$$

We can then generalize the formulas (10.110) and (10.111) and express the Laplace transform of the n-th derivative in terms of the Laplace transform of the function itself and the *initial conditions*:

$$\begin{aligned}
\hat{L}\left(\frac{d^n f}{dt^n}\right) &\\
&= p^n F(p) - p^{n-1} f(0) - p^{n-2}\frac{df}{dt}\Big|_{t=0} - \cdots \\
&\quad - \frac{d^{n-1} f}{dt^{n-1}}\Big|_{t=0}. \tag{10.112}
\end{aligned}$$

The same approach can be used for *antiderivatives*, that is, for indefinite integrals:

$$\hat{L}\left(\int dt\, f(t)\right) = \hat{L}\left(\int_0^t dt'\, f(t')\right)$$

$$= \int_0^\infty dt\, e^{-pt} \cdot \int_0^t dt'\, f(t') = -\frac{1}{p}\int_0^\infty dt\, \frac{d}{dt}(e^{-pt})\int_0^t dt'\, f(t')$$

$$= -\frac{1}{p}e^{-pt}\int_0^t dt'\, f(t')\Big|_0^\infty + \frac{1}{p}\int_0^\infty dt\, e^{-pt}\frac{d}{dt}(\int_0^t dt'\, f(t'))$$

$$= \frac{1}{p}\int_0^\infty dt\, e^{-pt} f(t) = \frac{F(p)}{p}. \tag{10.113}$$

"Technical" properties of the Laplace transform

Our immediate goal is to develop technical skills for handling both direct and inverse Laplace transforms and, then, to apply this newly gained knowledge in solving specific problems. For these purposes, two further properties of the Laplace transform will be of great use in applications.

a) *p - Shifting Theorem*: if

$$\hat{L}(f(t)) = F(p), \tag{10.114}$$

then

$$\hat{L}(e^{-at}f(t)) = F(p+a), \tag{10.115}$$

provided $Re(p+a) > R_0$, the abscissa of absolute convergence.

To prove this theorem, it is sufficient to use the definition of the Laplace transform:

$$\hat{L}(e^{-at}f(t)) = \int_0^\infty dt\, e^{-pt}e^{-at}f(t)$$

$$= \int_0^\infty dt\, e^{-(p+a)t}f(t) = F(p+a). \tag{10.116}$$

Of course, the statement of Eq. (10.115) implies that, conversely,

$$\hat{L}^{-1}(F(p)) = e^{-at}\hat{L}^{-1}(F(p-a)). \tag{10.117}$$

b) *t - Shifting Theorem*: if, again,

$$\hat{L}(f(t)) = F(p), \tag{10.118}$$

then

$$\hat{L}(f(t-\tau)u(t-\tau)) = e^{-\tau p}F(p). \tag{10.119}$$

Again, to prove the statement of Eq. (10.119) we need nothing more than the basic definition:

$$\hat{L}(f(t-\tau) \cdot u(t-\tau)) = \int_0^\infty dt\, e^{-pt} f(t-\tau) u(t-\tau)$$

$$= \int_\tau^\infty dt\, e^{-p(t-\tau)} e^{-p\tau} f(t-\tau)$$

$$= \int_0^\infty dt'\, e^{-p\tau} e^{-pt'} f(t') = e^{-\tau p} F(p), \qquad (10.120)$$

where we changed the variable $t \to t' = t - \tau$.

Two useful formulas follow as corollaries from (10.119):

$$\hat{L}(f(t)u(t-\tau)) = e^{-p\tau} \hat{L}(f(t+\tau)) \qquad (10.121)$$

and

$$\hat{L}^{-1}(e^{-p\tau} F(p)) = f(t-\tau)u(t-\tau). \qquad (10.122)$$

10.11 BASIC LAPLACE TRANSFORMS

Now we attempt to find Laplace transforms of the most important and most commonly-encountered functions and to summarize them eventually in the form of a table. We already know that the Laplace transform of the step-function,

$$\hat{L}(u(t)) = \frac{1}{p}. \qquad (10.123)$$

Then, making use of the p-Shifting Theorem, we can immediately write down:

$$\hat{L}(e^{-at}) = \frac{1}{p+a}. \qquad (10.124)$$

This, in turn, allows us to easily obtain Laplace transforms of the basic trigonometric functions. For instance:

$$\begin{aligned}
\hat{L}(\cos(\alpha t)) &= \hat{L}\left(\frac{1}{2}(e^{i\alpha t} + e^{-i\alpha t})\right) \\
&= \frac{1}{2}\left(\hat{L}(e^{i\alpha t}) + \hat{L}(e^{-i\alpha t})\right) \\
&= \frac{1}{2}\left(\frac{1}{p+i\alpha} + \frac{1}{p-i\alpha}\right) = \frac{p}{p^2 + \alpha^2}. \qquad (10.125)
\end{aligned}$$

BASIC LAPLACE TRANSFORMS

We, of course, could with equal ease find the Laplace Transform of $\sin(\alpha t)$, but we, just for curiosity, prefer to use another approach:

$$\sin(\alpha t) = -\frac{1}{\alpha} \cdot \frac{d}{dt}(\cos(\alpha t)), \tag{10.126}$$

thus, by virtue of Eq. (10.110),

$$\hat{L}(\sin(\alpha t)) = -\frac{1}{\alpha}\hat{L}\left(\frac{d}{dt}(\cos(\alpha t))\right)$$
$$= -\frac{1}{\alpha}(p\hat{L}(\cos(\alpha t)) - \cos(0)) = -\frac{1}{\alpha}\left(\frac{p^2}{p^2+\alpha^2} - 1\right). \tag{10.127}$$

So,

$$\hat{L}(\sin(\alpha t)) = \frac{\alpha}{p^2+\alpha^2}. \tag{10.128}$$

To explore another family of related Laplace transforms, let us start with the function $f(t) = t$. Its Laplace transform is obtained as

$$\hat{L}(t) = \int_0^\infty dt\, t e^{-pt} = -\frac{\partial}{\partial p}\left(\int_0^\infty dt\, e^{-pt}\right)$$
$$= -\frac{\partial}{\partial p}(\hat{L}(u(t))) = -\frac{\partial}{\partial p}\left(\frac{1}{p}\right) = \frac{1}{p^2}. \tag{10.129}$$

In the same way,

$$\hat{L}(t^2) = -\frac{\partial}{\partial p}(\hat{L}(t)) = -\frac{\partial}{\partial p}\left(\frac{1}{p^2}\right) = \frac{2}{p^3}. \tag{10.130}$$

Straightforward generalization gives the following important formula:

$$\hat{L}(t^n) = \frac{n!}{p^{n+1}}. \tag{10.131}$$

(Note that this formula is actually valid even for the case $f(t) = t^0$, which is just our initial case $f(t) = u(t)$, Eq. (10.123)).

For yet another important case, $f(t) = \sqrt{t}$, we need a somewhat different consideration:

$$\hat{L}(\sqrt{t}) = \int_0^\infty dt\, \sqrt{t}e^{-pt} = 2\int_0^\infty d(\sqrt{t})\, t e^{-pt}$$
$$= 2\int_0^\infty dz\, z^2 e^{-pz^2} = -2\frac{\partial}{\partial p}\left(\int_0^\infty dz\, e^{-pz^2}\right), \tag{10.132}$$

where we changed the dummy variable, $t \to z = \sqrt{t}$. The final integral in (10.132) looks very similar to that we dealt with when studying Fourier transform. The only difference is that p now is a *complex number*,

$$p = Re(p) + iIm(p) = |p|e^{i \cdot arg(p)}. \tag{10.133}$$

Therefore,

$$\int_0^\infty dz \, e^{-pz^2} = \frac{1}{\sqrt{p}} \int_0^{e^{iarg(p)/2} \cdot \infty} d\tilde{z} \, e^{-\tilde{z}^2}, \tag{10.134}$$

and the path of integration in the complex $\tilde{z}$ plane is not the positive half of the real axis but a ray having angle $arg(p)/2$ with this axis. One easily sees, however, that the sector of the $\tilde{z}$ plane between the $Re(\tilde{z})$ axis and the ray going from $z = 0$ at the angle $arg(p)/2$ to this axis does not contain any singular points of the integrand function. This means that we may rotate the integration path back to the real axis and get back to the Poisson integral:

$$\int_0^{e^{iarg(p)/2} \cdot \infty} d\tilde{z} \, e^{-\tilde{z}^2} = \int_0^\infty d\tilde{z} \, e^{-\tilde{z}^2} = \frac{\sqrt{\pi}}{2}. \tag{10.135}$$

Making use of (10.134) and (10.135) in (10.132), we eventually get

$$\hat{L}(\sqrt{t}) = -2 \frac{\partial}{\partial p} \left(\frac{\sqrt{\pi}}{2\sqrt{p}} \right) = \frac{\sqrt{\pi}}{2} p^{-3/2}. \tag{10.136}$$

Then, as in the previous case of integer powers, we can generalize formula (10.136) as follows:

$$\begin{aligned} \hat{L}(t^{\frac{2n+1}{2}}) &= \hat{L}(t^n \sqrt{t}) \\ &= (-1)^n \frac{\partial^n}{\partial p^n} \left(\hat{L}(\sqrt{t}) \right) \\ &= (-1)^n \frac{\sqrt{\pi}}{2} \frac{\partial^n}{\partial p^n} \left(p^{-3/2} \right) \\ &= \frac{\sqrt{\pi}}{2} \left(\frac{3}{2} \cdot \frac{5}{2} \cdot \frac{7}{2} \cdot \ldots \cdot \left(\frac{3}{2} + n \right) \right) p^{-\frac{3}{2} - n} \\ &= \frac{\sqrt{\pi}}{2^{n+1}} (3 \cdot 5 \cdot 7 \cdot \ldots \cdot (3 + 2n)) p^{-\frac{3}{2} - n}. \end{aligned} \tag{10.137}$$

10.12 THE LAPLACE TRANSFORM TABLE

We now summarize our achievements in the form of a table, which we will use in the following for both direct and inverse Laplace transformations without any regard to the methods we used to obtain the formulas.

$$\hat{L}(t^n) = \frac{n!}{p^{n+1}} \quad (n \geq 0). \tag{10.138}$$

$$\hat{L}(e^{-at}) = \frac{1}{p+a}. \tag{10.139}$$

$$\hat{L}(\cos(\alpha t)) = \frac{p}{p^2 + \alpha^2}. \tag{10.140}$$

$$\hat{L}(\sin(\alpha t)) = \frac{\alpha}{p^2 + \alpha^2}. \tag{10.141}$$

$$\hat{L}(t^{n+\frac{1}{2}}) = \frac{\sqrt{\pi}}{2^{n+1}}(3 \cdot 5 \cdot 7 \cdot \ldots \cdot (3+2n))p^{-\frac{3}{2}-n}. \tag{10.142}$$

The set of formulas (10.138)–(10.142) is quite sufficient to solve most, even very complicated, problems in the realm of the Laplace transform.

10.13 SAMPLE PROBLEMS II

To demonstrate the use of the Laplace Transform Table and the properties of the Laplace transform, we consider more complicated examples.

I. Find the inverse Laplace transform of the function:

$$F(p) = \frac{2p+5}{p^2+4p+13}. \tag{10.143}$$

First, we rewrite the denominator in the complete-square form:

$$p^2 + 4p + 13 = (p+2)^2 + 9 = (p+2)^2 + 3^2. \tag{10.144}$$

Then, we represent $F(p)$ as a function of the binomial $p + 2$:

$$\begin{aligned} F(p) &= \frac{2(p+2)+1}{(p+2)^2+3^2} \\ &= 2\frac{p+2}{(p+2)^2+3^2} + \frac{1}{3} \cdot \frac{3}{(p+2)^2+3^2}. \end{aligned} \tag{10.145}$$

Now, we have two terms with known inverse transforms,

$$\begin{aligned} &\hat{L}^{-1}(F(p)) \\ &= 2\hat{L}^{-1}\left(\frac{p+2}{(p+2)^2+3^2}\right) + \frac{1}{3}\hat{L}^{-1}\left(\frac{3}{(p+2)^2+3^2}\right). \end{aligned} \tag{10.146}$$

By using the p-shifting Theorem and the formulas (10.140) and (10.141), we come to

$$\hat{L}^{-1}(F(p))$$
$$= 2e^{-2t}\hat{L}^{-1}\left(\frac{p}{(p)^2 + 3^2}\right) + \frac{1}{3}e^{-2t}\hat{L}^{-1}\left(\frac{3}{(p)^2 + 3^2}\right)$$
$$= e^{-2t}(2\cos(3t) + \frac{1}{3}\sin(3t)). \qquad (10.147)$$

II. Find the Laplace transform of

$$e^{-3t}\int_0^t dt' t' \sin(2t'). \qquad (10.148)$$

Solution
According to p-shifting theorem and to the formula for the transform of derivatives (Section 10.10),

$$\hat{L}\left(e^{-3t}\int_0^t dt' t' \sin(2t')\right) = \Phi(p+3), \qquad (10.149)$$

where

$$\Phi(p) = \hat{L}\left(\int_0^t dt' \, t' \sin(2t')\right) = \frac{1}{p}\hat{L}(t\sin(2t)). \qquad (10.150)$$

Then,

$$\hat{L}(t\sin(\gamma t)) = \hat{L}\left(-\frac{\partial}{\partial \gamma}\cos(\gamma t)\right) = -\frac{\partial}{\partial \gamma}(\hat{L}_s(\cos(\gamma t)))$$
$$= -\frac{\partial}{\partial \gamma}\left(\frac{p}{p^2 + \gamma^2}\right) = \frac{2\gamma p}{(p^2 + \gamma^2)^2}. \qquad (10.151)$$

Thus, for our particular case, $\gamma = 2$, we have:

$$\hat{L}(t\sin(2t)) = \frac{4p}{(p^2 + 4)^2}. \qquad (10.152)$$

Finally, the Laplace transform of the initial function (10.148) is

$$\hat{L}\left(e^{-3t}\int_0^t dt' t' \sin(2t')\right)$$
$$= \frac{1}{p+3} \cdot \frac{4(p+3)}{((p+3)^2 + 4)^2} = \frac{4}{((p+3)^2 + 4)^2}. \qquad (10.153)$$

10.14 LAPLACE TRANSFORM OF THE CONVOLUTION PRODUCT

Yet another important formula, which is widely used in various applications of Laplace transforms, concerns the convolution product. We define the convolution product of two functions $f(t)$ and $g(t)$ in the same way we did when we considered the Fourier transform. Now, however, we take into account that $f(t) = g(t) = 0$ at $t < 0$. Thus, the definition takes the form:

$$f * g = \int_{-\infty}^{+\infty} d\tau f(t-\tau)g(\tau) = \int_0^t d\tau f(t-\tau)g(\tau). \tag{10.154}$$

For the Laplace transform of the convolution product we have

$$\hat{L}(f * g) = \int_0^{+\infty} dt\, e^{-pt} \int_0^t d\tau f(t-\tau)g(\tau)$$

$$= \int_{-\infty}^{+\infty} d\tau\, g(\tau) \int_0^{+\infty} dt\, e^{-pt} f(t-\tau), \tag{10.155}$$

provided $f(t)$ and $g(t)$ are Laplace-transformable functions and $Re(p) \geq \max\{R_f, R_g\}$. Then,

$$\int_0^{+\infty} dt\, e^{-pt} f(t-\tau) = \int_{-\tau}^{+\infty} dt'\, e^{-pt'} e^{-p\tau} f(t'), \tag{10.156}$$

where the new variable $t' = t - \tau$. As $f(t') = 0$ at $t' < 0$, we rewrite the last integral in (10.156) as

$$e^{-p\tau} \int_0^{+\infty} dt'\, e^{-pt'} f(t') = e^{-p\tau} F(p) \tag{10.157}$$

and substitute this result into (10.155) to obtain

$$\hat{L}(f * g) = \int_{-\infty}^{+\infty} d\tau\, g(\tau)e^{-p\tau} F(p)$$

$$= F(p) \int_0^{+\infty} d\tau\, g(\tau)e^{-p\tau} = F(p)G(p). \tag{10.158}$$

268 FOURIER AND LAPLACE TRANSFORMS

So, the result of the Fourier transform theory is still valid for the Laplace transform:
$$\hat{L}(f * g) = \hat{L}(f)\hat{L}(g). \tag{10.159}$$

10.15 APPLICATIONS OF THE LAPLACE TRANSFORM TO SOLVING DIFFERENTIAL EQUATIONS

Probably the most illustrious application of the Laplace transform is its use in solving linear differential equations with constant coefficients. A typical problem of this kind has the form:
$$a_n \frac{d^n y}{dt^n} + a_{n-1} \frac{d^{n-1} y}{dt^{n-1}} + \ldots + a_0 y = f(t), \tag{10.160}$$

with the initial conditions,
$$y(0) = y_0; \quad \left.\frac{dy}{dt}\right|_{t=0} = y_1; \quad \left.\frac{d^2 y}{dt^2}\right|_{t=0} = y_2; \quad \ldots \tag{10.161}$$

Making use of the formulas for transforms of derivatives, we write the Laplace transform of Eq.(10.160) as
$$a_0 Y(p) + \sum_{j=1}^{n} a_j \left(p^j Y(p) - \sum_{k=0}^{j-1} y_k p^{j-1-k} \right) = F(p). \tag{10.162}$$

This is an algebraic equation, which we easily solve to obtain:
$$Y(p) = \frac{1}{\sum_{j=0}^{n} a_j p^j} \left(F(p) + \sum_{j=1}^{n} \sum_{k=0}^{j-1} a_j y_k p^{j-1-k} \right). \tag{10.163}$$

Inverting $Y(p)$, we get the solution of the initial problem in the form:
$$y(t) = \hat{L}^{-1} \left(\frac{1}{\sum_{j=0}^{n} a_j p^j} \left(F(p) + \sum_{j=1}^{n} \sum_{k=0}^{j-1} a_j y_k p^{j-1-k} \right) \right). \tag{10.164}$$

Thus, the initial problem is reduced to the evaluation of integrals (and, in fact, very simple ones). We illustrate this method of solving differential equations on the following example.

10.16 SAMPLE PROBLEM III

Find the solution of the differential equation:

$$x'' + 2x' + x = te^{-t}, \qquad (10.165)$$

with initial conditions:

$$x(0) = 1; \quad \left.\frac{dx}{dt}\right|_{t=0} = -2. \qquad (10.166)$$

We first write the Laplace transform of both sides of the equation,

$$\left(p^2 X(p) - px(0) - \left.\frac{dx}{dt}\right|_{t=0}\right)$$
$$+ 2(pX(p) - x(0)) + X(p) = \frac{1}{(p+1)^2}. \qquad (10.167)$$

Then, we substitute in this equation the initial conditions, and obtain:

$$(p^2 X(p) - p + 2) + 2(pX(p) - 1) + X(p) = \frac{1}{(p+1)^2}, \qquad (10.168)$$

which is reduced to the equation,

$$(p^2 + 2p + 1)X(p) = \frac{1}{(p+1)^2} + p. \qquad (10.169)$$

The obvious solution of Eq. (10.169) is

$$\begin{aligned} X(p) &= \frac{1}{(p+1)^4} + \frac{p}{(p+1)^2} = \frac{1}{(p+1)^4} + \frac{p+1-1}{(p+1)^2} \\ &= \frac{1}{(p+1)^4} - \frac{1}{(p+1)^2} + \frac{1}{(p+1)}. \end{aligned} \qquad (10.170)$$

We know the inverse Laplace transform of all the three terms in Eq. (10.170) (see The Laplace Transform Table, Section 10.12), so, we can write at once:

$$x(t) = \frac{t^3 e^{-t}}{3!} - te^{-t} + e^{-t}. \qquad (10.171)$$

10.17 SUMMARY OF CHAPTER 10

1) The Fourier transform is obtained by passing to the limit in the Fourier series for a symmetric interval when the length of this interval goes to infinity.

270   FOURIER AND LAPLACE TRANSFORMS

2) The Fourier transform is a bijection mapping between the Hilbert space of square-integrable functions of the real variable x defined on the entire real axis and the Hilbert space of square-integrable functions of the real variable ω also defined on the entire real axis.

3) The Fourier transform is a linear mapping. It preserves the inner product and norms of functions.

4) The delta-function $\delta(x)$ is zero at all the points x except for $x = 0$, where its value is infinitely large. It is not a function in the usual interpretation. An integral of $\delta(x)$ with a continuous function $f(x)$ over an interval containing $x = 0$ produces the value $f(0)$.

5) The response of a linear system is given by the convolution product of an input signal and the system kernel. The Fourier transform of a convolution product is just equal to the product of the Fourier transforms of the signal and the kernel. Thus, a linear system works as a signal filter.

6) The Residue Theorem provides a very powerful technique for evaluating integrals encountered in direct and inverse Fourier transforms.

7) The Laplace transform is derived from the Fourier transform by implementing the complex variable p instead of the real variable ω. This transform works beyond the applicability of the Fourier transform, specifically, for functions of x which do not vanish when x goes to infinity.

8) The properties of the Laplace transform often allow one to reduce the transform of a given function to a combination of known cases.

9) The Laplace transform is especially helpful in solving linear differential equations.

10.18   PROBLEMS

I. Consider a linear system with "Gaussian-window" kernel,

$$K(t) = e^{-\alpha t^2}.$$

Find the corresponding frequency filter.

II.
Find inverse Fourier transform of the function

$$F(\omega) = \frac{e^{-4\pi i\omega}}{(i\omega + 3)^2}.$$

III. Find the Laplace transform of the function

$$f(t) = \frac{\sin^2(\pi t)}{t}.$$

IV. Find the Laplace transform of the function

$$\frac{\exp(-3t)\sin(2t)}{t}.$$

V. Find the inverse Laplace transform of the function

$$\frac{1}{(p^2+4)^2}.$$

VI. Find the inverse Laplace transform of the function

$$F(p) = \ln\left(\frac{p^2+2}{p^2+3p}\right).$$

VII. Find inverse Laplace transform of the function

$$F(p) = 1 - \exp\left(\frac{3}{p}\right).$$

VIII. Solve the following differential equation by means of Laplace transform:

$$2\frac{d^2y}{dt^2} + 2y = \sin t + \sqrt{3}\cos t; \quad t \geq 0; \quad y(0) = 0; \quad \left.\frac{dy}{dt}\right|_{t=0} = 2.$$

IX. Solve the following differential equation by means of Laplace transform:

$$\frac{d^2y}{dt^2} + 2\frac{dy}{dt} + 10y = e^{-t}\sin(3t); \quad t \geq 0; \quad y(0) = -1; \quad \left.\frac{dy}{dt}\right|_{t=0} = 2.$$

10.19 FURTHER READING

The Fourier transform — basics:
R. N. Bracewell, *The Fourier Transform and Its Applications*, McGraw-Hill, New York, 1978;

D. C. Champeney, *Fourier Transforms and Their Physical Applications*, Academic Press, London, 1973;

V. M. Cartwrite, *Fourier Methods for Mathematicians, Scientists, and Engineers*, Ellis Horwood, New York, 1990.

The Fourier transform — advanced topics:

J. S. Walker, *Fourier Analysis*, Oxford University Press, New York, 1988;

H. J. Weaver, *Theory of Discrete and Continuous Fourier Analysis*, Wiley, New York, 1989;

A. Papoulis, *The Fourier Integral and Its Applications*, McGraw-Hill, New York, 1962;

E. C. Titchmarsh, *Introduction to the Theory of Fourier Integrals*, Clarendon Press, Oxford, 1937;

C. S. Rees, S. M. Shah, C. V. Stanojevich, *Theory and Applications of Fourier Analysis*, Marcel Dekker, New York, 1981.

Computational aspects:

E. O. Brigham, *The Fast Fourier Transform and Its Applications*, Prentice-Hall, Englewood Cliffs, NJ, 1988;

R. Tolimieri, M. An, C. Lu, *Algorithms for Discrete Fourier Transform and Convolution*. Springer-Verlag, New York, 1989;

H. J. Nussbauer, *Fast Fourier Transform and Convolution Algorithms*, Springer-Verlag, New York, 1982;

W. L. Briggs, Van Emden Hensen, *The Discrete Fourier Transform*, SIAM, Philadelphis, 1995.

The Laplace transform:

P. K. F. Kuhfittig, *Introduction to the Laplace Transform*, Plenum Press, New York, 1978;

R. E. Bellman, *The Laplace Transform*, World Scientific, Singapore, 1984;

J. Shiff, *The Laplace Transform: Theory and Applications*, Springer, New York, 1999;

R. Beals, *Advanced Mathematical Analysis*, Springer-Verlag, New York, 1973;

E. J. Muth, *Transform Methods*, Prentice-Hall, Englewood Cliffs, NJ, 1977.

Applications to signal processing:

L. J. Geis, *Transform Analysis and Filters*, Prentice-Hall, Englewood Cliffs, NJ, 1989;

D. G. Myers, *Digital Signal Processing: Efficient Convolution and Fourier Transform Techniques*, Prentice-Hall, New York, 1990.

11
Partial Differential Equations

11.1 A SHORT INTRODUCTION

This chapter, the last in this book, differs from the preceding ones in both content and style. In the previous ten chapters, we have completed the exposition of the essential mathematical structures we see as necessary for any engineer. As the reader might notice, these mathematical structures are deeply interconnected. Taken together, they, as multi-color smalti of a bristling mosaic, constitute a harmonious pattern of modern engineering mathematics we hope the reader will retain during his/her future career. In this chapter we intend to demonstrate some holistic application of all these concepts. In this sense, the chapter is optional, since it is a case study. The choice of partial differential equations (PDEs) was made for three reasons. First, this topic is really universal, PDEs arise in virtually any field of analytical engineering except computer architecture. Second, the topic is wide and deep enough to require for its comprehension the entire toolbox of mathematical methods we have considered. Last, but not least, the authors like the topic personally, after many years of intensive and (hopefully) successful use of PDEs in their particular domains.

Although the theme is virtually inexhaustible, our goal in the following sections is quite humble. We will study how to handle a PDE, based on the general concepts of linear space and mappings. We will detail the ways to reduce any PDE to one of the most general (*canonical*) forms, demonstrate how to find solutions to these canonical equations and elucidate the important general features of these solutions.

11.2 DEFINITIONS

First of all, in what follows we consider only the two-dimensional case with independent variables x_1 and x_2. The generalization to higher dimensions will be obvious. In what follows, either both x_1 and x_2 will be spatial variables or one of them will be a spatial variable and the other time. Then, we will consider only equations of *second* order, because they invariably arise in a wide variety of applications. The most general form of such an equation is

$$\mathcal{F}\left(x_1,\, x_2,\, f,\, \frac{\partial f}{\partial x_1},\, \frac{\partial f}{\partial x_2},\, \frac{\partial^2 f}{\partial x_1^2},\, \frac{\partial^2 f}{\partial x_2^2},\, \frac{\partial^2 f}{\partial x_1 \partial x_2}\right) = 0, \tag{11.1}$$

where $f(x_1, x_2)$ is an unknown function. Of course, it is very difficult to seriously address this general equation. So, we make another leap in our restrictions, and commit ourselves to the *linear case*,

$$a\frac{\partial^2 f}{\partial x_1^2} + 2b\frac{\partial^2 f}{\partial x_1 \partial x_2} + c\frac{\partial^2 f}{\partial x_2^2}$$
$$+ 2d\frac{\partial f}{\partial x_1} + 2e\frac{\partial f}{\partial x_2} + gf = 0 \tag{11.2}$$

with *constant* coefficients a, b, c, d, e, g. (We will shortly see the convenience of introducing the "2-factors" in some of the coefficients.)

11.3 CANONICAL FORMS

Now we intend to maximally simplify the general form of linear equation of the second order in (11.2). In route to this simplification we will also see the proper way to *classify* the equation into one of several basic cases (the so-called *canonical forms*), depending on the values of the coefficients $a, b, c, \ldots$, or, more specifically, on the relation between these coefficients.

Transformations
Generally speaking, we have at hand only two ways of transforming equation (11.2):
1) Transformation of variables, that is, introducing new variables:

$$\begin{cases} \tilde{x}_1 = \tilde{x}_1(x_1, x_2) \\ \tilde{x}_2 = \tilde{x}_2(x_1, x_2) \end{cases}. \tag{11.3}$$

2) Transformation of the unknown function, that is, introducing a new unknown function,

$$\mathcal{F}(x_1, x_2) = \mathcal{F}(x_1, x_2, f(x_1, x_2)), \tag{11.4}$$

CANONICAL FORMS   275

which would simplify the equation and from which we could restore the original unknown function $f(x_1, x_2)$.

We will subsequently make use of both of these methods.

Simplification of the principal part
Let us first try to simplify the *principal part* of Eq. (11.2), that is, the terms containing second derivatives:

$$a\frac{\partial^2 f}{\partial x_1^2} + 2b\frac{\partial^2 f}{\partial x_1 \partial x_2} + c\frac{\partial^2 f}{\partial x_2^2} = \hat{m}(f), \qquad (11.5)$$

where the operator $\hat{m}$ is defined as

$$\hat{m} = a\frac{\partial^2}{\partial x_1^2} + 2b\frac{\partial^2}{\partial x_1 \partial x_2} + c\frac{\partial^2}{\partial x_2^2}. \qquad (11.6)$$

We can rewrite $\hat{m}$ in *matrix form*,

$$\hat{m} = \left(\frac{\partial}{\partial x_1}, \frac{\partial}{\partial x_2}\right) \begin{pmatrix} a & b \\ b & c \end{pmatrix} \begin{pmatrix} \frac{\partial}{\partial x_1} \\ \frac{\partial}{\partial x_2} \end{pmatrix}$$

$$= \left(\frac{\partial}{\partial x_1}, \frac{\partial}{\partial x_2}\right) \hat{M} \begin{pmatrix} \frac{\partial}{\partial x_1} \\ \frac{\partial}{\partial x_2} \end{pmatrix} \qquad (11.7)$$

(clearly, upon carrying out the matrix multiplication, the expression of Eq. (11.7) becomes that of Eq. (11.6), and we see the utility of introducing one of the "2-factors"). We choose as our next task that of constructing the variable transformation, which will make the matrix of coefficients diagonal. This should be a linear transformation, which we write in general form as:

$$\begin{cases} \tilde{x}_1 = r_{11} \cdot x_1 + r_{12} \cdot x_2 \\ \tilde{x}_2 = r_{21} \cdot x_1 + r_{22} \cdot x_2 \end{cases} \qquad (11.8)$$

or, in the matrix form,

$$\begin{pmatrix} \tilde{x}_1 \\ \tilde{x}_2 \end{pmatrix} = \hat{R} \begin{pmatrix} x_1 \\ x_2 \end{pmatrix}, \quad \hat{R} = \begin{pmatrix} r_{11} & r_{12} \\ r_{21} & r_{22} \end{pmatrix}. \qquad (11.9)$$

Let us first see, what will be the transformation of the derivatives. We express the derivatives with respect to x_1 and x_2 in terms of those with respect to $\tilde{x}_1$ and $\tilde{x}_2$:

$$\begin{cases} \dfrac{\partial}{\partial x_1} = \dfrac{\partial \tilde{x}_1}{\partial x_1} \cdot \dfrac{\partial}{\partial \tilde{x}_1} + \dfrac{\partial \tilde{x}_2}{\partial x_1} \cdot \dfrac{\partial}{\partial \tilde{x}_2} \\[2mm] \dfrac{\partial}{\partial x_2} = \dfrac{\partial \tilde{x}_1}{\partial x_2} \cdot \dfrac{\partial}{\partial \tilde{x}_1} + \dfrac{\partial \tilde{x}_2}{\partial x_2} \cdot \dfrac{\partial}{\partial \tilde{x}_2} \end{cases} \qquad (11.10)$$

Taking the partial derivatives of the new variables with respect to the old ones in Eq. (11.8) and substituting in Eq. (11.10), we get:

$$\begin{cases} \dfrac{\partial}{\partial x_1} = r_{11} \cdot \dfrac{\partial}{\partial \tilde{x}_1} + r_{21} \cdot \dfrac{\partial}{\partial \tilde{x}_2} \\ \dfrac{\partial}{\partial x_2} = r_{12} \cdot \dfrac{\partial}{\partial \tilde{x}_1} + r_{22} \cdot \dfrac{\partial}{\partial \tilde{x}_2} \end{cases} \tag{11.11}$$

$$\begin{pmatrix} \dfrac{\partial}{\partial x_1} \\ \dfrac{\partial}{\partial x_2} \end{pmatrix} = \begin{pmatrix} r_{11} & r_{21} \\ r_{12} & r_{22} \end{pmatrix} \begin{pmatrix} \dfrac{\partial}{\partial \tilde{x}_1} \\ \dfrac{\partial}{\partial \tilde{x}_2} \end{pmatrix} = \hat{R}^T \begin{pmatrix} \dfrac{\partial}{\partial \tilde{x}_1} \\ \dfrac{\partial}{\partial \tilde{x}_2} \end{pmatrix}. \tag{11.12}$$

Applying this result to Eq. (11.7), we rewrite the operator m in the new variables as

$$\tilde{m} = \left(\dfrac{\partial}{\partial \tilde{x}_1}, \dfrac{\partial}{\partial \tilde{x}_2} \right) \hat{R}\hat{M}\hat{R}^T \begin{pmatrix} \dfrac{\partial}{\partial \tilde{x}_1} \\ \dfrac{\partial}{\partial \tilde{x}_2} \end{pmatrix}. \tag{11.13}$$

So, we need to choose the elements of matrix $\hat{R}$ in such a way that the matrix

$$\hat{\tilde{M}} = \hat{R}\hat{M}\hat{R}^T \tag{11.14}$$

will be diagonal. But we know (see Section 5.11) the regular way to diagonalize the matrix:

$$\hat{M} = \hat{S} \begin{pmatrix} \lambda_1 & 0 \\ 0 & \lambda_2 \end{pmatrix} \hat{S}^{-1}. \tag{11.15}$$

where λ_1 and λ_2 are the eigenvalues of the original matrix, and where the diagonalizing matrix $\hat{S}$ has the corresponding eigenvectors as its columns. The matrix under consideration is a symmetric one. Thus, both its eigenvalues are real numbers, and the eigenvectors are orthogonal (see Section 7.12). Now, if we choose these eigenvectors to be *normalized*, then

$$\hat{S}^{-1} = \hat{S}^T, \tag{11.16}$$

and

$$\hat{M} = \hat{S} \begin{pmatrix} \lambda_1 & 0 \\ 0 & \lambda_2 \end{pmatrix} \hat{S}^T. \tag{11.17}$$

Comparing Eq. (11.17) with Eq. (11.14), we come to conclusion that the matrix $\hat{R}$ which performs the coordinate transformation in Eq. (11.9) must be the *transposed* matrix of eigenvectors of the matrix of coefficients, $\hat{M}$.

Correspondingly, the diagonal form of the principal part is expressed as

$$\left(\frac{\partial}{\partial \tilde{x}_1}, \frac{\partial}{\partial \tilde{x}_2}\right) \begin{pmatrix} \lambda_1 & 0 \\ 0 & \lambda_2 \end{pmatrix} \begin{pmatrix} \dfrac{\partial}{\partial \tilde{x}_1} \\ \dfrac{\partial}{\partial \tilde{x}_2} \end{pmatrix} = \lambda_1 \frac{\partial^2}{\partial \tilde{x}_1{}^2} + \lambda_2 \frac{\partial^2}{\partial \tilde{x}_2{}^2}, \qquad (11.18)$$

where λ_1, λ_2 are the eigenvalues of the original coefficient matrix.

Classification

The diagonal form of the principal part provides a natural way to classify a PDE accordingly to the relation between λ_1 and λ_2. To do this classification, let us first write the explicit expressions for these eigenvalues. The characteristic equation:

$$\begin{vmatrix} a - \lambda & b \\ b & c - \lambda \end{vmatrix} = (a - \lambda)(c - \lambda) - b^2 = 0; \qquad (11.19)$$

$$\lambda^2 - \lambda(a + c) + ac - b^2 = 0. \qquad (11.20)$$

Its solutions are given by the formula,

$$\lambda_{1,2} = \frac{a+c}{2} \pm \sqrt{\left(\frac{a+c}{2}\right)^2 + [b^2 - ac]}. \qquad (11.21)$$

We can rewrite the expression under the radical as

$$\left(\frac{a-c}{2}\right)^2 + b^2. \qquad (11.22)$$

Since this number is always positive both eigenvalues are always real numbers (as we have already mentioned, this is due to the matrix of coefficients being symmetric). Then, these real numbers λ_1 and λ_2 having the same or different signs is determined by the sign of the combination $b^2 - a \cdot c$. So, we classify an equation as follows:
1) If $b^2 - a \cdot c > 0$, λ_1 and λ_2 are of different sign. This is *the hyperbolic case*.
2) If $b^2 - a \cdot c < 0$, λ_1 and λ_2 are of the same sign. This is *the elliptic case*.
3) If $b^2 - a \cdot c = 0$, $\lambda_2 = 0$. This degenerated situation is called *the parabolic case*.
(The names were coined by analogy of the canonical forms corresponding to these cases, Eqs. (11.54)–(11.57) of Section 11.5, with the analytical formulas for conical sections, hyperbola, ellipse, and parabola, in cartesian coordinates.)

This classification is not that superficial as it might seem. In fact, it is very essential, because it forecasts cardinally different responses of the solutions to variations of the variables.

Coefficients of the first derivatives

Let us trace the transformation of the terms containing first derivatives in Eq. (11.2)

278 PARTIAL DIFFERENTIAL EQUATIONS

under the transformation of the variables of Eq. (11.9). By virtue of (11.12), we have

$$2d\frac{\partial f}{\partial x_1} + 2e\frac{\partial f}{\partial x_2} = 2(d, e)\begin{pmatrix} \frac{\partial f}{\partial x_1} \\ \frac{\partial f}{\partial x_2} \end{pmatrix}$$

$$= 2(d, e)\hat{R}^T \begin{pmatrix} \frac{\partial f}{\partial \tilde{x}_1} \\ \frac{\partial f}{\partial \tilde{x}_2} \end{pmatrix} = 2(\tilde{d}, \tilde{e})\begin{pmatrix} \frac{\partial f}{\partial \tilde{x}_1} \\ \frac{\partial f}{\partial \tilde{x}_2} \end{pmatrix}, \quad (11.23)$$

where

$$\begin{pmatrix} \tilde{d} \\ \tilde{e} \end{pmatrix} = \hat{R}\begin{pmatrix} d \\ e \end{pmatrix} = \hat{S}^T \begin{pmatrix} d \\ e \end{pmatrix}. \quad (11.24)$$

Normalization

Upon completing the diagonalization of the principal part, we have reduced the initial PDE to the form:

$$\lambda_1 \frac{\partial^2 f}{\partial \tilde{x}_1{}^2} + \lambda_2 \frac{\partial^2 f}{\partial \tilde{x}_2{}^2} + 2\tilde{d}\frac{\partial f}{\partial \tilde{x}_1} + 2\tilde{e}\frac{\partial f}{\partial \tilde{x}_2} + gf = 0. \quad (11.25)$$

In our pursuit of transforming the equation towards the most general form, we now get rid of the numeric factors λ_1 and λ_2. We do this by normalizing the variables:

$$\bar{x}_1 = \frac{\tilde{x}_1}{\sqrt{|\lambda_1|}}; \quad \bar{x}_2 = \frac{\tilde{x}_2}{\sqrt{|\lambda_2|}}. \quad (11.26)$$

(Of course, the second formula makes sense in the elliptic and hyperbolic cases only. So, we will treat the parabolic case separately.)

Having normalized the variables in this way, we obtain from (11.25):

$$\frac{\partial^2 f}{\partial \bar{x}_1{}^2} \pm \frac{\partial^2 f}{\partial \bar{x}_2{}^2} + 2\bar{d}\frac{\partial f}{\partial \bar{x}_1} + 2\bar{e}\frac{\partial f}{\partial \bar{x}_2} + gf = 0, \quad (11.27)$$

where the positive sign corresponds to the elliptic case, and the negative sign corresponds to the hyperbolic case. The transformed coefficients $\bar{d}$ and $\bar{e}$ are expressed as

$$\bar{d} = \frac{\tilde{d}}{\sqrt{|\lambda_1|}}; \quad \bar{e} = \frac{\tilde{e}}{\sqrt{|\lambda_2|}}. \quad (11.28)$$

CANONICAL FORMS 279

Eliminating first derivatives

We still have intact our second weapon of transformation, the transformation of the unknown function $f(\bar{x}_1, \bar{x}_2)$. We now use it to eliminate the first-order derivatives in Eq. (11.27). Specifically, to eliminate the $\bar{x}_1$-derivative we represent the unknown function as

$$f(\bar{x}_1, \bar{x}_2) = \chi(\bar{x}_1) \cdot Y(\bar{x}_1, \bar{x}_2), \tag{11.29}$$

where $Y(\bar{x}_1, \bar{x}_2)$ is a new unknown function and $\chi(\bar{x}_1)$ is the auxiliary function we will choose to cancel the first-order derivative. From Eq. (11.29), the partial derivatives of the function $f(\bar{x}_1, \bar{x}_2)$ with respect to $\bar{x}_1$ are found as

$$\frac{\partial f}{\partial \bar{x}_1} = \chi \frac{\partial Y}{\partial \bar{x}_1} + Y \frac{\partial \chi}{\partial \bar{x}_1}; \tag{11.30}$$

$$\frac{\partial^2 f}{\partial \bar{x}_1^2} = \chi \frac{\partial^2 Y}{\partial \bar{x}_1^2} + 2 \frac{\partial \chi}{\partial \bar{x}_1} \frac{\partial Y}{\partial \bar{x}_1} + Y \frac{\partial^2 \chi}{\partial \bar{x}_1^2}. \tag{11.31}$$

Then, the $\bar{x}_1$-derivative-related terms in Eq. (11.27) produce

$$\chi \frac{\partial^2 Y}{\partial \bar{x}_1^2} + 2 \frac{\partial \chi}{\partial \bar{x}_1} \frac{\partial Y}{\partial \bar{x}_1} + Y \frac{\partial^2 \chi}{\partial \bar{x}_1^2} + 2\bar{d}\chi \frac{\partial Y}{\partial \bar{x}_1} + 2\bar{d}Y \frac{\partial \chi}{\partial \bar{x}_1}. \tag{11.32}$$

To make the coefficient of $\partial Y/\partial \bar{x}_1$ be zero requires

$$2\frac{\partial \chi}{\partial \bar{x}_1} + 2\bar{d}\chi = 0. \tag{11.33}$$

Calculating $\chi(\bar{x}_1)$ from this equation yields

$$\chi(\bar{x}_1) = e^{-\bar{d}\bar{x}_1}. \tag{11.34}$$

We then substitute (11.32), (11.33) and (11.34) in Eq. (11.27), and obtain

$$e^{-\bar{d}\bar{x}_1}\left(\frac{\partial^2 Y}{\partial \bar{x}_1^2} \pm \frac{\partial^2 Y}{\partial \bar{x}_2^2}\right) + 2\bar{e}e^{-\bar{d}\bar{x}_1} \frac{\partial Y}{\partial \bar{x}_2}$$
$$+ Y(\bar{x}_1, \bar{x}_2)(ge^{-\bar{d}\bar{x}_1} - 2\bar{d}\bar{d}e^{-\bar{d}\bar{x}_1} + (\bar{d})^2 e^{-\bar{d}\bar{x}_1}) = 0. \tag{11.35}$$

For finite $\bar{d}\bar{x}_1$ we can eliminate $e^{-\bar{d}\bar{x}_1}$, and Eq. (11.35) takes the form,

$$\frac{\partial^2 Y}{\partial \bar{x}_1^2} \pm \frac{\partial^2 Y}{\partial \bar{x}_2^2} + 2\bar{e}\frac{\partial Y}{\partial \bar{x}_2} + (g - \bar{d}^2)Y(\bar{x}_1, \bar{x}_2) = 0. \tag{11.36}$$

We can use the same procedure to eliminate the first-x_2-derivative terms. We write

$$Y(\bar{x}_1, \bar{x}_2) = \chi_1(\bar{x}_2) \cdot \psi(\bar{x}_1, \bar{x}_2), \tag{11.37}$$

and find the auxiliary function $\chi_1(\bar{x}_2)$ as

$$\chi_1(\bar{x}_2) = e^{\pm \bar{e}\bar{x}_2}. \tag{11.38}$$

Then, Eq. (11.36) yields:

$$\frac{\partial^2 \psi}{\partial \bar{x}_1{}^2} \pm \frac{\partial^2 \psi}{\partial \bar{x}_2{}^2} + (g - \bar{d}^2 \mp \bar{e}^2)\psi = 0. \qquad (11.39)$$

The eventual result, the so-called *canonical form* of elliptic or hyperbolic equations is commonly written as

$$\frac{\partial^2 \psi}{\partial \bar{x}_1{}^2} \pm \frac{\partial^2 \psi}{\partial \bar{x}_2{}^2} + k\psi = 0, \qquad (11.40)$$

with the coefficient

$$k = g - \bar{d}^2 \mp \bar{e}^2, \qquad (11.41)$$

and the cumulative transformation leading from Eq. (11.27) to Eq. (11.39) is

$$f(\bar{x}_1, \bar{x}_2) = e^{-\bar{d}\bar{x}_1 \mp \bar{e}\bar{x}_2} \psi(\bar{x}_1, \bar{x}_2). \qquad (11.42)$$

In the formulas (11.40)–(11.42), the upper sign corresponds to the elliptic case, and the lower sign to the hyperbolic case.

11.4 THE PARABOLIC CASE

In this degenerate case Eq. (11.25) takes the form:

$$\lambda_1 \frac{\partial^2 f}{\partial \tilde{x}_1{}^2} + 2\tilde{d}\frac{\partial f}{\partial \tilde{x}_1} + 2\tilde{e}\frac{\partial f}{\partial \tilde{x}_2} + gf = 0. \qquad (11.43)$$

Now, the most natural choice of the normalized variables is

$$\bar{x}_1 = \frac{\tilde{x}_1}{\sqrt{|\lambda_1|}}; \quad \bar{x}_2 = \frac{\tilde{x}_2}{2\tilde{e}}. \qquad (11.44)$$

Using these, we get Eq. (11.43) as

$$\frac{\partial^2 f}{\partial \bar{x}_1{}^2} + 2\bar{d}\frac{\partial f}{\partial \bar{x}_1} + \frac{\partial f}{\partial \bar{x}_2} + gf = 0, \qquad (11.45)$$

with the normalizing coefficient,

$$\bar{d} = \frac{\tilde{d}}{\sqrt{|\lambda_1|}}. \qquad (11.46)$$

Then, exactly as previously, we eliminate the first-order-$\bar{x}_1$-derivative term by transforming the unknown function as

$$f(\bar{x}_1, \bar{x}_2) = e^{-\bar{d}\bar{x}_1} Y(\bar{x}_1, \bar{x}_2), \qquad (11.47)$$

to obtain the equation,

$$\frac{\partial^2 Y}{\partial \bar{x}_1{}^2} + \frac{\partial Y}{\partial \bar{x}_2} + (g - \bar{d}^2)Y = 0, \tag{11.48}$$

and stop, because we have no means to eliminate $\partial Y/\partial \bar{x}_2$, having no higher $\bar{x}_2$-derivative. What we can do in compensation is to eliminate the zero-order derivative by means of the first-order one! We take:

$$Y(\bar{x}_1, \bar{x}_2) = \chi_1(\bar{x}_2)\psi(\bar{x}_1, \bar{x}_2), \tag{11.49}$$

and obtain in Eq. (11.48)

$$\chi_1(\bar{x}_2)\frac{\partial^2 \psi}{\partial \bar{x}_1{}^2} + \chi_1(\bar{x}_2)\frac{\partial \psi}{\partial \bar{x}_2}$$
$$+\psi(\bar{x}_1, \bar{x}_2)\left(\frac{\partial \chi_1}{\partial \bar{x}_2} + (g - \bar{d}^2) \cdot \chi_1(\bar{x}_2)\right) = 0. \tag{11.50}$$

By analogy to (11.33), we require

$$\frac{\partial \chi_1}{\partial \bar{x}_2} + (g - \bar{d}^2)\chi_1(\bar{x}_2) = 0, \tag{11.51}$$

which gives

$$\chi_1(\bar{x}_2) = e^{-(g-\bar{d}^2)\bar{x}_2}. \tag{11.52}$$

Thus, in the parabolic case the canonical form is

$$\frac{\partial^2 \psi}{\partial \bar{x}_1{}^2} + \frac{\partial \psi}{\partial \bar{x}_2} = 0, \tag{11.53}$$

with the cumulative transformation leading from Eq. (11.48) to Eq. (11.53),

$$f(\bar{x}_1, \bar{x}_2) = e^{-\bar{d}\bar{x}_1 - (g-\bar{d}^2)\bar{x}_2}\psi(\bar{x}_1, \bar{x}_2). \tag{11.54}$$

11.5 CANONICAL FORMS — A SUMMARY

We have obtained three maximally simplified generic forms of linear second-order PDE's,

$$\frac{\partial^2 \psi}{\partial \bar{x}_1{}^2} + \frac{\partial^2 \psi}{\partial \bar{x}_2{}^2} + k\psi = 0 \quad \text{elliptic case;} \tag{11.55}$$

$$\frac{\partial^2 \psi}{\partial \bar{x}_1{}^2} - \frac{\partial^2 \psi}{\partial \bar{x}_2{}^2} + k\psi = 0 \quad \text{hyperbolic case;} \tag{11.56}$$

$$\frac{\partial^2 \psi}{\partial \bar{x}_1{}^2} + \frac{\partial \psi}{\partial \bar{x}_2} = 0 \quad \text{parabolic case},\tag{11.57}$$

and developed straightforward algorithms for reducing a linear second-order PDE to one of these canonical forms. To proceed further, a natural approach would be to try solving these three equations, (11.55)–(11.57), in their general form and to discuss the important properties of these general solutions and their dependencies, say, on the initial or/and boundary conditions.

Such an approach, however, is beyond the scope of this book. We will merely consider typical examples in each of the cases and develop basic methods and skills which are sufficient for handling the most commonly-encountered PDEs.

11.6 SAMPLE PROBLEM I

Classify the following partial differential equation and find the simplest transformation of variables $x, y,$ and the unknown function ψ which puts it into canonical form:

$$\frac{\partial^2 \phi}{\partial x^2} + 2\frac{\partial^2 \phi}{\partial x \partial y} + 2\frac{\partial \phi}{\partial y} = 0.\tag{11.58}$$

Solution
1) To diagonalize the matrix of coefficients of the principal part,

$$\hat{A} = \begin{pmatrix} 1 & 1 \\ 1 & 0 \end{pmatrix},\tag{11.59}$$

we start with the characteristic equation:

$$\begin{vmatrix} 1-\lambda & 1 \\ 1 & -\lambda \end{vmatrix} = \lambda \cdot (\lambda - 1) - 1 = \lambda^2 - \lambda - 1 = 0.\tag{11.60}$$

The solutions are given as

$$\lambda_{1,2} = \frac{1}{2} \pm \sqrt{\frac{1}{4} + 1} = \frac{1 \pm \sqrt{5}}{2}.\tag{11.61}$$

$$\lambda_1 = \frac{1+\sqrt{5}}{2} > 0; \quad \lambda_2 = \frac{1-\sqrt{5}}{2} < 0.\tag{11.62}$$

So, we have the *hyperbolic* case.
The eigenvector for λ_1 satisfies the equation

$$\begin{pmatrix} 1-\lambda_1 & 1 \\ 1 & -\lambda_1 \end{pmatrix} \begin{pmatrix} x_1 \\ x_2 \end{pmatrix} = 0.\tag{11.63}$$

$$\begin{pmatrix} \dfrac{1-\sqrt{5}}{2} & 1 \\ 1 & -\dfrac{1+\sqrt{5}}{2} \end{pmatrix} \begin{pmatrix} x_1 \\ x_2 \end{pmatrix} = 0. \tag{11.64}$$

From the first equation of this system we obtain

$$x_2 = \frac{\sqrt{5}-1}{2} x_1. \tag{11.65}$$

Thus, the normalized eigenvector is

$$\vec{x}^{(1)} = \gamma \cdot \begin{pmatrix} 1 \\ \dfrac{\sqrt{5}-1}{2} \end{pmatrix}, \tag{11.66}$$

where

$$\gamma = \sqrt{\frac{2}{5-\sqrt{5}}} = 5^{-1/4}\sqrt{\frac{2}{\sqrt{5}-1}}. \tag{11.67}$$

The eigenvector for λ_2 satisfies the equation

$$\begin{pmatrix} \dfrac{1+\sqrt{5}}{2} & 1 \\ 1 & \dfrac{\sqrt{5}-1}{2} \end{pmatrix} \begin{pmatrix} x_1 \\ x_2 \end{pmatrix} = 0. \tag{11.68}$$

Here, it is more convenient to determine the eigenvector from the second equation. In this way, it is obtained in a form very similar to that of the first eigenvector:

$$x_1 = -\frac{\sqrt{5}-1}{2} x_2, \tag{11.69}$$

so that the normalized eigenvector is

$$\vec{x}^{(2)} = \gamma \cdot \begin{pmatrix} -\dfrac{\sqrt{5}-1}{2} \\ 1 \end{pmatrix}. \tag{11.70}$$

Finally, the diagonalizing matrix is

$$\hat{S} = \gamma \cdot \begin{pmatrix} 1 & -\dfrac{\sqrt{5}-1}{2} \\ \dfrac{\sqrt{5}-1}{2} & 1 \end{pmatrix}. \tag{11.71}$$

284   PARTIAL DIFFERENTIAL EQUATIONS

2) The corresponding coordinate transformation is given as

$$\begin{pmatrix} \tilde{x} \\ \tilde{y} \end{pmatrix} = \hat{S}^T \begin{pmatrix} x \\ y \end{pmatrix} = \gamma \cdot \begin{pmatrix} 1 & \frac{\sqrt{5}-1}{2} \\ -\frac{\sqrt{5}-1}{2} & 1 \end{pmatrix} \begin{pmatrix} x \\ y \end{pmatrix}. \quad (11.72)$$

Upon normalization, the new coordinates are obtained as

$$\begin{aligned}
\bar{x} &= \sqrt{\frac{2}{\sqrt{5}+1}} \cdot \gamma \cdot \left(x + \frac{\sqrt{5}-1}{2} y \right) \\
&= 5^{-1/4} \left(x + \frac{\sqrt{5}-1}{2} y \right); \quad (11.73)
\end{aligned}$$

$$\begin{aligned}
\bar{y} &= \sqrt{\frac{2}{\sqrt{5}-1}} \cdot \gamma \cdot \left(-\frac{\sqrt{5}-1}{2} x + y \right) \\
&= 5^{-1/4} \left(-x + \frac{2}{\sqrt{5}-1} \cdot y \right). \quad (11.74)
\end{aligned}$$

3) The coefficients at first-order partial derivatives are being transformed as

$$\begin{aligned}
\begin{pmatrix} \tilde{d} \\ \tilde{e} \end{pmatrix} &= \hat{S}^T \begin{pmatrix} d \\ e \end{pmatrix} \\
&= \gamma \cdot \begin{pmatrix} 1 & \frac{\sqrt{5}-1}{2} \\ -\frac{\sqrt{5}-1}{2} & 1 \end{pmatrix} \begin{pmatrix} 0 \\ 1 \end{pmatrix} \\
&= \gamma \cdot \begin{pmatrix} \frac{\sqrt{5}-1}{2} \\ 1 \end{pmatrix} \quad (11.75)
\end{aligned}$$

Upon normalization, we get

$$\begin{pmatrix} \bar{d} \\ \bar{e} \end{pmatrix} = \begin{pmatrix} \frac{\tilde{d}}{\sqrt{|\lambda_1|}} \\ \frac{\tilde{e}}{\sqrt{|\lambda_2|}} \end{pmatrix} = 5^{-1/4} \begin{pmatrix} \frac{\sqrt{5}-1}{2} \\ \frac{2}{\sqrt{5}-1} \end{pmatrix}. \quad (11.76)$$

Thus, steps 1) – 3) of the coordinate transformation result in

$$\frac{\partial^2 \phi}{\partial \bar{x}^2} - \frac{\partial^2 \phi}{\partial \bar{y}^2}$$

$$+2 \cdot 5^{-1/4} \frac{\sqrt{5}-1}{2} \cdot \frac{\partial \phi}{\partial \bar{x}} + 2 \cdot 5^{-1/4} \frac{2}{\sqrt{5}-1} \cdot \frac{\partial \phi}{\partial \bar{y}} = 0. \quad (11.77)$$

To eliminate the first-order derivatives, we represent the unknown function in the form:

$$\phi(\bar{x},\bar{y}) = \exp\left(-5^{-1/4} \frac{\sqrt{5}-1}{2} \bar{x} + 5^{-1/4} \frac{2}{\sqrt{5}-1} \bar{y}\right) \psi(\bar{x},\bar{y})$$
$$= \exp\left[\left(\frac{5^{-1/4}}{2}\right)\left(-(\sqrt{5}-1)\bar{x} + (\sqrt{5}+1)\bar{y}\right)\right] \psi(\bar{x},\bar{y}), \quad (11.78)$$

so that the equation for $\Psi(\bar{x},\bar{y})$ turns out to be:

$$\frac{\partial^2 \psi}{\partial \bar{x}^2} - \frac{\partial^2 \psi}{\partial \bar{y}^2} + k\psi = 0, \quad (11.79)$$

where

$$k = -\bar{d}^2 + \bar{e}^2$$
$$= 5^{-1/2}\left(\left(-\frac{\sqrt{5}-1}{2}\right)^2 + \left(\frac{2}{\sqrt{5}-1}\right)^2\right) = 1. \quad (11.80)$$

Thus, the canonical form of the equation is

$$\frac{\partial^2 \psi}{\partial \bar{x}^2} - \frac{\partial^2 \psi}{\partial \bar{y}^2} + \psi = 0. \quad (11.81)$$

11.7 THE WAVE EQUATION. D'ALEMBERT'S FORMULA

As a typical example of the hyperbolic case we consider the so called *wave equation*,

$$\frac{1}{c^2} \cdot \frac{\partial^2 f}{\partial t^2} = \frac{\partial^2 f}{\partial x^2}; \quad t > 0; \quad -\infty < x < \infty. \quad (11.82)$$

Here, the x and t are spatial and temporal variables, respectively. The factor $1/c^2$ is added for the sake of dimensionality. This equation (11.82) describes the evolution of small perturbations of a continuous medium when there exists a mechanism that tends to restore the medium to its normal or *equilibrium* state. In such circumstances, the disturbance travels, or *propagates*, without the transport of matter. Examples of waves are the sound you hear, visible light, elastic waves in strings and membranes, radio signals of wireless communication, ripples on a pond, ocean waves, and so on. Moreover, on a *microscopic* scale the charge carriers in electronic devices also must be described as some specific waves, and their wave behavior essentially determines the characteristics of modern devices.

The disturbance created by a wave is represented by a wave function; in our case this is a function of two variables, $f(t, x)$. Correspondingly, Eq. (11.82) is called the wave equation, and the constant c has the physical meaning of the wave velocity.

Wave propagation is described in (11.82) by means of the second derivative with respect to t. This means that we need two initial conditions to uniquely specify the solution. We will not go into detailed discussion on this matter, but just take these conditions as

$$f(0, x) = \phi(x); \quad \left.\frac{\partial f}{\partial t}\right|_{t=0} = \psi(x). \tag{11.83}$$

The complication in solving Eq. (11.82) compared to *ordinary* differential equations is that it contains partial derivatives in both the t and x variables. There is, however, a simple method to get rid of "extra" derivatives. This method is just the Fourier transform with respect to the most suitable variable. The initial conditions (11.83) dictate the single choice of this transformable variable, this is x. Recalling that the Fourier transform is

$$F(t, k) = \int_{-\infty}^{+\infty} f(t, x) e^{-2\pi i k x} \, dx \tag{11.84}$$

and making use of the formulas (10.52) for the transform of derivatives, we easily obtain from Eq. (11.82):

$$\frac{\partial^2 F}{\partial t^2} = c^2 (2\pi i k)^2 F. \tag{11.85}$$

Equation (11.85) contains k only as a parameter. This means that we can treat it as an *ordinary* differential equation in t with solutions depending on the parameter k. This equation has the obvious solution:

$$F(t, k) = A e^{2\pi i c k t} + B e^{-2\pi i c k t}, \tag{11.86}$$

where constants A and B now can (and will) be functions of the parameter k, $A(k)$ and $B(k)$. To find these functions, we transform the initial conditions (11.83) into k-space:

$$F(0, k) = \int_{-\infty}^{+\infty} \phi(x) e^{-2\pi i k x} \, dx = \Phi(k); \tag{11.87}$$

$$\left.\frac{dF}{dt}\right|_{t=0} = \int_{-\infty}^{+\infty} \psi(x) e^{-2\pi i k x} \, dx = \Psi(k). \tag{11.88}$$

Then,

$$\frac{dF}{dt} = 2\pi i c k (A(k) e^{2\pi i c k t} - B(k) e^{-2\pi i c k t}). \tag{11.89}$$

Thus, we obtain from (11.86) and (11.89) for the initial instant:

$$\begin{cases} A(k) + B(k) = \Phi(k) \\ A(k) - B(k) = \dfrac{\Psi(k)}{2\pi i c k} \end{cases} \qquad (11.90)$$

Solving this system for A and B gives us

$$\begin{cases} A(k) = \dfrac{1}{2}\left(\Phi(k) + \dfrac{\Psi(k)}{2\pi i c k}\right) \\ B(k) = \dfrac{1}{2}\left(\Phi(k) - \dfrac{\Psi(k)}{2\pi i c k}\right) \end{cases} \qquad (11.91)$$

From the formulas (11.86) and (11.91) we eventually get the expression for the function $F(t, k)$,

$$\begin{aligned} F(t,k) &= \frac{1}{2}\Phi(k)(e^{2\pi i c k t} + e^{-2\pi i c k t}) \\ &\quad + \frac{\Psi(k)}{4\pi i c k}(e^{2\pi i c k t} - e^{-2\pi i c k t}) \\ &= \Phi(k)\cos(2\pi c k t) + \frac{\Psi(k)}{2\pi c k}\sin(2\pi c k t). \end{aligned} \qquad (11.92)$$

To find the inverse Fourier transform of the expression (11.92), we first note that for any function $g(x)$ and its Fourier transform $G(k)$,

$$\int_{-\infty}^{+\infty} dk\, e^{2\pi i k x} G(k) \frac{1}{2\pi i k}$$

$$= \int dx \int_{-\infty}^{+\infty} dk\, e^{2\pi i k x} G(k) = \int dx\, g(x). \qquad (11.93)$$

Therefore, for the function $f(t,x)$ we obtain

$$\begin{aligned} f(t,x) &= \int_{-\infty}^{\infty} dk\, e^{2\pi i k x}\left(\Phi(k)\cos(2\pi c k t) + \frac{\Psi(k)}{2\pi c k}\sin(2\pi c k t)\right) \\ &= \int_{-\infty}^{\infty} dk\, \Big(\frac{1}{2}\Phi(k)(e^{2\pi i k \cdot (x+ct)} \\ &\quad + e^{2\pi i k(x-ct)}) + \frac{\Psi(k)}{4\pi i c k}(e^{2\pi i k(x+ct)} - e^{-2\pi i k(x-ct)})\Big) \\ &= \frac{1}{2}(\phi(x+ct) + \phi(x-ct)) \end{aligned}$$

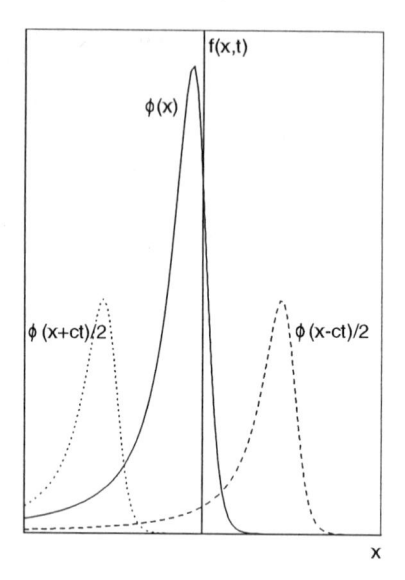

Fig. 11.1 To the general solution of the wave equation.

$$+ \frac{1}{c} \int dx' \psi(x') \bigg|_{x'=x+ct} - \frac{1}{c} \int dx' \psi(x') \bigg|_{x'=x-ct} \Bigg)$$

$$= \frac{1}{2} \left(\phi(x+ct) + \phi(x-ct) + \frac{1}{c} \int_{x-ct}^{x+ct} ds\, \psi(s) \right). \quad (11.94)$$

The result obtained for the solution of the wave equation (11.82) with the initial conditions (11.83),

$$f(t,x) = \frac{1}{2} \left(\phi(x+ct) + \phi(x-ct) \right) + \frac{1}{2c} \int_{x-ct}^{x+ct} ds\, \psi(s), \quad (11.95)$$

is surprisingly simple and fundamentally important. Due to its widespread applicability, the formula (11.95) has been given a proper name; it is known as the *D'Alembert's Formula*. In a particular clear case when $\psi = 0$, the evolution of the initial perturbation of the media is described as the superposition of two perturbations, each of which conserves the shape of the initial one and moves it with constant velocity c in the positive (argument $x - ct$) or negative (argument $x + ct$) directions with respect to the x-axis (see Fig. 11.1).

11.8 THE DIFFUSION EQUATION. TRANSFORM METHODS

As a typical example of a parabolic PDE we consider the equation that is often used to describe the evolution of a disturbance in a continuous medium when some

THE DIFFUSION EQUATION. TRANSFORM METHODS

mechanisms of *dissipation* are involved. Examples are the process of diffusion of a scent in air or of an ink drop in water. More tangible examples are heat propagation in various heating and cooling systems, charge carrier propagation in semiconductor electronic devices on a *macroscale*, and even the spread of a flu epidemic. Yet another important engineering example is propagation of telegraph signals in a coaxial cable.

The general form of the one-dimensional diffusion equation is

$$\frac{\partial f}{\partial t} = \kappa \frac{\partial^2 f}{\partial x^2}. \tag{11.96}$$

For historical reasons this equation differs somewhat from our canonical form, Eq. (11.57). The coefficient κ is added for dimensionality matching. In diffusion problems it is known as the *diffusion coefficient*. For our purposes we pick the simplest case from a vast variety of diffusion problems. We suppose the unknown function $f(t, x)$ to be defined on the entire x-axis ($x \in (-\infty, \infty)$), and we impose initial conditions in the form $f(0, x) = f_0(x)$, where $f_0(x)$ is a given function of the spatial variable. (Equation (11.96) is of first order in t, so we need only one initial condition.) Thus, the problem is formulated as

$$\frac{\partial f}{\partial t} = \kappa \frac{\partial^2 f}{\partial x^2}; \quad f = f(t, x), \quad x \in (-\infty, \infty); \quad f(0, x) = f_0(x). \tag{11.97}$$

For the same reasons as described in the previous section, we again use the Fourier transform with respect to variable x:

$$F(t, k) = \int_{-\infty}^{\infty} dx \, e^{-2\pi i k x} f(t, x); \tag{11.98}$$

$$F_0(k) = \int_{-\infty}^{\infty} dx \, e^{-2\pi i k x} f_0(x). \tag{11.99}$$

Then, the inverse Fourier transform gives

$$f(t, x) = \int_{-\infty}^{\infty} dk \, e^{2\pi i k x} F(t, k) \tag{11.100}$$

and we substitute the latter expression into the original equation (11.97):

$$\int_{-\infty}^{\infty} dk \, e^{2\pi i k x} \frac{\partial F}{\partial t} = \kappa \int_{-\infty}^{\infty} dk \, e^{2\pi i k x} (-(2\pi k)^2) F(t, k). \tag{11.101}$$

Thus, we obtain the equation for the Fourier transform, $F(t, k)$, in the form:

$$\frac{\partial F}{\partial t} = -\kappa (2\pi k)^2 F(t, k). \tag{11.102}$$

290   PARTIAL DIFFERENTIAL EQUATIONS

This equation is, in fact, *ordinary* differential equation, because it contains only derivative with respect to the *t*-variable. The solution of this ordinary differential equation of first order is obvious:

$$F(t,k) = F(0,k)e^{-\kappa(2\pi k)^2 t} = F_0(k)e^{-\kappa(2\pi k)^2 t}. \qquad (11.103)$$

We then get the solution of the initial problem by means of the inverse transform:

$$\begin{aligned}
f(t,x) &= \int_{-\infty}^{\infty} dk\, e^{2\pi ikx} F_0(k) e^{-\kappa(2\pi k)^2 t} \\
&= \int_{-\infty}^{\infty} dk\, e^{2\pi ikx - \kappa(2\pi k)^2 t} \int_{-\infty}^{\infty} dx'\, f_0(x') e^{-2\pi kx'} \\
&= \int_{-\infty}^{\infty} dx'\, f_0(x') \int_{-\infty}^{\infty} dk\, e^{2\pi ik(x-x') - \kappa(2\pi k)^2 t}. \qquad (11.104)
\end{aligned}$$

To evaluate the integral over *k* in this expression, we make a perfect square in the exponent:

$$\begin{aligned}
&2\pi ik(x-x') - \kappa(2\pi k)^2 t \\
&= -\kappa t \left((2\pi k)^2 - \frac{i(x-x')}{\kappa t}(2\pi k) \right. \\
&\left. \quad + \left(\frac{i(x-x')}{2\kappa t}\right)^2 - \left(\frac{i(x-x')}{2\kappa t}\right)^2 \right) \\
&= -\kappa t \left(2\pi k - \frac{i(x-x')}{\kappa t}\right)^2 - \frac{(x-x')^2}{4\kappa t}. \qquad (11.105)
\end{aligned}$$

Then, we choose a new dummy variable,

$$\eta = 2\pi k - \frac{i(x-x')}{\kappa t}, \qquad (11.106)$$

and obtain

$$\begin{aligned}
\int_{-\infty}^{\infty} dk\, e^{2\pi ik(x-x') - \kappa(2\pi k)^2 t} &= \frac{1}{2\pi} \int_{-\infty}^{\infty} d\eta\, e^{-\kappa t \eta^2} e^{-\frac{(x-x')^2}{4\kappa t}} \\
&= \frac{1}{2\pi\sqrt{\kappa t}} e^{-\frac{(x-x')^2}{4\kappa t}} \int_{-\infty}^{\infty} d(\eta\sqrt{\kappa t}) e^{-(\eta\sqrt{\kappa t})^2} \\
&= \frac{1}{2\pi\sqrt{\kappa t}} e^{-\frac{(x-x')^2}{4\kappa t}} \sqrt{\pi}
\end{aligned}$$

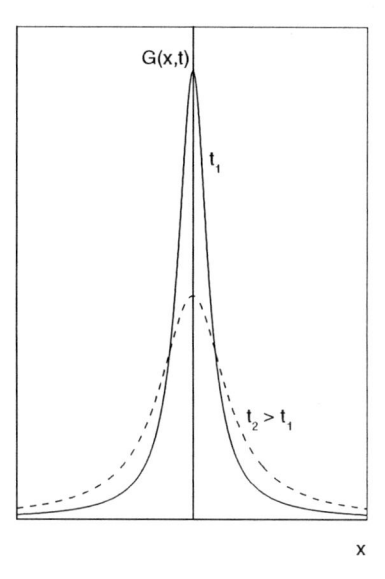

Fig. 11.2 The fundamental solution (Green's function) of the diffusion equation.

$$= \frac{1}{\sqrt{4\pi\kappa t}} e^{\frac{-(x-x')^2}{4\kappa t}}. \tag{11.107}$$

This function,

$$U(t, x) = \frac{1}{\sqrt{4\pi\kappa t}} e^{\frac{-x^2}{4\kappa t}} \tag{11.108}$$

is known as the *solution operator* or the *fundamental solution* of the diffusion equation. This function (Fig. 11.2) has two significant properties. First, at any instant,

$$\int_{-\infty}^{\infty} dx\, U(t, x) = 1, \tag{11.109}$$

which can easily be checked by direct calculation. Second, when $t \to 0$ the function $U(t, x)$ goes to 0 at any point on the x-axis except $x = 0$, at which point the function goes to ∞. Thus, in this limit the fundamental solution satisfies the properties of the δ-function (Section 10.3),

$$\lim_{t \to 0}(U(t, x)) = \delta(x). \tag{11.110}$$

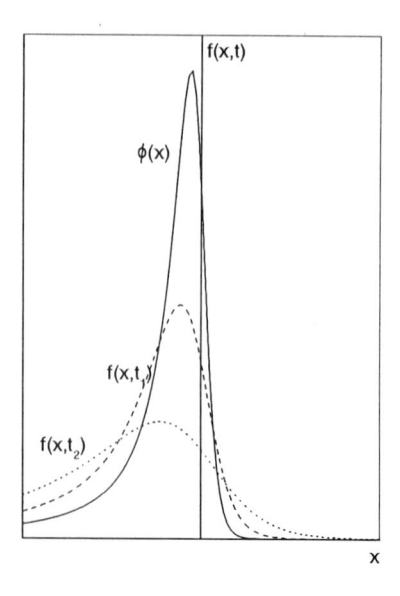

Fig. 11.3 To the general solution of the diffusion equation.

We have succeeded in expressing the evolution of an arbitrary initial distribution $f_0(x)$ in the handy form of a convolution integral. At any moment of time,

$$f(t, x) = \int_{-\infty}^{\infty} dx' \, U(t, x - x') f_0(x') \qquad (11.111)$$

(see illustration in Fig. 11.3). Note, that, in view of this equation, the limiting property of Eq. (11.110) is a necessary attribute of the function $U(t, x)$. Indeed, at the initial instant the solution, $f(t, x)$ must become the initial distribution, $f_0(x)$. An additional observation we can make here is that $U(t, x)$ is itself a solution of (11.96) with the initial distribution given as $f_0(x) = \delta(x)$, that is, with an initial "point source". In general, the approach leading to expressions like that of Eq. (11.111) is known as the Green's function approach, and the function $U(t, x - x')$ is referred to as the *Green's function of the diffusion equation*.

11.9 THE ELLIPTIC CASE: POISSON'S EQUATION

Following the previous cases, we will not consider here the general canonical form of the elliptic equation (11.40), but that formulation which is in most use in engineering (and scientific) applications. Moreover, we will need to break the tradition of all the previous narration in this chapter to consider just two independent variables. The reason to abandon this convenient "minimal" form of a PDE is that solving the two-dimensional elliptic equation faces some essential (and very specific) difficulties.

Although quite surmountable, these difficulties would inevitably distract the reader's attention from the basic ideas we are pursuing here. Thus, in this section we reluctantly turn to considering the *three-dimensional* situation, with the independent variables, x, y, and z.

The application-oriented equation of the elliptic case is the so-called *Laplace equation*,

$$\frac{\partial^2 f}{\partial x^2} + \frac{\partial^2 f}{\partial y^2} + \frac{\partial^2 f}{\partial z^2} = 0. \tag{11.112}$$

This equation is also frequently referred to as the *potential equation*, the name it has acquired in electrostatic field theory. in this case the function $f(x, y, z)$ has the meaning of the electrostatic potential in a two-dimensional domain. The same equation describes steady-state two-dimensional flow of an *incompressible* fluid, with the function $f(x, y, z)$ taking the meaning of the velocity potential, which determines the local velocity of the fluid as

$$\mathbf{v} = -\frac{\partial f}{\partial x} e_x - \frac{\partial f}{\partial y} e_y - \frac{\partial f}{\partial z} e_z, \tag{11.113}$$

e_x, e_y, and e_z being the unit vectors in the x, y, and z directions, respectively. Moreover, Eq. (11.112) can be treated as a limiting case of the three-dimensional *diffusion equation* (cf. Eq. (11.96) which corresponds to the steady-state situation. Then, $f(x, y, z)$ is a temperature distribution in a bulk volume.

Since now x, y, and z are all spatial variables, the additional condition required to completely specify the function $f(x, y, z)$ is not the *initial condition* but the *boundary condition*. Thus, a typical problem related to the elliptic case is to find the function $f(x, y, z)$, which is the solution of Eq. (11.112), in some three-dimensional domain when the function $f(x, y, z)$, or its *normal* derivative, or a combination of these two, are given on the surface of the domain. In adherence to our spirit of briefness, we will only consider the problem where the function $f(x, y, z)$ is to be defined in an *infinite* volume, that is, in the entire (x, y, z) space. Then, the natural additional conditions on the function $f(x, y, z)$ are that this function should vanish when $|x| \to \infty$, or $|y| \to \infty$, or $|z| \to \infty$.

The complication, however, is that the only solution of the equation (11.112) which satisfies these additional conditions is the function $f(x, y)$ being identically zero throughout the space! To have a non-trivial solution in such situation, the equation (11.112) must have some non-zero function $\phi(x, y, z)$ in its right-hand side. The equation

$$\frac{\partial^2 f}{\partial x^2} + \frac{\partial^2 f}{\partial y^2} + \frac{\partial^2 f}{\partial z^2} = \phi(x, y, z) \tag{11.114}$$

with the right-hand side being an arbitrary function of the variables x, y, and y is known as *Poisson's equation*. This equation, for instance, determines the electrostatic potential induced by a given charge distribution $\phi(x, y, z)$.

In solving the Poisson equation, we will assume the function $\phi(x, y, z)$ to be Fourier transformable (i.e., square integrable) with respect to all the three variables.

294 PARTIAL DIFFERENTIAL EQUATIONS

The symmetry of the variables in Eq. (11.114) makes the choice of the variable for the Fourier transform a hard problem. To avoid procrastination, we take the transform with respect to *all* variables,

$$F(\omega_x, \omega_y, \omega_z) = \int_{-\infty}^{\infty} dx \int_{-\infty}^{\infty} dy \int_{-\infty}^{\infty} dz\, e^{-2\pi i \omega_x x} e^{-2\pi i \omega_y y}$$
$$\times e^{-2\pi i \omega_z z} f(x, y, z). \tag{11.115}$$

Then, the function $F(\omega_x, \omega_y \omega_z)$ satisfies the algebraic equation,

$$(2\pi i \omega_x)^2 F + (2\pi i \omega_y)^2 F + (2\pi i \omega_z)^2 F = \Phi(\omega_x, \omega_y, \omega_z), \tag{11.116}$$

where, of course,

$$\Phi(\omega_x, \omega_y, \omega_z) = \int_{-\infty}^{\infty} dx \int_{-\infty}^{\infty} dy \int_{-\infty}^{\infty} dz\, e^{-2\pi i \omega_x x} e^{-2\pi i \omega_y y}$$
$$\times e^{-2\pi i \omega_z z} \phi(x, y, z). \tag{11.117}$$

The obvious solution of Eq. (11.116) is

$$F(\omega_x, \omega_y \omega_z) = \frac{\Phi(\omega_x, \omega_y, \omega_z)}{(2\pi i \omega_x)^2 + (2\pi i \omega_y)^2 + (2\pi i \omega_z)^2}. \tag{11.118}$$

To return to the (x, y, z) realm, we intend to make use of the concept of Green's function introduced in the previous section. Thus, we write the inverse Fourier transform as

$$f(x, y, z) = \int_{-\infty}^{\infty} d\omega_x \int_{-\infty}^{\infty} d\omega_y \int_{-\infty}^{\infty} d\omega_z\, e^{2\pi i \omega_x x} e^{2\pi i \omega_y y} e^{2\pi i \omega_z z} F(\omega_x, \omega_y, \omega_z)$$
$$= \int_{-\infty}^{\infty} d\omega_x \int_{-\infty}^{\infty} d\omega_y \int_{-\infty}^{\infty} d\omega_z\, \frac{e^{2\pi i \omega_x x} e^{2\pi i \omega_y y} e^{2\pi i \omega_z z} \Phi(\omega_x, \omega_y, \omega_z)}{(2\pi i \omega_x)^2 + (2\pi i \omega_y)^2 + (2\pi i \omega_z)^2}.$$

Now, we substitute in Eq. (11.119) the expression for $F(\omega_x, \omega_y, \omega_z)$ from Eq. (11.117), and obtain

$$f(x, y, z) = \int_{-\infty}^{\infty} d\omega_x \int_{-\infty}^{\infty} d\omega_y \int_{-\infty}^{\infty} d\omega_z \frac{e^{2\pi i \omega_x x} e^{2\pi i \omega_y y} e^{2\pi i \omega_z z}}{(2\pi i \omega_x)^2 + (2\pi i \omega_y)^2 + (2\pi i \omega_z)^2}$$
$$\times \int_{-\infty}^{\infty} dx' \int_{-\infty}^{\infty} dy' \int_{-\infty}^{\infty} dz'\, e^{-2\pi i \omega_x x'} e^{-2\pi i \omega_y y'} e^{-2\pi i \omega_z z'} \phi(x', y', z').$$

THE ELLIPTIC CASE: POISSON'S EQUATION   295

Just as in the preceding section, we change the order of integrations,

$$f(x,y,z) = \int_{-\infty}^{\infty} dx' \int_{-\infty}^{\infty} dy' \int_{-\infty}^{\infty} dy' \phi(x',y',z')$$

$$\times \int_{-\infty}^{\infty} d\omega_x \int_{-\infty}^{\infty} d\omega_y \int_{-\infty}^{\infty} d\omega_z \frac{e^{2\pi i \omega_x (x-x')} e^{2\pi i \omega_y (y-y')} e^{2\pi i \omega_z (z-z')}}{(2\pi i \omega_x)^2 + (2\pi i \omega_y)^2 + (2\pi i \omega_z)^2}$$

$$= \int_{-\infty}^{\infty} dx' \int_{-\infty}^{\infty} dy' \int_{-\infty}^{\infty} dz' \phi(x',y',z') G(x-x',y-y',z-z'), \quad (11.119)$$

where the Green's function is given by the expression,

$$G(x-x',y-y',z-z')$$

$$= \int_{-\infty}^{\infty} d\omega_x \int_{-\infty}^{\infty} d\omega_y \int_{-\infty}^{\infty} d\omega_z \frac{e^{2\pi i \omega_x (x-x')} e^{2\pi i \omega_y (y-y')} e^{2\pi i \omega_z (z-z')}}{(2\pi i \omega_x)^2 + (2\pi i \omega_y)^2 + (2\pi i \omega_z)^2}.$$

However, in computing the Green's function, we face unexpected problems. Indeed, let us start this computation by integrating over ω_z,

$$\int_{-\infty}^{\infty} d\omega_z \frac{e^{2\pi i \omega_z (z-z')}}{(2\pi i \omega_x)^2 + (2\pi i \omega_y)^2 + (2\pi i \omega_z)^2}$$

$$= -\frac{1}{4\pi^2} \int_{-\infty}^{\infty} d\omega_z \frac{e^{2\pi i \omega_z (z-z')}}{\omega_z^2 + \omega_x^2 + \omega_y^2}. \quad (11.120)$$

The poles of the integrand are

$$\omega_z = \pm i \sqrt{\omega_x^2 + \omega_y^2}, \quad (11.121)$$

so that the integral is equal to

$$-\frac{1}{4\pi^2}(2\pi i)\frac{1}{2i\sqrt{\omega_x^2 + \omega_y^2}} e^{-2\pi (z-z')\sqrt{\omega_x^2 + \omega_y^2}} \quad (11.122)$$

for $(z-z') > 0$, and

$$-\frac{1}{4\pi^2}(2\pi i)(-1)\frac{1}{(-2i\sqrt{\omega_x^2 + \omega_y^2})} e^{2\pi (z-z')\sqrt{\omega_x^2 + \omega_y^2}} \quad (11.123)$$

for $(z-z') < 0$. Combining the formulas (11.122) and (11.123), we get

$$-\frac{1}{4\pi \sqrt{\omega_x^2 + \omega_y^2}} e^{-2\pi |z-z'| \sqrt{\omega_x^2 + \omega_y^2}}. \quad (11.124)$$

Upon substituting the result of Eq. (11.124) in Eq. (11.120), the latter formula turns out to be

$$G(x-x', y-y', z-z')$$

$$= -\frac{1}{4\pi} \int_{-\infty}^{\infty} d\omega_x \int_{-\infty}^{\infty} d\omega_y \frac{1}{\sqrt{\omega_x^2 + \omega_y^2}}$$

$$\times e^{2\pi i \omega_x (x-x')} e^{2\pi i \omega_y (y-y')} e^{-2\pi |z-z'| \sqrt{\omega_x^2 + \omega_y^2}}.$$

The square root in Eq. (11.125) makes any further use of the Residue Theorem highly problematic. Thus, we need some other approach to compute the integrals over ω_x and ω_y. The approach which looks both natural and promising, is switching to *polar coordinates* in the plane ω_x, ω_y,

$$\omega_x = \omega \cos\theta; \qquad \omega_y = \omega \sin\theta. \tag{11.125}$$

In these coordinates,

$$\sqrt{\omega_x^2 + \omega_y^2} = \omega;$$
$$\omega_x(x-x') + \omega_y(y-y') = \omega((x-x')\cos\theta + (y-y')\sin\theta);$$
$$d\omega_x d\omega_y = \omega d\omega d\theta. \tag{11.126}$$

So, the double integral in Eq. (11.125) yields

$$G(x-x', y-y', z-z')$$

$$= -\frac{1}{4\pi} \int_0^{2\pi} d\theta \int_0^{\infty} d\omega \, \omega$$

$$\times \frac{1}{\omega} e^{2\pi i \omega((x-x')\cos\theta + (y-y')\sin\theta)} e^{-2\pi |z-z'|\omega}$$

$$= -\frac{1}{4\pi} \int_0^{2\pi} d\theta \int_0^{\infty} d\omega \, e^{2\pi i \omega((x-x')\cos\theta + (y-y')\sin\theta - |z-z'|)}. \tag{11.127}$$

In Eq. (11.127), the integral over ω is the trivial integral of an exponential. So, we immediately get

$$G(x-x', y-y', z-z')$$

$$= \frac{1}{8\pi^2} \int_0^{2\pi} \frac{d\theta}{|z-z'| - i((x-x')\cos\theta + (y-y')\sin\theta)}. \tag{11.128}$$

The calculation of the remaining integral over θ is a good exercise in the theory of functions of a complex variable. First of all, we transform the complicated expression in the denominator of the integrand:

$$(x - x') \cos\theta + (y - y') \sin\theta$$

$$= \sqrt{(x-x')^2 + (y-y')^2} \left(\frac{(x-x')}{\sqrt{(x-x')^2 + (y-y')^2}} \cos\theta \right.$$

$$\left. + \frac{(y-y')}{\sqrt{(x-x')^2 + (y-y')^2}} \sin\theta \right)$$

$$= \sqrt{(x-x')^2 + (y-y')^2} (\cos\theta' \cos\theta + \sin\theta' \sin\theta)$$

$$= \sqrt{(x-x')^2 + (y-y')^2} \cos(\theta - \theta'), \qquad (11.129)$$

where the auxiliary angle θ' is introduced by the condition

$$\tan^{-1}\theta' = \frac{y - y'}{x - x'}. \qquad (11.130)$$

Then, due to the periodicity of trigonometric functions,

$$\frac{1}{8\pi^2} \int_0^{2\pi} \frac{d\theta}{|z - z'| - i\sqrt{(x-x')^2 + (y-y')^2} \cos(\theta - \theta')}$$

$$= \frac{1}{8\pi^2} \int_0^{2\pi} \frac{d(\theta - \theta')}{|z - z'| - i\sqrt{(x-x')^2 + (y-y')^2} \cos(\theta - \theta')}$$

$$= \frac{1}{8\pi^2} \int_0^{2\pi} \frac{d\tilde{\theta}}{|z - z'| - i\sqrt{(x-x')^2 + (y-y')^2} \cos\tilde{\theta}}, \qquad (11.131)$$

with $\tilde{\theta} = \theta - \theta'$. Thus, the Green's function is now expressed as

$$G(x - x', y - y', z - z')$$

$$= \frac{i}{8\pi^2 \sqrt{(x-x')^2 + (y-y')^2}} \int_0^{2\pi} \frac{d\tilde{\theta}}{ia + \cos\tilde{\theta}}, \qquad (11.132)$$

where

$$a = \frac{|z - z'|}{\sqrt{(x-x')^2 + (y-y')^2}}. \qquad (11.133)$$

To make use of complex variable theory, we rewrite the integral in Eq. (11.132) in complex form:

$$\int_0^{2\pi} \frac{d\tilde\theta}{ia + \cos\tilde\theta} = 2\int_0^{2\pi} \frac{d\tilde\theta}{2ia + e^{i\tilde\theta} + e^{-i\tilde\theta}}$$

$$= 2\int_0^{2\pi} \frac{e^{i\tilde\theta}\,d\tilde\theta}{2iae^{i\tilde\theta} + e^{2i\tilde\theta} + 1} = -2i\int_0^{2\pi} \frac{d(e^{i\tilde\theta})}{2iae^{i\tilde\theta} + (e^{i\tilde\theta})^2 + 1}$$

$$= -2i\int_C \frac{d\zeta}{2ia\zeta + \zeta^2 + 1}. \qquad (11.134)$$

Thus, it is now an integral of a complex variable ζ over a closed contour which is the unit circle in the ζ-plane. According to the Residue Theorem, the value of this integral is equal to

$$2\pi i \sum \text{Res}\,\frac{1}{2ia\zeta + \zeta^2 + 1} \qquad (11.135)$$

over all the poles lying inside the circle. The poles are determined by the equation

$$\zeta^2 + 2ia\zeta + 1 = 0, \qquad (11.136)$$

which has the solutions,

$$\zeta_\pm = -ia \pm \sqrt{(-ia)^2 - 1} = i(\pm\sqrt{a^2 + 1} - a). \qquad (11.137)$$

It is easy to check that ζ_+ always lies inside the circle, while ζ_- always lies outside the circle. Then,

$$\text{Res}\,\frac{1}{2ia\zeta + \zeta^2 + 1} = \frac{1}{\zeta_+ - \zeta_-} = \frac{1}{2i\sqrt{a^2 + 1}}. \qquad (11.138)$$

Thus, the integral of Eq. (11.134) is equal to

$$\int_0^{2\pi} \frac{d\tilde\theta}{ia + \cos\tilde\theta} = -2i(2\pi i)\frac{1}{2i\sqrt{a^2 + 1}} = \frac{-2\pi i}{\sqrt{a^2 + 1}}. \qquad (11.139)$$

Substituting this result in Eq. (11.132) and taking the parameter a from Eq. (11.133), we obtain

$$G(x - x', y - y', z - z')$$

$$= \frac{i}{8\pi^2\sqrt{(x - x')^2 + (y - y')^2}}$$

$$\times \frac{-2\pi i\sqrt{(x - x')^2 + (y - y')^2}}{\sqrt{(z - z')^2 + (x - x')^2 + (y - y')^2}}$$

$$= \frac{1}{4\pi\sqrt{(z - z')^2 + (x - x')^2 + (y - y')^2}}. \qquad (11.140)$$

Thus, the general solution of Poisson's equation (11.114) with arbitrary left-hand-side function $\phi(x, y, z)$ is obtained as

$$f(x, y, z) = \int_{-\infty}^{\infty} dx' \int_{-\infty}^{\infty} dy' \int_{-\infty}^{\infty} dz' \phi(x', y', z')$$
$$\times G(x - x', y - y', z - z'), \qquad (11.141)$$

with the Green's function,

$$G(x - x', y - y', z - z')$$

$$= \frac{1}{4\pi} \frac{1}{\sqrt{(z - z')^2 + (x - x')^2 + (y - y')^2}}. \qquad (11.142)$$

Those readers who remember the theory of electromagnetic fields can recognize in Eqs. (11.141), (11.142), Poisson's formula for the electrostatic field of a distributed electric charge. Note also, that substitution of the Green's function of Eq. (11.142) in Poisson's equation (11.114) produces a function $\phi(x - x', y - y', z - z')$ which is the *three-dimensional* delta-function.

11.10 A VERY BRIEF COMMENT IN DEFENSE OF THE LAPLACE TRANSFORM

As the reader has seen, in all cases of second-order PDE's we successfully employed Fourier transform. It might seem, then, that this transform is especially convenient for solving PDE's when compared to the Laplace transform. This is not the case. The real reasons for the use of the Fourier transform in these examples stemmed not in the equations themselves, but from the *initial (or boundary) conditions*. In contrast, there can be (and really are) such boundary conditions which naturally invoke the use of the Laplace transform for solving the corresponding PDE.

An example of such initial conditions is the so-called *problem of a string with a moving end*, where the form of the string at any instant is determined by the wave equation (11.82) with the boundary condition,

$$f(0, t) = \phi(t), \qquad (11.143)$$

$\phi(t)$ being a given function of t, and the initial condition,

$$f(x, 0) = \left.\frac{\partial f}{\partial t}\right|_{t=0} = 0. \qquad (11.144)$$

Another example is the propagation of a signal in a coaxial cable which is governed by Eq. (11.96). However, the boundary conditions in this case are quite different

300 PARTIAL DIFFERENTIAL EQUATIONS

from those of Eq. (11.97). Now, they are given by Eq. (11.143), the signal being induced at the end of the cable. The initial condition is, naturally,

$$f(x, 0) = 0. \tag{11.145}$$

In both of these examples, the problem can be easily solved by making use of the Laplace transform with respect to the temporal variable.

11.11 SAMPLE PROBLEM II

Solve the following partial differential equation for the function $\phi(x, t)$ (a diffusion problem with time-dependent diffusion coefficient) by means of the Fourier transform:

$$\frac{\partial \phi}{\partial t} - \sin(4t) \frac{\partial^2 \phi}{\partial x^2} = 0; \qquad \phi(x, 0) = e^{-x^2}.$$

Hint. Poisson's integral:

$$\int_{-\infty}^{\infty} d\xi \, e^{-\xi^2} = \sqrt{\pi}$$

Solution
We use the formula for the Fourier transform of the second derivative,

$$\hat{F}\left(\frac{\partial^2 \phi}{\partial x^2}\right) = (2\pi i \omega)^2 \Phi(\omega, t), \tag{11.146}$$

where $\Phi(\omega, t)$ is the Fourier transform of the function $\phi(x, t)$ with respect to variable x, and obtain an equation for $\Phi(\omega, t)$:

$$\frac{\partial \Phi}{\partial t} - \sin(4t)(2\pi i \omega)^2 \Phi(\omega, t) = 0, \tag{11.147}$$

with initial condition:

$$\Phi(\omega, 0) = \int_{-\infty}^{\infty} dx \, e^{-2\pi i \omega x} \phi(x, 0). \tag{11.148}$$

For the given function $\phi(x, 0)$,

$$\begin{aligned}
\Phi(\omega, 0) &= \int_{-\infty}^{\infty} dx \, e^{-2\pi i \omega x} e^{-x^2} \\
&= \int_{-\infty}^{\infty} dx \, e^{-x^2 - 2x(\pi i \omega) - (\pi i \omega)^2 + (\pi i \omega)^2} \\
&= e^{-(\pi \omega)^2} \int_{-\infty}^{\infty} dx \, e^{-(x + \pi i \omega)^2} = \sqrt{\pi} e^{-(\pi \omega)^2}. \tag{11.149}
\end{aligned}$$

SAMPLE PROBLEM II   301

We solve the equation for $\Phi(w, t)$ as follows:

$$\frac{d\Phi}{\Phi} = -4(\pi w)^2 \sin(4t)\,dt, \qquad (11.150)$$

$$\ln \Phi = -(\pi w)^2 \int d(4t)\sin(4t) + const$$
$$= (\pi w)^2 \cos(4t) + const; \qquad (11.151)$$

$$\Phi(w, t) = C(w) e^{(\pi w)^2 \cos(4t)}. \qquad (11.152)$$

The constant $C(\omega)$ is found from the initial condition:

$$\Phi(\omega, 0) = C(\omega) e^{(\pi w)^2} = \sqrt{\pi} e^{-(\pi w)^2}, \qquad (11.153)$$

$$C(\omega) = \sqrt{\pi} e^{-2(\pi w)^2}. \qquad (11.154)$$

Thus,

$$\Phi(\omega, t) = \sqrt{\pi} e^{-(\pi w)^2 \cdot (2 - \cos(4t))}. \qquad (11.155)$$

Then, we find the function $\phi(x, t)$ as the result of the inverse Fourier transform:

$$\phi(x, t) = \int_{-\infty}^{\infty} d\omega\, e^{2\pi i \omega x} \Phi(\omega, t)$$

$$= \int_{-\infty}^{\infty} d\omega\, e^{2\pi i \omega x} \sqrt{\pi} e^{-(\pi w)^2 \cos(4t)}$$

$$= \sqrt{\pi} \cdot \frac{1}{\pi\sqrt{(2 - \cos(4t))}}$$

$$\times \int_{-\infty}^{\infty} d\xi\, e^{-\xi^2 + 2\xi \frac{ix}{\sqrt{(2 - \cos(4t))}}}, \qquad (11.156)$$

where $\xi = \pi\omega\sqrt{(2 - \cos(4t))}$.

We complete the integral with the same perfect-square method as we used in Eq. (11.107):

$$\int_{-\infty}^{\infty} d\xi\, \exp\left(-\xi^2 + 2\xi \frac{ix}{\sqrt{(2 - \cos(4t))}}\right)$$

$$
\begin{aligned}
&= \int_{-\infty}^{\infty} d\xi \, \exp\left(-\xi^2 + 2\xi \frac{ix}{\sqrt{(2-\cos(4t))}}\right.\\
&\qquad\left. - \left(\frac{ix}{\sqrt{(2-\cos(4t))}}\right)^2 + \left(\frac{ix}{\sqrt{(2-\cos(4t))}}\right)^2\right)\\
&= \exp\left(-\frac{x^2}{(2-\cos(4t))}\right)\\
&\qquad \times \int_{-\infty}^{\infty} d\xi \, \exp\left(-\left(\xi - \frac{ix}{\sqrt{(2-\cos(4t))}}\right)^2\right)\\
&= \sqrt{\pi} \exp\left(\frac{x^2}{(2-\cos(4t))}\right). \qquad\qquad (11.157)
\end{aligned}
$$

We get the function $\phi(x,t)$, which is the solution of the original equation (11.146):

$$
\phi(x,t) = \frac{1}{\sqrt{(2-\cos(4t))}} \exp\left(-\frac{x^2}{(2-\cos(4t))}\right). \qquad (11.158)
$$

11.12 SUMMARY OF CHAPTER 11

1) A linear partial differential equation of second order with constant coefficients always can be reduced to one of three *canonical forms*: elliptic, hyperbolic, or parabolic. This reduction is achieved by a linear transformation of the independent variables and the use of an exponential multiplier for the unknown function.

2) The solutions to PDEs behave very differently in these three cases.

3) A typical example of a hyperbolic PDE is the wave equation. Its general solution is given by the D'Alembert formula. The initial disturbance conserves its form while propagating with constant velocity in the positive and negative directions on the spatial axis.

4) A typical example of a parabolic PDE is the diffusion equation. Its solution for arbitrary initial distribution is obtained as a convolution integral of the initial distribution and the *fundamental solution*. The initial disturbance spreads over the spatial axis while dissipating throughout.

5) A typical example of an elliptic PDE is Poisson's equation, an equation in three spatial variables with non-zero right-hand side. Its solution is given as the three-dimensional convolution integral of this right-hand side and Green's function.

11.13 PROBLEMS

I. Classify the following partial differential equation, find the transformations of variables, x, y, and unknown function, ϕ, which put the equation into canonical form, and write down this canonical form:

$$\frac{\partial^2 \phi}{\partial x^2} - 12\frac{\partial^2 \phi}{\partial x \partial y} + \frac{\partial^2 \phi}{\partial y^2} + \frac{\partial \phi}{\partial x} + \frac{\partial \phi}{\partial y} = 0.$$

II. Classify the following partial differential equation and find the simplest transformation of variables x, y which put it into canonical form:

$$\frac{1+\sqrt{3}}{4}\frac{\partial^2 \phi}{\partial x^2} + \frac{1-\sqrt{3}}{2}\frac{\partial^2 \phi}{\partial x \partial y} + \frac{1+\sqrt{3}}{4}\frac{\partial^2 \phi}{\partial y^2} = 0.$$

III. Solve the following partial differential equation by means of the Fourier transform:

$$\frac{\partial \psi}{\partial t} + 2t\frac{\partial \psi}{\partial x} = 3t^2\frac{\partial^2 \psi}{\partial x^2}; \quad \psi(x, t=0) = \exp\left(\frac{-x^2}{2}\right).$$

IV. Find the general solution of the *two-dimensional* Poisson's equation,

$$\frac{\partial^2 \psi}{\partial x^2} + \frac{\partial^2 \psi}{\partial y^2} = f(x, y),$$

by means of the Fourier transform, provided $f(x, y)$ is a Fourier-transformable function with respect to both variables. Discuss the difference between this two-dimensional case and three-dimensional case considered in Section 11.9.

11.14 FURTHER READING

Basic-level texts:
    P. V. O'Neil, *Beginning Partial Differential Equations*, Wiley, New York, 1999;
    S. J. Farlow, *Partial Differential Equations for Scientists and Engineers*, Wiley, New York, 1982;
    G. L. Lamb, Jr., *Introductory Applications of Partial Differential Equations*, Wiley, New York, 1995;
    E. Zauderer, *Partial DIfferential Equations of Applied Mathematics*, 2nd ed., Wiley, New York, 1989;
    S. Shu, *Boundary Value Problems of Linear Partial Differential Equations for Engineers and Scientists*, World Scientific, Singapore, 1987.

Advanced texts:

G. B. Folland, *Introduction to Partial Differential Equations*, 2nd ed., Princeton University Press, Princeton, NJ, 1995;

M. A. Pinsky, *Partial Differential Equations and Boundary-Value Problems With Applications*, 2nd ed., McGraw-Hill, New York, 1991;

J. Wloka, *Partial Differential Equations*, Cambridge University Press, Cambridge,1987;

K. E. Gustafson, *Introduction to Partial Differential Equations and Hilbert Space Methods*, 2nd ed., Wiley, New York,1987.

Computational methods:

D. G. Duffy, *Transform Methods for Solving Partial Differential Equations*, CRC Press, Boca Raton, FL, 1994;

M. Pikering, *An Introduction to Fast Fourier Transform Methods for Partial Differential Equations*, Wiley, New York, 1986;

E. H. Twizell, *Computational Methods for Partial Differential Equations*, Ellis Horwood, Chichester, 1984;

J. C. Strikweda, Finite Difference Schemes and Partial Differential Equations, Wadsworth & Brooks, Pacific Grove, CA, 1989;

V. G. Ganzha, E. V. Vorozhtsov, *Computer-Aided Analysis of Difference Schemes for Partial Differential Equations*, Wiley, New York, 1996;

G. D. Smith, *Numerical Solutions of Partial Differential Equations: Finite Difference Methods*, 3rd Ed., Clarendon Press, Oxford, 1985.

Index

Abelian groups, 71
Abscissa of absolute convergence, 258
Algebra of complex numbers, 73
Algebraic structures, 68
Amplitude of a complex number, 76
Argument of a complex number, 76
Atomic statements, 42
Banach space, 148
Basis, 82
Bessel's Inequality, 180
Bijection, 31
Binary relation, 28
Binomial coefficients, 12
Block-diagonal matrix, 119
Boolean algebra, 58
Bromwich path, 258
Cancellation laws, 71
Canonical forms, 274
Canonical forms of PDE, 282
Cauchy-Schwarz inequality, 155
Cauchy criterion, 134
Cauchy sequences, 134
Causality, 249
Characteristic equation, 110
Closed interval, 5
Closed loop, 253
Closed set, 139
Closure point, 139
Codomain, 29
Combinations, 12

Complementation, 60
Complete orthonormal sequence, 187
Complete set, 139
Complex conjugation, 75
Complex plane, 75
Constants of an algebra, 68
Continuous mapping, 140
Contraction mapping, 141
Contraction Mapping Theorem, 202
Contradiction, 49
Contraposition, 55
Convolution product, 249
Cross product, 22
D'Alembert's Formula, 288
De Morgan laws, 53
Decomposition, 83
Delta-function, 239
Diffusion coefficient, 289
Diffusion equation, 289
Dimension, 84
Direct image, 31
Distance, 127
Domain, 29
Echelon matrix, 105
Eigenvalue problem, 201
Eigenvalues, 110
Eigenvectors, 110
Element (member) of a set, 3
Elliptic case, 277
Empty set, 10

Euclidian space, 130
Euler's formula, 76
Extended matrix, 226
Finite-dimensional linear space, 82
Fourier coefficients, 191
Fourier decomposition, 191
Fourier pair, 237
Fourier transforms, 237
Frequency filter, 250
Frequency shifting, 245
Function of a matrix, 113
Fundamental solution, 291
Gaussian elimination, 104
General contradiction, 56
General tautology, 56
Generalized Fourier Series, 184
Generalized functions, 241
Gram-Schmidt orthonormalization, 176
Green's function, 292
Group, 70
Group of symmetries, 72
Hermitian conjugate, 158
Hermitian matrices, 158
Hermitian operator, 158
Hilbert space, 153
Hyperbolic case, 277
Identity element, 69
Identity matrix, 100
Imaginary axis, 75
Imaginary part, 73
Imaginary unity, 74
Incompletely specified equations, 222
Induced metric, 148
Infinite-dimensional linear space, 83
Injections, 30
Inner product, 152
Input–output relation, 249
Input signal, 248
Integral operator, 209
Interior point, 139
Intersection of sets, 9
Inverse Fourier transform, 241
Inverse image, 32
Inverse mapping, 32
Inversion, 70
Isolated poles, 253
Jordan block, 120
Jordan form, 119
Kernel, 248
Laplace equation, 293
Laplace transform, 258
Laplace Transform Table, 265
Lattice, 72
Laurent expansion, 253
Least-Square Theorem, 223
Limit of a sequence, 133

Linear combination, 80
Linear mappings, 80
Linear signal processing, 250
Linear space, 77
Linear spaces of functions, 79
Linear subspace, 80
Linear systems, 246
Linearly dependent sets, 82
Lipschitz condition, 141
Lipschitz constant, 141
Logical connectives, 2
Logical object, 41
Logical statements, 41
Mapping, 30
Matrix, 97
Matrix algebra, 100
Matrix diagonalization, 112
Matrix multiplication, 97
Mean square deviation, 225
Measurement errors, 221
Metric mapping, 128
Metric space, 129
Minimal residual approximation, 216
Modus ponens, 55
Modus tollens, 55
Monoid, 70
Newton's binomial formula, 12
Non-singular matrices, 100
Norm mapping, 148
Normed space, 148
Nullity, 103
Nullspace, 103
One-to-one mappings, 30
Onto mapping, 31
Open ball, 134
Open interval, 5
Operations, 13
Operator, 200
Operator equation, 200
Ordered n-tuple, 20
Ordinary differential equations, 207
Orthogonal complement, 160
Orthogonal subsets, 160
Orthogonality, 160
Orthonormal basis, 176
Orthonormal sequence, 173
Orthonormal set, 171
Output signal, 248
Parabolic case, 277
Parseval's Identity, 188
Partial differential equations, 273
PDE classification, 277
Phasor, 245
Picard sequence, 202
Poisson's equation, 293
Power set, 11

308 INDEX

Predicate function, 42
Projection Decomposition, 181
Proof formats, 54
Pseudo-inverse operator, 228
Quantifiers, 44
Range, 31
Rank, 103
Real axis, 75
Real part, 73
Reductio ad absurdum, 56
Relation, 19
Residual, 215
Residual Theorem, 216
Residue Theorem, 253
Root-finding problem, 200
Russel's paradox, 5
Scalar, 78
Semigroup, 69
Set of integer numbers, 4
Set of natural numbers, 4
Set of real numbers, 4
Shifting theorem, 261
Signal processing, 246

Signature, 68
Singular point, 253
Solution operator, 291
Space of polynomials, 178
Span, 80
Square-integrable functions, 131
Subalgebra, 69
Subset, 7
Successive approximations, 204
Surjection, 31
Tautology, 49
The Projection Theorem, 163
The triangle axiom, 128
Time-invariant systems, 248
Time shifting, 245
Transpose matrix, 101
Trial function, 220
Truth-value function, 43
Truth values, 43
Union of sets, 8
Unit-step function, 251
Universal set, 6
Upper-triangular matrix, 105
Vector-by-matrix product, 97
Vector, 78